Advances in Intelligent Systems and Computing

Volume 539

Series editor

Janusz Kacprzyk, Polish Academy of Sciences, Warsaw, Poland
e-mail: kacprzyk@ibspan.waw.pl

About this Series

The series "Advances in Intelligent Systems and Computing" contains publications on theory, applications, and design methods of Intelligent Systems and Intelligent Computing. Virtually all disciplines such as engineering, natural sciences, computer and information science, ICT, economics, business, e-commerce, environment, healthcare, life science are covered. The list of topics spans all the areas of modern intelligent systems and computing.

The publications within "Advances in Intelligent Systems and Computing" are primarily textbooks and proceedings of important conferences, symposia and congresses. They cover significant recent developments in the field, both of a foundational and applicable character. An important characteristic feature of the series is the short publication time and world-wide distribution. This permits a rapid and broad dissemination of research results.

Advisory Board

More information about this series at http://www.springer.com/series/11156

Jerzy Świątek · Jakub M. Tomczak
Editors

Advances in Systems Science

Proceedings of the International Conference
on Systems Science 2016 (ICSS 2016)

 Springer

Editors
Jerzy Świątek
Wroclaw University of Science
and Technology
Wroclaw
Poland

Jakub M. Tomczak
Wroclaw University of Science
and Technology
Wroclaw
Poland

ISSN 2194-5357   ISSN 2194-5365 (electronic)
Advances in Intelligent Systems and Computing
ISBN 978-3-319-48943-8   ISBN 978-3-319-48944-5 (eBook)
DOI 10.1007/978-3-319-48944-5

Library of Congress Control Number: 2016955520

Printed on acid-free paper

This Springer imprint is published by Springer Nature
The registered company is Springer International Publishing AG
The registered company address is: Gewerbestrasse 11, 6330 Cham, Switzerland

Preface

This year we had a great pleasure to organize 19th edition of the International Conference on Systems Science (ICSS) held in Wrocław (Poland). ICSS is the series of international conferences, jointly organized on a rotational basis among three universities, namely Wrocław University of Science and Technology (Poland), University of Nevada Las Vegas (the USA), and Coventry University (UK). The conference covered major topics in systems science and was divided into four workshops: Workshop on Applications of Machine Learning, Workshop on Uncertain Systems, Workshop on Cloud Computing, and Workshop on Transportation and Multi-Robot Systems.

Workshop on Applications of Machine Learning. Machine Learning (ML) is currently one of the fastest growing disciplines in computer science. The ML algorithms are widely applicable in many different areas including (but not limited to) automatic drug design, business intelligence, computer vision, image processing, information retrieval, natural language processing, online advertising, recommendation systems, social networks, speech recognition, systems biology, text mining, bioinformatics, biomedicine, credit scoring, economy, and spam detection.

The Workshop on Applications of Machine learning organized as a part of the 19th edition of the International Conference on Systems Science gathered outstanding researchers that presented valuable applications of various machine learning methods including medical image processing and diagnosis, power load prediction, voice recognition or character recognition. All the accepted and presented papers are included in this book.

Workshop on Cloud Computing. Cloud computing has emerged as one of the most widespread used paradigms for on-demand resource provisioning and application development. Due to its popular characteristics such as resource pooling, rapid elasticity, broad network access, or pay-as-you-go pricing models, it has been widely adopted for a variety of application scenarios and use cases. Those unique attributes have led to a number of new research topics. Possibilities offered by modern cloud infrastructures gave researchers and developers an opportunity to design new application architectures, develop algorithms and methods for Big Data processing in the cloud, introduce new methods for web service management and resource management, propose new service composition methods, apply new business models for cloud services' delivery, and develop methods for management of methods supporting Internet of Things data aggregation and processing in the cloud.

All these research topics are closely related and can be implemented in scenarios utilizing cloud computing in the Internet of Things (IoT). Applications in the scope of the IoT, such as telemedicine, smart homes, smart cities, or applications with industrial background such as factory automation, logistics, or automotive are often based on a variety of heterogeneous sensor nodes and sensor networks collecting data about the environment. Both, amount and diversity of sensor nodes and resulting data streams are rapidly increasing. Thus, IoT applications can benefit from the ability to elastically provide computing, network, and storage resources offered by the cloud. However, the cloud computing and Internet of Things domains show divergent characteristics in terms of their underlying resources. While IoT developers often have to be aware of resource constraints, location and semantics of sensor nodes, non-functional and infrastructure management requirements, cloud computing is perceived as a rather homogenous and endless resource being accessible within seconds and without limits. Therefore, it is necessary to investigate and develop appropriate concepts that allow leveraging the advantages of the cloud computing for the challenging application scenarios of the Internet of Things domain.

Workshop on Transportation and Multi-Robot Systems. Development and usage of autonomous devices capable of solving spatially distributed problems in a timely fashion is an ever-growing field of interest. This workshop set the study of movement in complex systems as its main field of interest. Fundamentally, problems arising from such considerations are on the perimeter of logistic, transportation, and robotics. It is therefore worthwhile to consider them in a jointly fashion.

The workshop covers a wide range of topics, from problems of multicriteria decision-making in logistics to task allocation in multi-robot environments. More specifically, tackled are problems of traffic modeling and transportation sharing, of service design for transportation grids and of sensing and control mechanisms in robotic systems. Finally, the workshop gives consideration towards social and environmental aspect of extensive automation.

Workshop on Uncertain Systems. Uncertain systems constantly attract much attention since they enable the better describing and understanding the reality. They are useful from the practical point of view, since the models of the actual processes and phenomenons are almost never exact and the precise values of their parameters are often not known in advance.

The aim of this workshop was to present the latest results concerning application of probabilistic and non-probabilistic descriptions of uncertainty to the systems science. The workshop presentations concerned the use of different uncertainty descriptions (i.e., fuzzy logics, grey systems, and uncertain variables) to many areas including economic, control systems, and even computer networking.

Jerzy Świątek

Jakub M. Tomczak

Editors

Adam Gonczarek

Maciej Zięba

AML Chairs

Adam Grzech

Paweł Świątek

CC Chairs

Grzegorz Filcek

Maciej Hojda

TMS Chairs

Dariusz Gąsior

US Chair

Contents

Applications of Machine Learning

Maximum Likelihood Estimation and Optimal Coordinates

P. Spurek$^{(\boxtimes)}$ and J. Tabor

Faculty of Mathematics and Computer Science,
Jagiellonian University, Łojasiewicza 6, 30-348 Kraków, Poland
przemyslaw.spurek@ii.uj.edu.pl

Abstract. We show that the MLE (maximum likelihood estimation) in the class of Gaussian densities can be understood as the search for the best coordinate system which "optimally" underlines the internal structure of the data. This allows in particular to the search for the optimal coordinate system when the origin is fixed in a given point.

Keywords: Maximum likelihood estimation · Cross-entropy · Gaussian distribution

1 Introduction

MLE (maximum likelihood estimation) is one the most important estimation methods in statistics [4,11]. In data engineering it plays the crucial role in particular in EM clustering [15], in information theory it can be "identified" with the cross-entropy, which jointly with the Kullback-Leibler divergence plays the basic role in computer science [6]. In this paper we discuss the MLE in the case when the considered density is Gaussian with the center belonging to a given set. We were inspired by the ideas presented by [5] and consider estimations in various subclasses of normal densities.

One of the crucial question in data analysis is how to choose the best coordinate system and define distance which "optimally" underlines the internal structure of the data [3,8,12,17,18,20]. A similar role is played by Mahalanobis distance in discrimination analysis [9]. In general, we first need to decide if we *allow or not the translation of the origin of coordinate system.* Next we usually consider one of the following:

- *no change in coordinates;*
- *possibly different change of scale separately in each coordinate;*
- *arbitrary coordinates.*

P. Spurek—The paper was supported by the National Centre of Science (Poland) Grant No. 2013/09/N/ST6/01178.

J. Tabor— The paper was supported by the National Centre of Science (Poland) Grant No. 2014/13/B/ST6/01792.

J. Świątek and J.M. Tomczak (eds.), *Advances in Systems Science*, Advances in Intelligent Systems and Computing 539, DOI 10.1007/978-3-319-48944-5_1

It occurs that the value of likelihood function, in the case when we restrict to the Gaussian densities, can be naturally interpreted as the measure of the fitness of the given coordinate system to the data. Thus in the paper we search for those coordinates in the above situations which best describe (with respect to MLE) the given dataset $\mathcal{Y} \subset \mathbb{R}^N$.

At the end of the introduction let us mention that our results can be also used in various density estimation and clustering problems which use Gaussian models [1,5], in particular in the case when we consider the model consisting of Gaussians with centers satisfying certain constraints.

2   Entropy and Gaussian Random Variables

Let X be a random variable with density f_X. The differential entropy

$$H(X) := \int - \ln(f_X(y)) f_X(y) dy \tag{1}$$

tells us what is the asymptotic expected amount of memory needed to code X [6], and thus the differential code-length optimized for X is given by $-\ln(f_X(x))$.

If we want to code Y (a continuous variable with density g_Y) with the code optimized for X we obtain the *cross-entropy* which was presented in [6,10] (we follow the notation from [16]):

$$H^{\times}(Y\|X) := \int g_Y(y) \cdot (-\ln f_X(y)) dy, \tag{2}$$

If A is a linear operator, then $H^{\times}(AY\|AX) = H^{\times}(Y\|X) + \ln|\det(A)|$. Since we consider X only from its density f_X point of view, we will commonly use the notation

$$H^{\times}(Y\|f_X) := \int g_Y(y) \cdot (-\ln f_X(y)) dy. \tag{3}$$

Roughly speaking, $H^{\times}(Y\|f)$ denotes (asymptotically) the memory needed to code random variable Y with the code optimized for the density f. In the case of given dataset $\mathcal{Y} \subset \mathbb{R}^N$ we interpret $\mathcal{Y}$ as an uniform discreet variable Y on $\mathcal{Y}$. Consequently, our formula is reduced to

$$H^{\times}(\mathcal{Y}\|f) := H^{\times}(Y\|f) = -\frac{1}{|\mathcal{Y}|} \sum_{x \in \mathcal{Y}} \ln(f(x)), \tag{4}$$

where $|\mathcal{Y}|$ denote cardinality of the set $\mathcal{Y}$.

In our investigations we are interested in (best) coding for Y by densities chosen from a set of densities $\mathcal{F}$, and thus we will need the following definition.

Definition 1. *By the* cross-entropy *of Y with respect to a family of coding densities $\mathcal{F}$ we understand*

$$H^{\times}(Y\|\mathcal{F}) := \inf_{f \in \mathcal{F}} H^{\times}(Y\|f). \tag{5}$$

Observe that the search for the density f with minimal cross entropy leads exactly to the maximum likelihood estimation. Thus in general the calculation of $H^\times(Y\|\mathcal{F})$ is nontrivial, as it is equivalent to finding ML estimator.

As is the case in many statistical or data-information problems, the basic role in our investigations is played by the Gaussian densities. We recall that the normal variable with mean m and a covariance matrix Σ has the density $\mathcal{N}_{(m,\Sigma)}(x) = \frac{1}{\sqrt{(2\pi)^N \det(\Sigma)}} e^{(-\frac{1}{2}\|x-m\|^2_\Sigma)}$, where $\|x-m\|_\Sigma$ is the Mahalanobis norm $\|x-m\|^2_\Sigma := (x-m)^T \Sigma^{-1}(x-m)$, see [13]. The differential entropy of Gaussian distribution is given by

$$H(\mathcal{N}_{(m,\Sigma)}) = \frac{N}{2}\ln(2\pi e) + \frac{1}{2}\ln(\det(\Sigma)).$$

From now on, if not otherwise specified, we assume that all the considered random variables have finite second moments and that they have values in $\mathbb{R}^N$. For a random variable Y, by $m_Y = E(Y)$ we denote its mean, and by Σ_Y its covariance matrix, that is $\Sigma_Y = E((Y-m_Y)\cdot(Y-m_Y)^T)$.

We will need the following result, which says that the cross-entropy of an arbitrary random variable Y versus normal can be computed just from the knowledge of covariance and mean of Y.

Theorem 1 *([4], Theorem 5.59). Let Y be a random variable with finite covariance matrix. Then for arbitrary m and positive-definite covariance matrix Σ we have*

$$H^\times(Y\|\mathcal{N}_{(m,\Sigma)}) = \tfrac{N}{2}\ln(2\pi) + \tfrac{1}{2}\|m-m_Y\|^2_\Sigma + \tfrac{1}{2}\mathrm{tr}(\Sigma^{-1}\Sigma_Y) + \tfrac{1}{2}\ln(\det(\Sigma)). \quad (6)$$

Remark 1. *Suppose that we are given a data set $\mathcal{Y}$. Then we usually understand the data as a sample realization of a random variable Y. Consequently as an estimator for the mean of Y we use the mean $m_\mathcal{Y} = \frac{1}{|\mathcal{Y}|}\sum_{y\in\mathcal{Y}} y$ of the data $\mathcal{Y}$ and as the covariance we use the ML estimator $\frac{1}{|\mathcal{Y}|}\sum_{y\in\mathcal{Y}}(y-m_\mathcal{Y})(y-m_\mathcal{Y})^T$.*

As a direct corollary we obtain the formula for the optimal choice of origin.

Proposition 1. *Let Y be a random variable and Σ be a fixed covariance matrix. Let M be a nonempty closed subset of $\mathbb{R}^N$. From all normal coding densities $\mathcal{N}_{(m,\Sigma)}$, where $m \in M$, the minimal cross-entropy is realized for that $m \in M$ which minimizes $\|m-m_Y\|_\Sigma$, and equals*

$$\inf_{m\in M} H^\times(Y\|\mathcal{N}_{(m,\Sigma)}) = \tfrac{1}{2}\big(d^2_\Sigma(m_Y;M) + \mathrm{tr}(\Sigma^{-1}\Sigma_Y) + \ln(\det(\Sigma)) + N\ln(2\pi)\big),$$

where d_Σ is a Mahalanobis distance.

Consequently, if $M = \mathbb{R}^N$ the minimum is realized for $m = m_Y$ and equals

$$\inf_{m\in\mathbb{R}^N} H^\times(Y\|\mathcal{N}_{(m,\Sigma)}) = \tfrac{1}{2}\big(\mathrm{tr}(\Sigma^{-1}\Sigma_Y) + \ln(\det(\Sigma)) + N\ln(2\pi)\big).$$

It occurs that our basic MLE problem, in the case when we restrict to the Gaussian densities, can be naturally interpreted as search for the optimal rescaling (optimal choice of coordinate system).

Remark 2. *Let us start from one dimensional space. In such a case, if we allow the translation of the origin of coordinate system, we usually apply the standarization/normalization: $s : Y \rightarrow (Y - \mathrm{m}_Y)/\sigma_Y$. In the multivariate case the normalization is given by the transformation $s : X \rightarrow \Sigma_Y^{-1/2}(Y - \mathrm{m}_Y)$. Then we obtain that the coordinates are uncorrelated, and the covariance matrix is identity. Taking the distance between the transformation of points x, y:*

$$\|sx - sy\|^2 = (sx - sy)^T(sx - sy) = (x - y)^T \Sigma^{-1}(x - y)$$

we arrive naturally at the Mahalanobis distance $\|x-y\|_\Sigma^2 = (x-y)^T\Sigma^{-1}(x-y)$. If we do not allow the translation of the origin, we usually only scale each coordinate by dividing it by its mean (then the unit-scale plays the normalizing role, as the mean of each coordinate is one), arriving in the case when the mean is one.

To study the question what is the optimal procedure, we need the criterion to compare different coordinate systems. Suppose that we are given a basis $\mathrm{v} = (v_1, \ldots, v_N)$ of $\mathbb{R}^N$ and an origin of coordinate system m. Then by $\mathcal{N}_{[\mathrm{m},\mathrm{v}]}$ we denote the "normalized" Gaussian density with respect to the basis v with center at m, that is

$$\mathcal{N}_{[\mathrm{m},\mathrm{v}]}(\mathrm{m} + x_1 v_1 + \ldots + x_N v_N) = \frac{1}{(2\pi)^{N/2}|\det(\mathrm{v})|} e^{-(x_1^2 + \ldots + x_N^2)/2}.$$

Then as a measure of fitness of the coordinate system $[\mathrm{m}, \mathrm{v}]$ we understand the cross-entropy $H^\times(Y\|\mathcal{N}_{[\mathrm{m},\mathrm{v}]})$.

3   Rescaling

Let us first consider the question how we should uniformly rescale the classical coordinates to optimally "fit" the data. *Assume that we have fixed an origin of the coordinate system at m and that we want to find how we should (uniformly) rescale the coordinates to optimally fit the data.* This means that we search for s such that $s \rightarrow H^\times(Y\|\mathcal{N}_{(\mathrm{m},s\mathrm{I})})$ attains minimum. Since

$$H^\times(Y\|\mathcal{N}_{(\mathrm{m},s\mathrm{I})}) = \tfrac{1}{2}([\mathrm{tr}(\Sigma_Y) + \|\mathrm{m} - \mathrm{m}_Y\|^2]s^{-1} + N\ln(s) + N\ln(2\pi)), \quad (7)$$

by the trivial calculations we obtain that the above function attains its minimum

$$\frac{N}{2}(\ln[\mathrm{tr}(\Sigma_Y) + \|\mathrm{m} - \mathrm{m}_Y\|^2] + \ln(2\pi e/N))$$

for $s = [\mathrm{tr}(\Sigma_Y) + \|\mathrm{m} - \mathrm{m}_Y\|^2]/N$. Thus we have arrived at the following theorem.

Theorem 2. *Let Y be a random variable with invertible covariance matrix and m be fixed. Then the minimum of $H^\times(Y\|\{\mathcal{N}_{(\mathrm{m},s\mathrm{I})}\}_{s>0})$ is realized for $s = (\mathrm{tr}(\Sigma_Y) + \|\mathrm{m} - \mathrm{m}_Y\|^2)/N$, and equals*

$$H^\times(Y\|\{\mathcal{N}_{(\mathrm{m},s\mathrm{I})}\}_{s>0}) = \tfrac{N}{2}(\ln[\mathrm{tr}(\Sigma_Y) + \|\mathrm{m} - \mathrm{m}_Y\|^2] + \ln(2\pi e/N)).$$

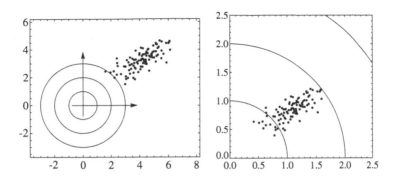

Fig. 1. The original data set with optimal coordinate system (the new "optimal" basis is marked by the bold arrows) in the case of the family $\{\mathcal{N}_{(m,sI)}\}_{s>0}$ (left figure). The data in the new basis (right figure).

Example 1. *Let $\mathcal{Y}$ be a realization of the normal random variable Y with $\mathrm{m}_Y = [3,4]^T$ and $\Sigma = \begin{bmatrix} 1 & 0.3 \\ 0.3 & 0.6 \end{bmatrix}$ and let $\mathrm{m} = [0,0]^T$. In Fig. 1 we present a sample $\mathcal{Y}$ with the coordinate system (marked by bold black segments) obtained by the Theorem 2 and data in the new basis.*

Observe that the above minimum depends only on the trace of the covariance matrix of Y and the Euclidean distance of m from m_Y. If we allow the change of the origin, we have to clearly put the origin it at m_Y:

Corollary 1. *Let Y be a random variable with invertible covariance matrix. Then $H^\times(Y\|\{\mathcal{N}_{(m,sI)}\}_{s>0,m\in\mathbb{R}^N})$ is realized for $\mathrm{m} = \mathrm{m}_Y$, $s = \frac{1}{N}\mathrm{tr}(\Sigma_Y)$, and equals*

$$H^\times(Y\|\{\mathcal{N}_{(m,sI)}\}_{s>0,m\in\mathbb{R}^N}) = \tfrac{N}{2}(\ln(\mathrm{tr}(\Sigma_Y)) + \ln(\tfrac{2\pi e}{N})).$$

Corollary 2. *Let $\mathcal{Y} = (y_1,\ldots,y_n)$ be a given data-set. Assume that we want to move the origin to m, and uniformly rescale the coordinates. Then*

$$s \to (s - \mathrm{m})/\sqrt{\tfrac{1}{N}(\mathrm{tr}(\Sigma_y) + \|\mathrm{m} - \mathrm{m}_Y\|^2)}$$

is the optimal rescaling, where Σ_y is a covariance of $\mathcal{Y}$. If we additionally allow the change of the origin, we should put $\mathrm{m} = \mathrm{m}_S$ and consequently the rescaling takes the form $s \to (s - \mathrm{m}_S)/\sqrt{\mathrm{tr}(\Sigma_y)/N}$.

Applying the above we obtain that in the one dimensional case the rescaling takes the form $s \to (s - \mathrm{m}_y)/\sigma_y$ (if we allow change of origin), and $s \to s/\sqrt{\mathrm{m}_y^2 + \sigma_y^2}$ (if we fix the origin at zero).

Example 2. *Let $\mathcal{Y}$ be a realization of the normal random variable Y from Example 1. In Fig. 2 we present a sample $\mathcal{Y}$ with the coordinate system obtained by the Corollary 1 and data in the new basis.*

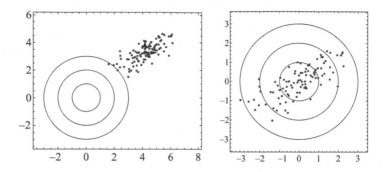

Fig. 2. The original data set with optimal coordinate system in the case of the family $\{\mathcal{N}_{(m,sI)}\}_{s>0,m\in\mathbb{R}^N}$ (left figure). The data in the new basis (figure on the right).

Now we consider the case when we allow to rescale each coordinate Y_i of $Y = (Y_1, \ldots, Y_N)$ separately. For simplicity we consider the case $N = 2$. Consider the splitting $\mathbb{R}^N = \mathbb{R}^{N_1} \times \mathbb{R}^{N_2}$. For densities f_1 and f_2 on $\mathbb{R}^{N_1}$ and $\mathbb{R}^{N_2}$, respectively, we define the product density $f_1 \otimes f_2$ on $\mathbb{R}^N = \mathbb{R}^{N_1} \times \mathbb{R}^{N_2}$ by the formula

$$(f_1 \otimes f_2)(x_1, x_2) := f_1(x_1) \cdot f_2(x_2),$$

for $(x_1, x_2) \in \mathbb{R}^{N_1} \times \mathbb{R}^{N_2}$. Given density families $\mathcal{F}_1$ and $\mathcal{F}_2$, we put $\mathcal{F}_1 \otimes \mathcal{F}_2 := \{f_1 \otimes f_2 : f_1 \in \mathcal{F}_1, f_2 \in \mathcal{F}_2\}$. Let $Y : (\Omega, \mu) \to \mathbb{R}^{N_1} \times \mathbb{R}^{N_2}$ be a random variable and let $Y_1 : \Omega \to \mathbb{R}^{N_1}$ and $Y_2 : \Omega \to \mathbb{R}^{N_2}$ denote the first and second coordinate of Y (observe that in general Y_1 and Y_2 are not independent random variables). On can easily observe that

Proposition 2. *Let $\mathcal{F}_1$ and $\mathcal{F}_2$ denote coding density families in $\mathbb{R}^{N_1}$ and $\mathbb{R}^{N_2}$, respectively, and let $Y : \Omega \to \mathbb{R}^{N_1} \times \mathbb{R}^{N_2}$ be a random variable. Then*

$$H^\times(Y\|\mathcal{F}_1 \otimes \mathcal{F}_2) = H^\times(Y_1\|\mathcal{F}_1) + H^\times(Y_2\|\mathcal{F}_2).$$

The above result means that if we allow to rescale coordinates, we can treat them as separate random variables. Thus we obtain the following theorem.

Theorem 3. *Let $\mathcal{Y}$ be a data set, and let $\mathcal{Y}_k$ denote the set containing its k-th coordinate. Then the optimal rescaling for each k-th coordinate is given by*

$$\mathcal{Y}_k \ni s \to (s - \mathrm{m}_{\mathcal{Y}_k})/\sigma_{\mathcal{Y}_k} \ (\text{if we allow change of origin}),$$
$$\mathcal{Y}_k \ni s \to s/\sqrt{\mathrm{m}_{\mathcal{Y}}^2 + \sigma_{\mathcal{Y}}^2} \ (\text{if we fix the origin at zero}).$$

Example 3. *Let $\mathcal{Y}$ be a realization of the normal random variable Y from Example 1. In Fig. 4 we present a sample $\mathcal{Y}$ and the coordinate system obtained by the Theorem 3 (if we fix the origin at zero) and data in the new basis. In Fig. 3 we present a sample $\mathcal{Y}$ and the coordinate system obtained by the Theorem 3 (when we allow change of origin) and data in the new basis.*

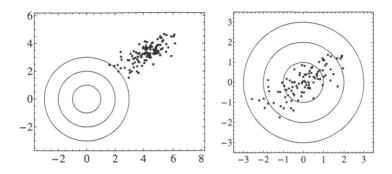

Fig. 3. The original data set with optimal coordinate system in the case of separated random variable when we allow change of origin (figure on the left) and the data in the new basis (figure on the right).

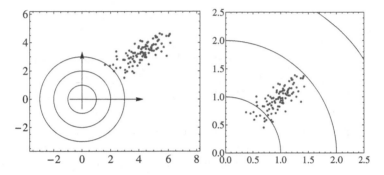

Fig. 4. The original data set with optimal coordinate system in the case of separated random variable when we do not allow change of origin (left hand side illustration) and data in the new basis (right hand side illustration).

4   Main Result

We find the optimal coordinate system in the general case by applying an approach similar to that from [19]. To do so, we need a simple consequence of the famous von Neuman trace inequality. Next we discuss the optimal rescaling if we move the coordinate to the mean of the data.

In most of our further results the following proposition will play an important role. In its proof we will use the well-known von Neumann trace inequality described by [7,14]:

Theorem [von Neumann trace inequality]. *Let E, F be complex $N \times N$ matrices. Then*

$$|\mathrm{tr}(EF)| \leq \sum_{i=1}^{N} s_i(E) \cdot s_i(F), \tag{8}$$

where $s_i(D)$ denote the ordered (decreasingly) singular values of matrix D.

We also need Sherman-Morrison formula [2]:

Theorem [Sherman-Morrison formula]. *Suppose A is an invertible square matrix and u, v are column vectors. Suppose furthermore that $1 + v^T A^{-1} u \neq 0$. Then the Sherman-Morrison formula states that*

$$(A + uv^T)^{-1} = A^{-1} - \frac{A^{-1} uv^T A^{-1}}{1 + v^T A^{-1} u}.$$

Let us recall that for the symmetric positive matrix its eigenvalues coincide with singular values. Given $\lambda_1, \ldots, \lambda_N \in \mathbb{R}$ by $S_{\lambda_1, \ldots, \lambda_N}$ we denote the set of all symmetric matrices with eigenvalues $\lambda_1, \ldots, \lambda_N$.

Proposition 3. *Let B be a symmetric nonnegative matrix with eigenvalues $\beta_1 \geq \ldots \geq \beta_N \geq 0$. Let $0 \leq \lambda_1 \leq \ldots \leq \lambda_N$ be fixed. Then*

$$\min_{A \in S_{\lambda_1, \ldots, \lambda_N}} \operatorname{tr}(AB) = \sum_i \lambda_i \beta_i.$$

Proof. Let e_i denote the orthogonal basis build from the eigenvectors of B, and let operator $\bar{A}$ be defined in this basis by $\bar{A}(e_i) = \lambda_i e_i$. Then trivially

$$\min_{A \in S_{\lambda_1, \ldots, \lambda_N}} \operatorname{tr}(AB) \leq \operatorname{tr}(\bar{A}B) = \sum_i \lambda_i \beta_i.$$

To prove the inverse inequality we will use the von Neumann trace inequality. Let $A \in S_{\lambda_1, \ldots, \lambda_N}$ be arbitrary. We apply the inequality (8) for $E = \lambda_N I - A$, $F = B$. Since E and F are symmetric nonnegatively defined matrices, their eigenvalues $\lambda_N - \lambda_i$ and β_i coincide with singular values, and therefore by (8)

$$\operatorname{tr}((\lambda_N I - A)B) \leq \sum_i (\lambda_N - \lambda_i)\beta_i = \lambda_N \sum_i \beta_i - \sum_i \lambda_i \beta_i. \tag{9}$$

Since

$$\operatorname{tr}((\lambda_N I - A)B) = \lambda_N \sum_i \beta_i - \operatorname{tr}(AB),$$

from inequality (9) we obtain that $\operatorname{tr}(AB) \geq \sum_i \lambda_i \beta_i$.

Now we proceed to the main result of the paper. Let $M \subset \mathbb{R}^N$ then by $\mathcal{G}_M$ we denote the set of Gaussians with mean $m \in M$.

Theorem 4. *Let $m \in \mathbb{R}^N$ be fixed and let $\mathcal{G}_{\{m\}}$ denote the set of Gaussians with mean m. Then $H^\times(Y \| \mathcal{G}_{\{m\}})$ equals*

$$\frac{1}{2} \left(\ln(1 + \|m - m_Y\|_{\Sigma_Y}^2) + \ln(\det(\Sigma_Y)) + N \ln(2\pi e) \right),$$

and is attained for $\Sigma = \Sigma_Y + (m - m_Y)(m - m_Y)^T$.

Proof. Let us first observe that by applying substitution

$$A = \Sigma_Y^{1/2} \Sigma^{-1} \Sigma_Y^{1/2}, v = \Sigma_Y^{-1/2}(\mathbf{m} - \mathbf{m}_Y),$$

we obtain

$$
\begin{aligned}
H^\times(Y\|\mathcal{N}_{(\mathbf{m},\Sigma)}) &= \tfrac{1}{2}(\operatorname{tr}(\Sigma^{-1}\Sigma_Y) + \|\mathbf{m} - \mathbf{m}_Y\|_\Sigma^2 + \ln(\det(\Sigma)) + N\ln(2\pi)) \\
&= \tfrac{1}{2}(\operatorname{tr}(\Sigma^{-1}\Sigma_Y) + (\mathbf{m} - \mathbf{m}_Y)^T \Sigma^{-1}(\mathbf{m} - \mathbf{m}_Y) \\
&\quad - \ln(\det(\Sigma^{-1}\Sigma_Y)) + \ln(\det(\Sigma_Y)) + N\ln(2\pi)) \\
&= \tfrac{1}{2}\big(\operatorname{tr}(A) + v^T A v - \ln(\det(A)) + \ln(\det(\Sigma_Y)) + N\ln(2\pi)\big).
\end{aligned}
\tag{10}
$$

Then A is a symmetric positive matrix. Contrary given a symmetric positive matrix A we can uniquely determine Σ by the formula

$$\Sigma = \Sigma_Y^{1/2} A^{-1} \Sigma_Y^{1/2}. \tag{11}$$

Thus finding minimum of (10) reduces to finding a symmetric positive matrix A which minimize the value of

$$\operatorname{tr}(A) + v^T A v - \ln(\det(A)). \tag{12}$$

Let us first consider $A \in S_{\lambda_1,\dots,\lambda_N}$, where $0 < \lambda_1 \le \dots \le \lambda_N$ are fixed. Our aim is to minimize

$$v^T A v = \operatorname{tr}(v^T A v) = \operatorname{tr}(A \cdot (vv^T)).$$

We fix an orthonormal basis such that $v/\|v\|$ is its first element, and then by applying von Neumann trace formula we obtain that the above minimizes when v is the eigenvector of A corresponding to λ_1, and thus the minimum equals $\lambda_1\|v\|^2$. Consequently we arrive at the minimization problem

$$\lambda_1(1 + \|v\|^2) + \sum_{i>1}\lambda_i - \sum_i \ln\lambda_i.$$

Now one can easily verify that the minimum of the above is realized for

$$\lambda_1 = 1/(1 + \|v\|^2), \lambda_i = 1 \text{ for } i > 1,$$

and then (12) equals $N + \ln(1 + \|\mathbf{m} - \mathbf{m}_Y\|_{\Sigma_Y}^2)$, while the formula for A minimizing it is given by $A = I - \frac{vv^T}{1+\|v\|^2}$. Consequently then the minimal value of (10) is

$$\frac{1}{2}\left(\ln(1 + \|\mathbf{m} - \mathbf{m}_Y\|_{\Sigma_Y}^2) + \ln(\det(\Sigma_Y)) + N\ln(2\pi e)\right).$$

and by (11) and Sherman-Morrison formula is attained for

$$\Sigma = \Sigma_Y^{1/2}\left(I - \frac{\Sigma_Y^{-1/2}(\mathbf{m}-\mathbf{m}_Y)(\mathbf{m}-\mathbf{m}_Y)^T\Sigma_Y^{-1/2}}{1+\|\mathbf{m}-\mathbf{m}_Y\|_{\Sigma_Y}^2}\right)^{-1}\Sigma_Y^{1/2} = \Sigma_Y + (\mathbf{m}-\mathbf{m}_Y)(\mathbf{m}-\mathbf{m}_Y)^T.$$

Example 4. *Let $\mathcal{Y}$ be a realization of the normal random variable Y from Example 1 and let $\mathbf{m}$ be fixed. In Fig. 5 is presented a sample $\mathcal{Y}$ and the coordinate system obtained by the Theorem 4 and data in the new basis.*

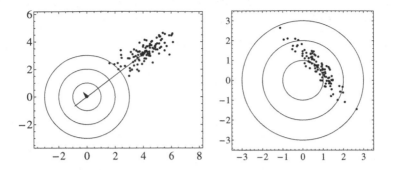

Fig. 5. The original data set with optimal coordinate system in the case of the family $\mathcal{G}_{\{(0,0)\}}$ (left hand side illustration) and data in the new basis (right hand side illustration).

5 Conclusion

In the paper we show that the MLE in the class of Gaussian densities can be understood equivalently as the search for the coordinates which best describe given dataset $\mathcal{Y} \subset \mathbb{R}^N$. The main result of the paper presents the formula of the optimal coordinate system in the case when the mean of the Gaussian density satisfies certain constrains.

Our work can be used in density estimation and clustering algorithms which use different Gaussian models.

References

1. Banfield, J.D., Raftery, A.E.: Model-based gaussian and non-gaussian clustering. Biometrics **49**(3), 803–821 (1993)
2. Bartlett, M.S.: An inverse matrix adjustment arising in discriminant analysis. Ann. Math. Stat. **22**(1), 107–111 (1951)
3. Borg, I., Groenen, P.: Modern multidimensional scaling: Theory and applications. Springer, Heidelberg (2005)
4. Van den Bos, A.: Parameter Estimation for Scientists and Engineers. Wiley Online Library, New York (2007)
5. Celeux, G., Govaert, G.: Gaussian parsimonious clustering models. Pattern Recogn. **28**(5), 781–793 (1995)
6. Cover, T., Thomas, J., Wiley, J., et al.: Elements of Information Theory. Wiley Online Library, New York (1991)
7. Grigorieff, R.: A note on von neumanns trace inequality. Math. Nachr. **151**, 327–328 (1991)
8. Han, J., Kamber, M., Pei, J.: Data Mining: Concepts and Techniques. Morgan Kaufmann, San Francisco (2011)
9. Krishnaiah, P.: Handbook of Statistics. North-Holland, New York (1988)
10. Kullback, S.: Information Theory and Statistics. Dover Publications, New York (1997)

11. Lehmann, E., Casella, G.: Theory of Point Estimation. Springer, New York (1998)
12. De Maesschalck, R., Jouan-Rimbaud, D., Massart, D.: The mahalanobis distance. Chemometr. Intell. Lab. Syst. **50**(1), 1–18 (2000)
13. Mahalanobis, P.C.: On the generalised distance in statistics. Proc. Nat. Inst. Sci. **2**, 49–55 (1936)
14. Mirsky, L.: A trace inequality of john von neumann. Monatshefte für mathematik **79**(4), 303–306 (1975)
15. Ng, S., Krishnan, T., McLachlan, G.: The em algorithm. In: Gentle, J.E., Härdle, W.K., Mori, Y. (eds.) Handbook of Computational Statistics Concepts and Methods. Springer Handbooks of Computational Statistics, pp. 139–172. Springer, Heidelberg (2004)
16. Nielsen, F., Nock, R.: Sided and symmetrized bregman centroids. IEEE Trans. Inf. Theory **55**(6), 2882–2904 (2009)
17. Raykov, T., Marcoulides, G.: An Introduction to Applied Multivariate Analysis. Routledge, London (2008)
18. Rencher, A.: Methods of Multivariate Analysis. Wiley Online Library, New York (1995)
19. Theobald, C.: An inequality with application to multivariate analysis. Biometrika **62**(2), 461–466 (1975)
20. Timm, N.: Applied Multivariate Analysis. Springer, New York (2002)

Domain Adaptation for Image Analysis: An Unsupervised Approach Using Boltzmann Machines Trained by Perturbation

Szymon Zaręba, Marcin Kocot, and Jakub M. Tomczak(✉)

Department of Computer Science, Faculty of Computer Science and Management,
Wrocław University of Science and Technology, Wrocław, Poland
{szymon.zareba,marcin.kocot,jakub.tomczak}@pwr.edu.pl

Abstract. In this paper, we apply Restricted Boltzmann Machine and Subspace Restricted Boltzmann Machine to domain adaptation. Moreover, we train these models using the Perturb-and-MAP approach to draw approximate sample from the Gibbs distribution. We evaluate our approach on domain adaptation task between two image corpora: MNIST and Handwritten Character Recognition dataset.

Keywords: Representation transfer · Deep learning · Boltzmann machine · Perturbation method · Low-dimensional perturbations

1 Introduction

Typically, in machine learning it is assumed that training and test samples are generated from the same underlying distribution. However, very often these two samples are drawn from the training (source) distribution that is similar (*e.g.*, in the sense of the Kullback-Leibler divergence) but not the same as the test (target) distribution). Furthermore, in many real-life applications, a source sample becomes out-dated and there is a need to adapt a model to the target domain. For example, in handwriting recognition there could be a vast of data from different users but the target distribution would correspond only to a specific user that possesses her own writing style. A recognition system, however, should quickly adapt to that user using a small amount of new data. It is worth stressing out that in the considered setting the source data is no longer available and we assume that the only information preserved from the source domain is the model itself. This assumption holds true in many real-life applications where we cannot afford to store and transform a large amount of training data, *e.g.*, on mobile devices.

The problem of domain adaptation becomes an important direction in current machine learning research [12]. The increasing interest in this issue is brought by modern applications, such as, natural language processing [3], sentiment analysis [7], text categorization [16] or image categorization [24]. Moreover, the ability to learn in an unsupervised manner and adapt to new problems using small samples is a distinctive trait of the human brain. Therefore, the domain adaptation seems to be one the most important means for formulating artificial intelligence [2].

© Springer International Publishing AG 2017
J. Świątek and J.M. Tomczak (eds.), *Advances in Systems Science*, Advances in Intelligent Systems and Computing 539, DOI 10.1007/978-3-319-48944-5_2

There are different approaches to the domain adaptation. One solution is to train a model using a sample from the target domain with a kind of regularizer utilizing the sample from the source domain, *e.g.*, by using Maximum Mean Discrepacy [16]. A different but similar approach aims at regularizing parameters of the model using the parameters estimated on the source domain. In [3,24] the model was regularized using ℓ_2-norm and in [14] it was further generalized and shown that from the Bayesian perspective the regularizer takes the form of the Kullback-Leibler divergence. In [1] conditions for learning a classifier for domain adaptation were provided that led to formulating an upper-bound combining source and target data. However, in many cases it remains unclear what kind of a regularization term should be proposed to adapt to new domains.

Another popular research direction is *instance weighting*. A simple modification of the risk functional for given loss function reveals that learning a model with domain adaptation relies on weighting instances from the target domain by the probability of these instances using the source distribution [12]. The manner these distributions are estimated leads to different methods. For example, one of them applies an optimization perspective and solves a minimax optimization problem [15]. Although, the instance weighting-based methods are theoretically well-motivated, they require maintaining the source sample and thus could be very time consuming.

Recently, a very promising approach is *representation adaptation* (or *representation transfer*) that utilizes hierarchical (deep) representation that contains information about different domains. First methods utilized probabilistic graphical models to combine source and target domains [5], however, these required fixed structure for the domains. More flexible approach was about augmenting feature representation for new domains [4]. Lately, *deep learning* gives the most promising results. The general idea is to apply deep neural networks that can share features among domains [6,7,26].

In this paper, we follow this line of thinking. However, we go beyond directed neural networks like auto-encoders [7] or convolutional neural networks [6,26] and utilize generative models, namely, Boltzmann Machines [11]. We aim at verifying whether adapting to new domain is possible using Restricted Boltzmann Machine [20] and its recently proposed modification, Subspace Restricted Boltzmann Machine [23], without any additional techniques for domain adaptation. In other words, we want to quantify to which extent these models provide representation transferable to new domains. Additionally, we apply new learning procedure for the two mentioned Restricted Boltzmann Machines using the Perturb-and-MAP (PM) approach [18]. The basic idea of the PM approach is to perturb the parameters of the model, and then, starting from the training data, to find the local optima of the energy function using and optimization method. This procedure produces approximate samples from the Gibbs distribution that can be further used to approximate the gradient.

The contribution of the paper is threefold:

– We apply Restricted Boltzmann Machine and Subspace Restricted Boltzmann machine to unsupervised domain adaptation task.

- We utilize the Perturb-and-MAP approach to learning Subspace Restricted Boltzmann Machine.
- We propose an evaluation metric for the domain adaptation in the fully unsupervised setting (*i.e.*, there are no labels available from source and target domains).

The paper is organized as follows. First, in Sect. 2 we formulate the problem of domain adaptation. Next, we outline Boltzmann machines used in the paper in Sect. 3. Further, we describe a new learning schema for the Subspace Restricted Boltzmann Machine basing on the Perturb-and-MAP approach in Sect. 4. The presented approaches are evaluated in Sect. 5 using MNIST dataset [13] and Handwritten Charactered Recognition dataset [25]. At the end, the conclusions are drawn in Sect. 6.

2   Unsupervised Domain Adaptation Problem

Let us define a data space $\mathcal{X}$ and a distribution $P(\mathbf{x})$ where $\mathbf{x} \in \mathcal{X}$. The objective of (unsupervised) machine learning is to learn a hypothesis (a model) $h(\mathbf{x})$ basing on N training data $\{x_n\}_{n=1}^N$. We define a *domain* as a pair $\mathbb{D} = (\mathcal{X}, P(\mathbf{x}))$. Further, we distinguish a *source domain* $\mathbb{D}_S$ and a *target domain* $\mathbb{D}_T$. We formulate the problem of *domain adaptation* as follows:

Given a source domain $\mathbb{D}_S$ *and a target domain* $\mathbb{D}_T$ *such that* $P_S(\mathbf{x}) \neq P_T(\mathbf{x})$, *and a model* $h_S(\mathbf{x})$ *trained using examples from* $\mathbb{D}_S$, *the* domain adaptation *aims to learn* $h(\mathbf{x})$ *using a sample from* $\mathbb{D}_T$ *that transfers knowledge from* $h_S(\mathbf{x})$ *and makes as small errors on both* $\mathbb{D}_S$ *and* $\mathbb{D}_T$ *as possible.*

There are different definitions of the domain adaptation (see, *e.g.*, [12,16]), however, in this paper we focus on a fully unsupervised problem statement. Moreover, we would like to emphasize that in our definition we have access to the source domain through the model $h_S(\mathbf{x})$ only. This forbids an application of a technique that takes advantage of examples from the source domain (see, *e.g.*, [16]). Therefore, a beneficial approach would be to transfer representation from source to target domain, *e.g.*, by applying deep learning [26].

3   Boltzmann Machines

A general Boltzmann machine (BM) is defined through a *Gibbs distribution* and an *energy function* that describe relationships among random variables.

Restricted Boltzmann Machine.   The binary Restricted Boltzmann Machine (RBM) is a bipartite BM that defines the joint distribution over binary visible and hidden units [21], where $\mathbf{x} \in \{0,1\}^D$ are the visibles and $\mathbf{h} \in \{0,1\}^M$ are the hiddens. The relationships among variables are specified through the energy function:

$$E(\mathbf{x}, \mathbf{h}|\Theta) = -\mathbf{x}^\top \mathbf{W} \mathbf{h} - \mathbf{b}^\top \mathbf{x} - \mathbf{c}^\top \mathbf{h}, \tag{1}$$

where $\boldsymbol{\theta} = \{\mathbf{W}, \mathbf{b}, \mathbf{c}\}$ is a set of parameters, $\mathbf{W} \in \mathbb{R}^{D \times M}$, $\mathbf{b} \in \mathbb{R}^D$, and $\mathbf{c} \in \mathbb{R}^M$.

We can train RBM using the maximum likelihood approach that seeks the maximum of the averaged log-likelihood. However, since the partition function (*i.e.*, normalizing constant) in the Gibbs distribution requires summing over exponential number of configurations, an application of a gradient-based optimization methods is troublesome. Therefore, RBM are trained using some kind of gradient approximations like in the *contrastive divergence* (CD) algorithm that applies T steps of the block Gibbs sampling [10].

Subspace Restricted Boltzmann Machine. The sRBM introduces third-order multiplicative interactions of one visible x_i and two types of hidden binary units, a gate unit g_j and a subspace unit s_{jk}. Each gate unit is associated with a group of subspace hidden units. The energy function of a joint configuration is defined as follows [23]:

$$E(\mathbf{x}, \mathbf{h}, \mathbf{S}|\boldsymbol{\theta}) = -\sum_{i=1}^{D}\sum_{j=1}^{M}\sum_{k=1}^{K} V_{ijk} x_i g_j S_{jk} - \sum_{i=1}^{D} b_i x_i - \sum_{j=1}^{M} c_j g_j - \sum_{j=1}^{M} g_j \sum_{k=1}^{K} D_{jk} S_{jk}. \quad (2)$$

where $\mathbf{x} \in \{0,1\}^D$ denotes a vector of visible variables, $\mathbf{g} \in \{0,1\}^M$ is a vector of gate units, $\mathbf{S} \in \{0,1\}^{M \times K}$ is a matrix of subspace units, the parameters are $\boldsymbol{\theta} = \{\mathbf{V}, \mathbf{b}, \mathbf{c}, \mathbf{D}\}$, $\mathbf{V} \in \mathbb{R}^{D \times M \times K}$, $\mathbf{b} \in \mathbb{R}^D$, $\mathbf{c} \in \mathbb{R}^M$, and $\mathbf{D} \in \mathbb{R}^{M \times K}$.

The sRBM can be seen as a mixture of many simple RBMs, where gate units allow to activate or deactivate subsets of hidden units (subspace units). We believe that these third-order relationships allow to better reflect domain adaptation by maintaining information about domains in different small RBMs.

4   Perturb-and-MAP Learning Algorithm

The idea of the *Perturb-and-MAP* (PM) approach is about first adding i.i.d. Gumbel perturbations to the energy function and next finding the MAP configuration that is a sample from the original Gibbs distribution [18]. Nevertheless, since the domain of visibles and hiddens in RBM and sRBM grows exponentially with the number of variables, it is troublesome to find the MAP assignment of the perturbed energy efficiently. Therefore, first order (low-dimensional) Gumbel perturbations are often employed to obtain an approximate sample [8,17,22]. Then the joint perturbation is fully decomposable into a sum of perturbations and it corresponds to perturbing unary potentials (*i.e.*, biases) only [8,9,18]. Further, we denote Gumbel perturbation for value z by $\gamma(z)$.

The manner of how the MAP configurations are found is crucial for learning process. Since finding MAP solutions is run for each example in a mini-batch, the method cannot be computationally complex. In [19] it was proposed to obtain samples from the RBM by first perturbing the unary potentials, and further, starting from a data point, applying block coordinate descent to optimize the energy function (Perturb-and-Descent, PD). The procedure for sampling from RBM is presented in Algorithm 1.[1]

---

[1] $\mathbb{I}[\cdot]$ denotes the indicator function, and $\odot$ is the element-wise multiplication.

Algorithm 1. Perturb-and-Descent for RBM

Input   : $\mathbf{x}^{(0)}$: training datum, T: number of optimization steps
Output: $\{\hat{\mathbf{x}}^{(T)}, \mathbf{h}^{(T)}\}$: approximate MAP solutions of the perturbed energy
for $i = 1, \ldots, D$, $j = 1, \ldots, M$ **do**
$\quad \tilde{b}_i = b_i + \gamma(x_i = 1) - \gamma(x_i = 0);$
$\quad \tilde{c}_j = c_j + \gamma(h_j = 1) - \gamma(h_j = 0);$
end
for $t = 1, \ldots, T$ **do**
$\quad \mathbf{x}^{(t)} = \mathbb{I}[\mathbf{W}\mathbf{h}^{(t-1)} + \tilde{\mathbf{b}} > \mathbf{0}];$
$\quad \mathbf{h}^{(t)} = \mathbb{I}[\mathbf{W}^\top \mathbf{x}^{(t)} + \tilde{\mathbf{c}} > \mathbf{0}];$
end
return $\{\hat{\mathbf{x}}^{(T)}, \mathbf{h}^{(T)}\};$

In the context of the sRBM the PD algorithm takes the similar form and is presented in Algorithm 2.

Algorithm 2. Perturb-and-Descent for sRBM

Input   : $\mathbf{x}^{(0)}$: training datum, T: number of optimization steps
Output: $\{\hat{\mathbf{x}}^{(T)}, \mathbf{g}^{(T)}, \mathbf{S}^{(T)}\}$: approximate MAP solutions of the perturbed
$\qquad\qquad$ energy
for $i = 1, \ldots, D$, $j = 1, \ldots, M$ and $k = 1, \ldots, K$ **do**
$\quad \tilde{b}_i = b_i + \gamma(x_i = 1) - \gamma(x_i = 0);$
$\quad \tilde{c}_j = c_j + \gamma(g_j = 1) - \gamma(g_j = 0);$
$\quad \tilde{D}_{jk} = D_{jk} + \gamma(S_{jk} = 1) - \gamma(S_{jk} = 0);$
end
for $t = 1, \ldots, T$ **do**
$\quad \mathbf{x}^{(t)} = \mathbb{I}[\sum_{j=1}^{M}\sum_{k=1}^{K} V_{\cdot jk} g_j^{(t-1)} S_{jk}^{(t-1)} + \tilde{\mathbf{b}} > \mathbf{0}];$
$\quad \mathbf{g}^{(t)} = \mathbb{I}[\sum_{i=1}^{D}\sum_{k=1}^{K} V_{i \cdot k} x_i^{(t)} S_{jk}^{(t-1)} + \tilde{\mathbf{c}} + \sum_{k=1}^{K} \tilde{D}_{\cdot k} \odot S_{\cdot k}^{(t-1)} > \mathbf{0}];$
$\quad \mathbf{S}^{(t)} = \mathbb{I}[\sum_{i=1}^{D}\sum_{j=1}^{M} V_{ij\cdot} x_i^{(t)} g_j^{(t)} + \sum_{j=1}^{M} g_j^{(t)} \tilde{D}_{j\cdot} \odot S_{j\cdot}^{(t-1)} > \mathbf{0}];$
end
return $\{\hat{\mathbf{x}}^{(T)}, \mathbf{g}^{(T)}, \mathbf{S}^{(T)}\};$

---

5   Experiments

Data and Training Details. We use two datasets in the experiment: Handwritten Character Recognition (HCR) [25], and MNIST [13]. MNIST contains handwritten digits of size 28×28, and HCR consists of images of digits and letters[2]. We split HCR into 24,000 training images, 8,134 test images, and remaining 8000 images formulate validation set. MNIST is divided into 50,000 training images, 10,000 validation images, and 10,000 test images (Fig. 1).

---

[2] The original HCR is scaled to fit MNIST images.

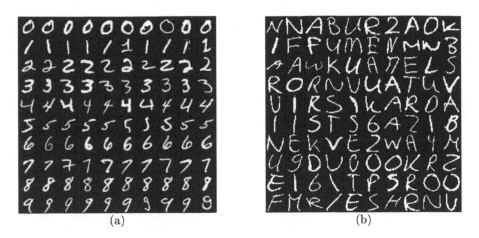

Fig. 1. Exemplary images from: (a) MNIST, and (b) HCR.

In the experiment we consider RBM with 500 hidden units and sRBM with 100 subspace units and 5 gate units to have comparable number of parameters. We consider learning rate in $\{0.1, 0.01\}$, weight decay in $\{0, 10^{-6}, 10^{-5}, 10^{-4}\}$, momentum equal 0.9 and optimization steps (or Gibbs sampler steps) in $\{1, 5, 10\}$. The hyperparameters are chosen using the validation set. Additionally, we apply early stopping with 30 look ahead steps. We train a model on first domain and once the training process converges, we switch the domain and proceed learning.

Evaluation Methodology. Our basic evaluation criterion is the negative *reconstruction error* (*cross-entropy*):

$$\varepsilon(\mathcal{D}) = \sum_{n=1}^{N} \sum_{d=1}^{D} \left(x_{n,d} \log \tilde{x}_{n,d} + (1 - x_{n,d}) \log(1 - \tilde{x}_{n,d}) \right), \tag{3}$$

where $\mathcal{D}$ is a dataset, $\mathbf{x}_n$ is an original image and $\tilde{\mathbf{x}}_n$ is its reconstruction. We measure the reconstruction error at different timestamps during learning: (i) after learning the first domain (*SWAP*), and (ii) at the end of training on the second domain (*END*). In order to measure the quality of a model on the domain adaptation task we define a domain adaptation accuracy as follows. The reconstruction error of a model that was first trained on $\mathcal{D}_i$ and further learned on $\mathcal{D}_j$ and eventually evaluated on $\mathcal{D}_i$ is denoted by $\varepsilon_{i \to j}(\mathcal{D}_i)$. In order to be invariant to specific domain error, we introduce a *baseline* model, multiple Bernoulli distribution with pixel probabilities calculated on the training data. The reconstruction error on dataset $\mathcal{D}_i$ for baseline is denoted by $\varepsilon_b(\mathcal{D}_i)$. The domain adaptation accuracy is then calculated as follows:

$$\alpha(\mathcal{D}_i \to \mathcal{D}_j) = \frac{1}{2} \left(\frac{\varepsilon_b(\mathcal{D}_i) - \varepsilon_{i \to j}(\mathcal{D}_i)}{\varepsilon_b(\mathcal{D}_i)} + \frac{\varepsilon_b(\mathcal{D}_j) - \varepsilon_{i \to j}(\mathcal{D}_j)}{\varepsilon_b(\mathcal{D}_j)} \right). \tag{4}$$

Table 1. Reconstruction errors and domain adaptation accuracy.

MNIST → HCR	SWAP		END		Domain adaptation accuracy
	HCR	MNIST	HCR	MNIST	
RBM-CD	159.1 ± 0.9	61.4 ± 0.2	32.9 ± 0.4	73.4 ± 0.4	76.6 % ± 0.1
RBM-PD	163.4 ± 2.0	41.8 ± 0.1	40.0 ± 0.1	45.6 ± 0.6	82.0 % ± 0.1
sRBM-CD	197.7 ± 8.9	42.2 ± 0.4	50.5 ± 0.6	55.3 ± 1.1	77.9 % ± 0.3
sRBM-PD	202.5 ± 4.0	47.3 ± 0.1	49.2 ± 0.8	55.3 ± 0.7	78.1 % ± 0.1
HCR → MNIST	SWAP		END		Domain adaptation accuracy
	HCR	MNIST	HCR	MNIST	
RBM-CD	29.6 ± 0.4	71.2 ± 0.2	109.6 ± 6.3	62.4 ± 0.1	65.9 % ± 0.1
RBM-PD	36.6 ± 0.4	42.6 ± 1.0	110.6 ± 2.4	44.3 ± 0.3	70.1 % ± 0.1
sRBM-CD	43.6 ± 0.2	52.0 ± 0.7	152 ± 10.2	43.1 ± 0.7	63.1 % ± 1.7
sRBM-PD	48.2 ± 0.3	55.3 ± 0.7	130.2 ± 9.9	47.3 ± 0.7	66.0 % ± 0.1

Results. The reconstruction errors and the domain adaptation accuracies are presented in Table 1.[3] The baseline achieves the following reconstruction errors: (i) HCR: 288.2, (ii) MNIST: 206.8. We present sampling capabilities of the considered methods in Fig. 2.

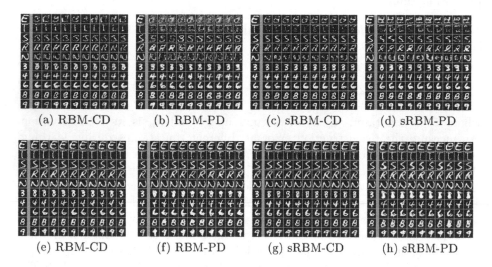

(a) RBM-CD	(b) RBM-PD	(c) sRBM-CD	(d) sRBM-PD

(e) RBM-CD	(f) RBM-PD	(g) sRBM-CD	(h) sRBM-PD

Fig. 2. Sampling using RBM and sRBM: (*top row*) HCR→MNIST, (*bottom row*) MNIST→HCR. In each figure the most left column contains real images.

---

[3] RBM and sRBM trained with contrastive divergence (CD) and PD are denoted by RBM-CD and sRBM-CD, and RBM-PD and sRBM-PD, respectively.

6   Discussion and Conclusion

In general, we notice that sRBM performs similarly to RBM, however, RBM has tendency to overfit to the first domain while the sRBM has problems with learning HCR (see Table 1). However, application of the PD training seems to help RBM to adapt to new domain by obtaining the best domain accuracy. This effect is less evident for sRBM. We hypothesize that sampling using PD approach allows to obtain more diverse examples during training and that is why the model does not drastically overfit to the first domain. We believe that similar result could be obtained for sRBM, however, in order to compare RBM with sRBM we chose the number of hidden units to be the same in both models but 5 gate and 100 subspace units could be too small amount to properly reconstruct images. Next, a closer inspection of the sampling capabilities of the considered approaches reveals that sampling with PD indeed results in more diverse sample. It seems that application of CD leads to copying training images while PD gives a model that generates more various images, *e.g.*, see letter S and digit 6 in Fig. 2e and f. Interestingly, sRBM trained with CD and PD allows to obtain more diverse examples than the ones from RBM-CD. Concluding, we notice that the learning procedure clearly matters in the domain adaptation using RBM, nevertheless, the obtained reconstructions are still imperfect. An open question is whether application of deeper BM is enough to handle the domain adaptation problem or other optimization techniques are needed. We leave investigating this issue for further research.

Acknowledgments. The work presented in the paper is partially co-financed by the Ministry of Science and Higher Education in Poland.

References

1. Ben-David, S., Blitzer, J., Crammer, K., Kulesza, A., Pereira, F., Vaughan, J.W.: A theory of learning from different domains. Mach. Learn. **79**(1–2), 151–175 (2010)
2. Bengio, Y.: Learning deep architectures for AI. Found. Trends Mach. Learn. **2**(1), 1–127 (2009)
3. Chelba, C., Acero, A.: Adaptation of maximum entropy capitalizer: little data can help a lot. Comput. Speech Lang. **20**(4), 382–399 (2006)
4. Daumé III., H., Frustratingly easy domain adaptation. arXiv preprint arXiv: 0907.1815 (2009)
5. Daume III, H., Marcu, D.: Domain adaptation for statistical classifiers. J. Artif. Intell. Res. **26**, 101–126 (2006)
6. Ganin, Y., Lempitsky, V.: Unsupervised domain adaptation by backpropagation. In: Proceedings of The 32nd International Conference on Machine Learning, pp. 1180–1189 (2015)
7. Glorot, X., Bordes, A., Bengio, Y.: Domain adaptation for large-scale sentiment classification: a deep learning approach. In: Proceedings of the 28th International Conference on Machine Learning (ICML 2011), pp. 513–520 (2011
8. Hazan, T., Jaakkola, T.: On the partition function and random maximum a-posteriori perturbations. In: ICML (2012)

9. Hazan, T., Maji, S., Jaakkola, T.: On sampling from the gibbs distribution with random maximum a-posteriori perturbations. In: NIPS, pp. 1268–1276 (2013)
10. Hinton, G.: Training products of experts by minimizing contrastive divergence. Neural Comput. **14**(8), 1771–1800 (2002)
11. Hinton, G.E., Sejnowski, T.J.: Learning and relearning in Boltzmann machines. Parallel Distrib. Process. Explor. Microstruct. Cogn. **1**, 282–317 (1986)
12. Jiang, J.: A literature survey on domain adaptation of statistical classifiers. Technical Report http://sifaka.cs.uiuc.edu/jiang4/domain_adaptation/survey/da_survey.pdf
13. LeCun, Y., Bottou, L., Bengio, Y., Haffner, P.: Gradient-based learning applied to document recognition. Proc. IEEE **86**(11), 2278–2324 (1998)
14. Li, X., Bilmes, J.: A bayesian divergence prior for classiffier adaptation. In: International Conference on Artificial Intelligence and Statistics, pp. 275–282 (2007)
15. Liu, A., Ziebart, B.: Robust classification under sample selection bias. In: Advances in Neural Information Processing Systems, pp. 37–45 (2014)
16. Long, M., Wang, J., Ding, G., Pan, S.J., Yu, P.S.: Adaptation regularization: a general framework for transfer learning. IEEE Trans. Knowl. Data Eng. **26**(5), 1076–1089 (2014)
17. Orabona, F., Hazan, T., Sarwate, A., Jaakkola, T.: On measure concentration of random maximum a-posteriori perturbations. In: ICML, pp. 432–440 (2014)
18. Papandreou, G., Yuille, A., Perturb-and-map random fields: using discrete optimization to learn and sample from energy models. In: ICCV, pp. 193–200 (2011)
19. Ravanbakhsh, S., Greiner, R., Frey, B.: Machine, Training Restricted Boltzmann by Perturbation. arXiv preprint arXiv: 1405.1436 (2014)
20. Sejnowski, T.: Higher-order Boltzmann machines. In: AIP Conference Proceedings, vol. 151, pp. 398–403 (1986)
21. Smolensky, P.: Information processing in dynamical systems: foundations of harmony theory. In: Parallel Distributed Processing: Explorations in the Microstructure of Cognition, vol. 1, pp. 194–281. MIT Press (1986)
22. Tomczak, J.M.: On some properties of the low-dimensional Gumbel perturbations in the Perturb-and-MAP model. Stat. Probab. Lett. **115**, 8–15 (2016). http://dx.doi.org/10.1016/j.spl.2016.03.019
23. Tomczak, J.M., Gonczarek, A.: Learning Invariant Features Using Subspace Restricted Boltzmann Machine. Neural Process. Lett. 1–10 (2016). doi:10.1007/s11063-016-9519-9
24. Tommasi, T., Orabona, F., Caputo, B.: Learning categories from few examples with multi model knowledge transfer. IEEE Trans. Pattern Anal. Mach. Intell. **36**(5), 928–941 (2014)
25. Van der Maaten, L.: A new benchmark dataset for handwritten character recognition, pp. 2–5. Tilburg University (2009)
26. Yosinski, J., Clune, J., Bengio, Y., Lipson, H.: How transferable are features in deep neural networks? In: Advances in Neural Information Processing Systems, pp. 3320–3328 (2014)

Relation Recognition Problems and Algebraic Approach to Their Solution

Juliusz L. Kulikowski[✉]

Nalecz Institute of Biocybernetics and Biomedical Engineering PAS.,
4 Ks. Trojdena Street, 02-109 Warsaw, Poland
juliusz.kulikowski@ibib.waw.pl

Abstract. Relation recognition as an extension of the well known pattern recognition problem is presented in the paper. Four types of such problems: simple, extended, matching and constructive relation recognition problems are considered. It is shown that such problems may arise in various application areas. There are presented possible approaches to the solution of the problems under consideration. It is shown that the extended algebra of relations is suitable as an universal tool to the relation recognition problems exact formulation and to description of the methods of their solution. Suitability of a concept of general (multi-aspect) similarity measure to the solution of constructive relation recognition problems based on a concept of covering by similarity spheres is also shown. Suggestions concerning desired future works in the domain of relation recognition theory and applications are given.

Keywords: Relation recognition · Pattern recognition · Extended algebra of relations · Learning systems · Artificial intelligence

1 Introduction

The notion of *relation* belongs to basic concepts of modern mathematics. It formally is defined as any subset of Cartesian product of a linearly ordered family of sets. It also plays a significant role in computer science as a formal model of data structures (Codd 1970), description of composite real objects or situations (Bagui 2005), description of program structures (Wirth 2002) etc. The well known *pattern recognition* (*PR*) problems are based on the concept of *similarity* being in fact a bi-variable reflexive, symmetrical and transitive relation described on a set of some (abstract or real) objects. An important property of similarity relation consists in partition of the given set of objects into mutually disjoint subsets called *similarity classes* (Kulikowski 2003). It was remarked in Kulikowski (1987) that *PR* can be interpreted as *checking* the fact that a given object x satisfies the relation of similarity to other objects belonging to a similarity class of objects. The pairs (in general – n-tuples) of elements satisfying a given relation Ξ are called *syndromes* of Ξ. The *PR* problem can thus be also interpreted as proving whether a given object x forms syndromes of similarity with the elements of some similarity classes established by the relation.

This brings to mind a more general concept of *relation recognition* (*RR*) consisting in identification of syndromes of a relation (Kulikowski 2002). In this case no

© Springer International Publishing AG 2017
J. Świątek and J.M. Tomczak (eds.), *Advances in Systems Science*, Advances in Intelligent Systems and Computing 539, DOI 10.1007/978-3-319-48944-5_3

homogeneity of the sets constituting the space U of n-tuples is assumed. This means that in general objects of different formal nature: arithmetical, algebraic, geometrical, Boolean, topological or symbolic denotations of qualitative features of real objects etc. may be linked together by the relations. Under such assumption several variants of *RR* problem can be formulated:

1st, it is given a relation Ξ described on a Cartesian product U of n non-empty sets and a n-tuple of objects x belonging to U; check whether Ξ is satisfied by x. This can be called a *simple RR* (*sRR*) problem.

2nd, it is given a relation Ξ and a finite subset $X \subset U$ of n-tuples; find in X all n-tuples being syndromes of Ξ. This can be called an *extended RR* (*eRR*) problem.

3rd, it is given a relation Ξ described on a Cartesian product U of n non-empty sets and its incomplete syndromes x, whose some components are unknown; restore their possible complete forms matching the relation. This can be called a *matching RR* (*mRR*) problem.

4th, it is given a Cartesian product U of n non-empty sets and a finite subset $X \subset U$ of n-tuples; find formal rules describing a relation Ξ in U containing X as a subset of its syndromes. This can be called a *constructive RR* (*cRR*) problem.

Some *RR* problems are closely connected with machine learning problems. This, for example, takes place in the *eRR* problem consisting in detection of tracks of particles in a series of snapshots (Sect. 3, Example 2) or in *mRR* problem consisting in meteorological prognosis based on a series of past observations. However, the paper does not present any extended example of *RR* application; it is rather aimed at showing the *RR* as a large and interesting area of investigations.

This paper is aimed at presentation of an approach to the solution of the above-mentioned classes of *RR* problems. However, the admitted heterogeneity of the space V of n-tuples limits the ability of a geometrical- or vector-space models to be used as a basis of the solution. That is why in this work an approach based on the extended algebra of relations (Kulikowski 1992) and on a general concept of similarity measure (Kulikowski 2001) is proposed.

The paper is organized as follows. Basic notions, concerning mainly the used in this work extended algebra of relations, are presented in Sect. 2. Section 3 presents solution of the *sRR* and *eRR* problems. An approach to the solution of *mRR* problems is presented in Sect. 4, while Sect. 5 presents a proposed solution of the *cRR* problem based on a concept of covering by similarity spheres. Concluding remarks and suggestions concerning future works are presented in Sect. 6.

2   Basic Notions

It will be taken into consideration a finite, linearly ordered family F of non-empty sets $[\Omega^{(i)}]$, $i = 1, 2, \ldots, n$. As mentioned above, no restrictions on formal nature of the elements of the sets $\Omega^{(i)}$ will be imposed. The Cartesian product of the sets $\Omega^{(i)}$ of F will be denoted by U and will be called an *universe*. In order not to suggest any connection with particular mathematical objects, the elements of the sets $\Omega^{(i)}$ are called *symptoms*, while the elements of U (strings of symptoms, n-tuples) are called *syndromes*.

Any subset $\varXi \subseteq U$ is called a *relation* described on U. The relation is called *trivial* if $\varXi \equiv U$ and *empty* if $\varXi \equiv \varnothing$ (empty set). For a given universe U the family of all possible relations described on U is denoted by $\varPhi$. In this family the well-known algebraic operations of sets sum ($\cup$), intersection ($\cap$) and asymmetrical (\\) and symmetrical ($\div$) difference can directly be applied to the relations. In particular, a *negation* of relation $\neg \varXi$ is defined as asymmetrical difference $U\backslash\varXi$. It directly follows that for any $\varXi$ it is $\varXi \cup (\neg\varXi) = U$ and $\varXi \cap (\neg\varXi) = \varnothing$.

If $\varXi$ is a relation described on U then any subset $S \subset \varXi$ is called a *sub-relation* of $\varXi$. On the other hand, if F' is a subfamily of sets, $F' \subset F$, (preserving their order in F), U' is a Cartesian product of the sets of F' (a *sub-universe* of U) and $\varXi$ is a relation described on U then a set $\varXi'$ consisting of all intersections of the syndromes of $\varXi$ and of U' will be called a *projection* of $\varXi$ onto U' and will be also denoted by $\varXi|_{U'}$.

For a given family F of consisting of n sets it is possible to extract from it 2^n-1 non-empty sub-families of sets (including F itself). On each such sub-family of selected sets a family of all described on them relations, called *partial relations* described in U, can be considered. Consequently, it is possible to take into consideration a family $\varPhi$ of all partial relations described in U (including also the relations described on U). An extension of the algebra of relations described on a fixed family of sets F on partial relations described on different subfamilies of F was given in Kulikowski (1992). If F' and F'' are any two subfamilies of F and $\varXi'$, $\varXi''$ are some relations described, respectively, on the Cartesian products U' of the sets of F' and U'' described on F'' then there can be established the following:

Extended algebraic operations:

(a) The sum $\varXi' \underline{\cup} \varXi''$ is a relation described in $U' \cup U''$, consisting of all syndromes whose projection on U' satisfies $\varXi'$ *or* the projection on U'' satisfies $\varXi''$.

(b) The intersection $\varXi' \underline{\cap} \varXi''$ is a relation described in $U' \cup U''$, consisting of all syndromes whose projection on U' satisfies $\varXi'$ *and* the projection on U'' satisfies $\varXi''$.

(c) The asymmetrical difference $\varXi' \underline{\backslash} \varXi''$ is a relation described in $U' \cup U''$, consisting of all syndromes whose projection on U' satisfies $\varXi'$ *and* the projection on U'' *does not* satisfy $\varXi''$.

Let us remark that $U' \cup U''$ is an universe constructed as a Cartesian product of the family of sets $F' \cup F''$ (not as $U' \times U''$). Then, on the basis of the above-given operations the following notions can also be defined:

(d) The difference $2^{(U' \cup U'')} \underline{\backslash} \varXi$ is called an *extended negation* of $\varXi$ and will be denoted by $\underline{\neg}\varXi$.

(e) The asymmetrical difference $\varXi' \underline{\div} \varXi''$ is a relation described in $U' \cup U''$, consisting of all syndromes satisfying $\varXi' \underline{\cup} \varXi''$ *and not* satisfying $\varXi' \underline{\cap} \varXi''$.

(f) For a relation $\varXi'$ described in a sub-universe U' and another relation $\varXi''$ described in a sub-universe U'' the projection $(\varXi' \underline{\cap} \varXi'')|_{U'}$ will be called a *relative relation* $\varXi'$ *assuming that* $\varXi''$ *holds*. It will be shortly denoted by $\varXi'/\varXi''$.

It can be shown that the quintuple $[U, \underline{\cup}, \underline{\cap}, \underline{\backslash}, \varnothing]$ constitutes a Boolean algebra (Rudeanu 2012) with U as its "unity" and $\varnothing$ as its "null relation" (no syndromes of any length); this will be called an *extended algebra of relations*. It thus introduces algebraic operations not only on the relations defined on the universe U but also on any their sub-relations and/or projections defined in U.

On the basis of the property (f) it can be established the following

Extension property:

For any relation Ξ' and Ξ'' described, respectively, in the universes U' and U'', the following property holds:

$$\Xi' \cap \Xi'' \equiv \Xi'/\Xi'' \cap \Xi''. \tag{1}$$

This property, seeming to be a trivial one, will be shown to be useful in solving some *RR* problems.

3 Solution of *Simple* and *Extended RR* Problems

Four basic methods can be used to description of relations:

(1) Analytical (functional) description,
(2) Logical description,
(3) Presentation by algebraic composition of other known (simpler) relations,
(4) Characterization by list of syndromes.

Analytical description takes place when syndromes of a relation should by definition satisfy some algebraic or analytical (functional, differential, integral etc.) equations. Solution of the equation provides all syndromes of the relation. On the other hand, an assumed solution can be proven as syndrome of the relation by substitution to the equation and checking whether it is satisfied.

Logical description of a relation may be given in the form of a formula:

$$\textit{if } T(x)\textit{then } x \in \Xi \tag{2}$$

where $T(x)$, is a logical predicate described in the universe U. The predicate can thus be used as a basis of construction of logical tests of the syndromes' validity. However, it does not deny existence of some other syndromes of Ξ, not satisfying $T(x)$.

Algebraic compositions of relations can be constructed on the basis of the above (in Sect. 2) described algebraic operations and concepts.

Characterization of a relation by a complete list of its syndromes is possible only in particular cases. In a more general case the list may be incomplete and it can be used as a rough representation of the relation.

Taking this into account we can more exactly formulate:

The *sRR* problem:

It is given an universe U and a relation Ξ described in it by some of the above-mentioned methods (1)–(2). It is also given an element $x \in U$. Check whether $x \in \Xi$.

Computer-aided solution of this problem depends on the method the relation has been described. In the case of analytical description it needs substitution of x into the corresponding equation and proving whether the equation is satisfied. This seems to be a rather trivial task. A less trivial task arises if the relation is described in a logical form (2). The predicate $T(x)$ may not directly characterize the relation, as it illustrates the following example.

Example 1. It is given an Euclidean plane E^2 with a system of Cartesian coordinates (u, v). It is taken into consideration an universe $U = E^2 \times E^2 \times E^2$ of triplets of points laying on E^2. It is also given a relation $\varXi \subset U$ described by the predicate:

"The triplet of points $[P_i, P_j, P_k]$ constitute vertices of an equilateral triangle",

Let it be given a triplet of points $x = [(u_i, v_i), (u_j, v_j), (u_k, v_k)]$. Check whether x is a syndrome of $\varXi$.

In this case a direct assessment of logical value of the statement:

"$[(u_i, v_i), (u_j, v_j), (u_k, v_k)]$ constitute vertices of an equilateral triangle"

is impossible. In this case it is necessary to use the definition of an equilateral triangle and to replace the statement by the following, semantically equivalent:

$$[P_i, P_j, P_k] \text{ constitute vertices of an equilateral triangle}$$
$$\textbf{\textit{if and only if}} \left[d(P_i, P_j) = d(P_j, P_k)\right] \wedge \left[d(P_j, P_k) = d(P_k, P_i)\right]$$

where $d(P_i, P_j)$ (similarly, $d(P_j, P_k)$ and $d(P_k, P_i)$) denotes an Euclidean distance between the points P_i, P_j. Now, the relation description leads to the following logical test of a triplet of points satisfying the relation:

$$\textbf{\textit{if}} \left[d(P_i, P_j) = d(P_j, P_k)\right] \wedge \left[d(P_j, P_k) = d(P_k, P_i)\right]$$
$$\textbf{\textit{then}} \left[P_i, P_j, P_k\right] \text{ satisfies the relation } \varXi'',$$
$$\textbf{\textit{otherwise}} \text{ it does not satisfy it.}$$

In fact, if the relations $\varXi$ are interpreted as formal characteristics of some classes of objects then the corresponding *sRR* problems become widely interpreted *PR* problems. However, the Example 1 shows that solution of some, apparently simple, *sRR* problems may need some wider knowledge about the application area in order to reformulate the primary logical description of the relation.

In some application problems (say, in management, medical treatment etc.) the *sRR* may take a more general form:

1st 'It is given a relation $\varXi$ described by an algebraic combination F of m partial relations $\varXi^{(1)}, ..., \varXi^{(m)}$ described on a Cartesian product U of n non-empty sets and a n-tuple of objects x belonging to U. Check whether $\varXi$ is satisfied by x.

Let us remind (see Sect. 2) that the partial relations can be defined on different subsets of variables. Their algebraic combination may be given in a canonical form, as an (extended) sum of products of selected partial relations or of their negations (denoted both by the brackets $\langle \rangle$):

$$\varXi = F(\varXi^{(1)}, \ldots, \varXi^{(m)}) = \underline{\cup \cap} \langle \varXi^{(v)} \rangle \tag{3}$$

This form of the relation $\varXi$ directly leads to a decomposition of the *sRR* problem into a finite set of simpler *sRR* sub-problems and then proving the validity of the corresponding composite logical predicate.

The calculation complexity of solution of a *sRR* problem depends linearly on the number of additive terms in expression (3). Moreover, each product-term $\cap \langle \varXi^{(v)} \rangle$ in this expression needs proving the validity of a subset of partial relations described on various subsets of x. This leads to an exponential complexity of calculation of the product-terms.

An *eRR* problem can be considered as simple extension of a corresponding *sRR* problem on several elements of the universe U; in such case its solution may consist of multiple solution of the *sRR* problem for assumed syndromes $x_1, x_2, \ldots$ etc. However, it also can exactly be formulated as follows:

The *eRR* problem:

It is given an universe U and two described on it relations: $\varXi$ described by the above-mentioned methods (1)–(2) and a finite relation χ described by method (4). Find $\varPsi = \varXi \cap \chi$

Really, $\varPsi$ is a finite set of n-tuples in U, constituting by definition a relation. We are looking in χ for all syndromes satisfying also the given relation $\varXi$. The following example should illustrate this idea.

Example 2. It is given a linearly ordered set of N photo snapshots. In any snapshot a finite set S_v, $v = 1, 2, \ldots, N$, of detected in it tracks $\omega_{v,p}$ of moving objects is given. Each track $\omega_{v,p}$ is presented by a sequence of m numerical and/or qualitative features describing its spatial coordinates and other (if any) its detected individual characteristics. Find the sequences $x = [\omega_{1,p}, \omega_{2,q}, \ldots, \omega_{N,s}]$, where $p, q, \ldots, s = 1, 2, \ldots, K$, denoting numbers assigned to the detected tracks of objects in the snapshots (K being a total number of objects under observation). The tracks illustrate position of selected real objects in consecutive phases of their motion. The example presents a simplified case of a more general problem of objects' action recognition and trajectories description (Wang et al. 2013). This may correspond in practice to monitoring (or back-monitoring) of selected individual vehicles in road traffic.

The universe has thus the form of a Cartesian product

$$U = S_1 \times S_2 \times \ldots \times S_N \tag{4}$$

consisting of K^N N-tuples and the relation $\varXi$ being to be recognized is described by the statement:

"$x \in \varXi$ *iff* x consists of elements representing the same real object"

Solution of the problem consists thus in finding in U all N-tuples ω satisfying a relation $\varXi$ of "representing the same real object". This needs, first of all, establishment the criteria of assigning tracks to fixed, individual objects Two such criteria are possible: 1st, the similarity of tracks if they present the appearance of real objects, 2nd, proximity of tracks in consecutive snapshots.

Two possible approaches to solution of the given eRR problem then can be taken into consideration:

(a) Solving the eRR problem by its consideration as multiple sRR problems. This means that the elements of U are separately taken into consideration and the property "representing the same real object" of consecutive pairs of tracks is proven. Any sequence $x = [\omega_{1,p}, \omega_{2,q}, \ldots, \omega_{N,s}]$ is thus subjected to a series of tests consisting in proving logical value of the statements:

"The pair $[\omega_{i,p}, \omega_{i+1,q}]$ represents the same real object"

for $i = 1, 2, \ldots, N–1$. First "$false$" statement in the series causes rejection of the given N-tuple x as not satisfying the relation $\varXi$. This, apparently simple approach does not guarantee uniqueness of the eRR problem's solution. The partial decisions being based on separately taken pairs of tracks neglecting wider context may lead to ambiguous assigning tracks to objects if the tracks are similar or laying close each to the other.

(b) Alternative approach to the eRR problem is based on the concept of conditional relation. For this purpose the following relations will be defined:

$\varXi'_i$: "The tracks $\omega_{1,p}, \ldots, \omega_{i,p}$ represent the same real object"

$\varXi''_i$: "The tracks $\omega_{i,p}, \ldots, \omega_{i+1,p}$ represent the same real object assuming that $\omega_{1,p}, \ldots, \omega_{i,p}$ represent the same real object"

for $2 \leq i \leq N–1$ and $1 \leq p \leq K$. Finally, $\varXi'_N$ contains K syndromes $\omega_1, \ldots, \omega_K$, each containing N tracks assigned to a given object.

The problem is then solved according to the following pseudo- procedure:

> **Input**: N lists of K tracks $\omega_{v,p}$, $1 \leq v \leq N$, $1 \leq p \leq K$;
> **Output**: $\varXi'_N = (\omega_1, \ldots, \omega_K)$;
>> **for** $i = 2$ **to** $N–1$ **do**
>>> 1. Find all syndromes of $\varXi'_i$;
>>> 2. Find all syndromes of $\varXi''_{i+1}$;
>>> 3. Create the syndromes of $\varXi'_{i+1} = \varXi'_i \cap \varXi''_{i+1}$;
>> **end for**

The procedure starts by finding K syndromes of $\varXi'_2$ consisting of ordered pairs $[\omega_{1p}, \omega_{2,q}]$ and K syndromes of $\varXi'_3$ consisting of ordered pairs $[\omega_{2q}, \omega_{3,r}]$ assigned to the same object. Next, the syndromes of $\varXi'$ are extended by adding tracks assigned to formerly fixed objects.

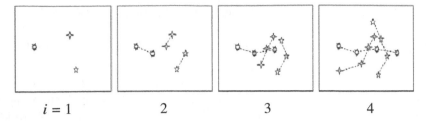

Fig. 1. Construction of moving particles' trajectories as a solution of the *eRR* problem for the relation Ξ = "assigned to the same real object".

At each step the similarity of tracks in the pairs $[\omega_{i,p}, \omega_{i+1,q}]$ is checked, however, it is established by selection of K most similar from $\frac{1}{2}K^2$ possible pairs of tracks in two consecutive snapshots. For this reason, this approach guarantees not worse solution than this provided by the approach (a). The way of step-wise construction of the trajectories of some moving particles is illustrated in Fig. 1.

4   Solution of *Matching RR* Problems

A *mRR* problem can be simply and exactly formulated in the extended relation algebra terms. Let us assume that it is given a family F of sets composed of two disjoint sub-families:

$$F = F' \cap F'', F' \cap F'' = \emptyset \tag{5}$$

and the described on them universe U and sub-universes U' and U''.

The *mRR* problem:

Let it be given a relation Ξ described on the universe U and a sub- relation $\chi \subset \Xi|_{U'}$ described by a list of syndromes in a sub-universe U'. Find the sub-relation $\Psi = \Xi \cap \chi$.

The sub-relation Ψ contains all syndromes satisfying both, Ξ and χ. Its projection $\Psi|_{U''}$ consists of the lacking elements in the syndromes of χ that transform them into syndromes of Ξ. There are lot of real situations leading to *mRR* problems.

Example 3. In a meteorological database data concerning observations of minimal and maximal air temperature, speed and direction of wind, type and intensity of falls, type and intensity of clouds, etc. in a certain geographical region are collected. The data have been daily recorded for several years. From this long time-interval shorter (say, $N = 10$ days long) sub-intervals have been cut out and the sequences of meteorological data in the sub-intervals as syndromes of a relation H describing admissible sequences of weather conditions have been extracted. Moreover, the relation H is presented as a sum of sub-relations:

$$H = H_w \cup H_{sp} \cup H_{su} \cup H_{au} \tag{6}$$

admitting various types of data to be recorded, correspondingly, in the winter, spring, summer and autumn seasons.

It is also given a sequence x of meteorological data recorded for the last (say, n') several days in a current season of the year. It is desired to forecast possible sequence of weather conditions in the next n'' days.

The task can be solved in three basic steps:

1. Construction, on the basis of the relation H, of a relation $\varXi$ describing admissible sequences of weather conditions for $n' + n''$ days in the given season of the year;
2. Finding a sub-relation $\chi \subset \varXi \mid_{U'}$ where U' denotes a sub-universe of formally possible sequences of weather conditions for n' days in the given season of the year; χ being reduced to the given single sequence x.
3. Finding the sub-relation $\varPsi = \varXi \cap \chi$.

The way of construction of the relation $\varXi$ (step 1) depends on the length $n' + n''$ with respect to N:

a. If $n' + n'' < N$ then $\varXi$ is given as a projection $H\mid_{U' \cup U''}$;
b. If $n' + n'' = N$ then $\varXi = H$;
c. If $n' + n'' > N$ then $\varXi$ should be given by an intersection of s, $s > N - (n' + n'')$, seasonal sub-relations of $\varXi$:

$$\varXi = \varXi^{(1)} \cap \varXi^{(2)} \cap \ldots \cap \varXi^{(s)} \tag{7}$$

corresponding to shifted overlapping time-intervals covering $n' + n''$ days. The solution of the problem is not unique; it provides a set of possible extensions of the given sequence x of weather conditions satisfying the relation $\varXi$. Let us also remark that the above-presented prognostic model is strongly deterministic. In fact, description of meteorological processes needs more sophisticated, randomized models which are not a subject of this paper.

5 Solution of *Constructive RR* Problems

Construction of a relation satisfied by a given set of syndromes belongs to a large class of knowledge discovery problems (Maimon and Rokach 2005). However, the problem needs to be more exactly formulated. If U is a Cartesian product of n non-empty sets and X, $X \subset U$, denotes a finite subset of n-tuples x in U then X, by definition, constitutes a "minimal" solution of the problem while U constitutes its "maximal" solution. From a practical point of view, both solutions are useless. On the other hand, the problem of reconstruction of an assumed "hidden" relation governing some social, economical, political etc. events on the basis of recorded symptoms preceding some other, similar-type events in the past is of high importance. It is thus necessary to put some constraints on the *cRR* problem in order to make it non-trivial and more useful.

One of possible ways to do it consists in introducing into the universe U a *similarity measure*. Lot of various similarity measures have been proposed in the literature (see, e.g. Sobecki (2009)). However, not all of them satisfy the requirement of ability to be used to multi-aspect assessment of similarity of composed objects. For this purpose it is here proposed using a similarity measure defined generally as a function:

$$\sigma : U \times U \rightarrow [0, 1] \tag{8}$$

satisfying the conditions: for any x', x'', $x''' \in U$ the following properties hold:

   I. $\sigma(x', x') = 1$,
   II. $\sigma(x', x'') = \sigma(x'', x')$,
   III. $\sigma(x', x'') \cdot \sigma(x'', x''') \leq \sigma(x', x''')$.

A very important property of so-defined similarity measure consists in its extensibility: if σ_1, σ_2 are any two similarity measures satisfying the conditions I – III, then their product $\sigma = \sigma_1 \cdot \sigma_2$ also satisfies the conditions and as such it constitutes a similarity measure (Kulikowski 2001). Consequently, any power σ^μ for $\mu \geq 1$ satisfies the conditions I – III of similarity measure, and this makes us able to modify some similarity measures so as to make them less or more sensible to some parameters of the objects. Moreover, taking into account that the above-given properties may be satisfied by a large class of mathematical objects: natural or real numbers, vectors, Boolean variables, etc., a similarity measure can be constructed for syndromes composed of various features of objects.

For a given universe U, described in it similarity measure σ, a fixed number d, $0 < d \leq 1$, and any element $u \in U$ the following relation can be defined:

"$x \in S(u,d)$ *if and only if* $\sigma(u, x) \geq d$"

The relation $S(u,d)$ can be called a *similarity sphere of center* u *and range* d. This is an analogue of a sphere in a metric space (existence of any metric in the universe U has not been assumed). Remark that the lower is d, the larger is the similarity sphere. The concept of similarity sphere makes us able to propose a solution of the following problem:

6 The *CRR* Problem

Let it be given an universe U, a described in it similarity measure σ, a finite set G, $G \subset U$, of syndromes and a constant δ such that $\delta_{min} \leq \delta \leq 1$ where δ_{min} denotes minimal similarity between the syndromes in G. Find a minimal relation $\Xi\ (G, d)$ such that for any $x \in G$ and any other $y \in \Xi\ (G, d)$ the inequality $\sigma\ (x, y) \leq \delta$ is satisfied.

We call G a *germ* of the relation $\Xi\ (G, d)$. For solution of the *cRR* problem there will be taken into consideration all unordered pairs (x_i, x_j) of different syndromes in G and their similarity measures $\delta_{i,j} = \sigma\ (x_i, x_j)$ will be calculated. Then for any x_i its minimal similarity to other syndromes in X:

$$d_i = min_{(j)}(\delta_{i,j}) \tag{9}$$

will be found. Let us take into consideration a relation:

$$\varXi = \cap_i S(x, di) \tag{10}$$

It can be shown that so-defined relation $\varXi$ is solution of the cRR problem for $\delta = \delta_{min} = min_{(i)}(d_i)$. Really, any $x \in \varXi$ is a syndrome belonging to all similarity spheres of syndromes in X. The largest similarity spheres are $S(x_*, \delta_{min})$ where x_* is one of (at least two) the less similar to any other syndrome in X. Therefore, if x belongs to $S(x^*, \delta_{min})$ then the more its similarity to any other syndrome in X is not less than δ_{min}. Geometrically, this situation is illustrated in Fig. 2.

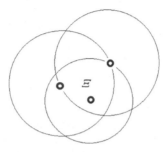

Fig. 2. Geometrical illustration of the solution of a cRR problem based on covering by similarity spheres.

The set X consists here of 3 syndromes denoted by **o**. The area of problem solution $\varXi$ lies in the intersection of three similarity circles. Its maximal range is thus limited by the intersection of the two largest similarity circles.

The spherical covering method of cRR problem solution is a cautious one. In case of a large number of syndromes in G and small δ it may happen that the intersection of similarity spheres is empty and the solution of the problem does not exist. This in particular may happen in learning cRR problems when the elements of the germ are step-by-step acquired. Then, instead of a single germ we have an increasing sequence of germs:

$$G^{(1)} \subset G^{(2)} \subset \ldots G^{(t)} \subset \ldots \tag{11}$$

Let $G^{(1)}$ be a maximal germ that provides a non-empty solution $\varXi^{(1)}$ of the cRR problem. Let us assume that a next germ, $G^{(2)}$ is acquired. It can be presented as a sum:

$$G^{(2)} = H^{(2,1)} \cup \Delta^{(2)} \tag{12}$$

where $H^{(2,1)}$ denotes a subset of syndromes such that a modified germ:

$$G^{(1,2)} = G^{(1)} \cup H^{(2,1)} \tag{13}$$

provides a still non-empty solution $\varXi^{(1,2)}$ of the cRR problem, while $\Delta^{(2)}$ contains the remaining syndromes of $G^{(2)}$. Then $\Delta^{(2)}$ can be used as a germ to construction of the next non–empty sub-relation $\varXi^{(2)}$ (or a set of non-empty sub-relations) which can be added to $\varXi^{(1,2)}$ giving the second approximation of the solution:

$$\varPsi^{(2)} = \varXi^{(1,2)} \cup \varXi^{(2)} \qquad (14)$$

Similar procedure can be continued in order to get next approximations of the solution of the learning cRR problem. The calculation complexity of the above-proposed method of the cRR problem solution is polynomial because each next syndrome acquired as a germ should be matched to the formerly acquired germs.

7 Conclusions

Relation recognition (RR) is a formal model of a large class of decision making problems. Its widely known example is any pattern recognition problem. However, the class of RR problems is much larger and it contains, in particular, the simple, extended, matching and constructive RR problems. Each type of the RR problems needs some specific methods of their solution. Both, the types of RR problems and the methods of their solution can be strongly described using the formalism of the extended algebra of relations. Also, the general concept of similarity measure are useful in some types of RR problems solution. Examples of such solution methods have been presented in the paper. However, they do not exhaust the RR area of investigation. In particular, the fuzzy RR problems (Rutkowski 2006), the problems following from using based on relations ontologies to describe real systems and processes (Abdoullayev 2008) and the problems of extension of the RR problems on the hyper-relation recognition problems (as it follows from the comparison of the Example 2 in this paper with a similar problem described in Kulikowski (2006) are worthy to be undertaken and deeply investigated.

Acknowledgement. This work was supported by the Nalecz Institute of Biocybernetics and Biomedical Engineering, Polish Academy of Sciences, in Warsaw.

References

Codd, C.J.: A relational model of data for large shared banks. Commun. ACM **13**, 377–387 (1970)
Bagui, S.: Extended entity relationship modeling. In: Rivero, L.C., et al. (eds.) Encyclopedia of Database Technologies and Applications, Idea Group Reference, Hershey, pp. 233–239 (2005)
Wirth, N.: Algorithms + Data Structures = Programs. Prentice Hall Inc., Englewood Cliffs (1976). (Polish translation WNT, Warsaw, 2002)
Kulikowski, J.L.: Rozpoznawanie obrazów. In: Nałęcz, M. (ed.) Biocybernetyka i Inżynieria Biomedyczna 2000, vol. 8, Obrazowanie Biomedyczne, AOW EXIT, Warsaw, Chap. 7, pp. 193–237 (2003). (in Polish)

Kulikowski, J.L.: Structural analysis of biomedical data. Biocybernetics Biomed. Eng. **7**(1–4), 143–154 (1987)

Kulikowski, J.L.: From pattern recognition to image interpretation. Biocybernetics Biomed. Eng. **22**(2–3), 177–197 (2002)

Kulikowski, J.L.: Relational approach to structural analysis of images. Mach. Graph. Vis. **1**(1/2), 299–309 (1992)

Rudeanu, S.: Sets and Ordered Structures, Sect. 4.2, pp. 121–138. Bentham Science Publishers, Oak Park (2012)

Wang, H., Kläser, A., Schmid, C., Liu, C.-L.: Dense trajectories and monitoring boundary descriptors for action recognition. Int. J. Comput. Vis. **103**, 594 (2013). SpringerLink

Maimon, O., Rokach, L.: The Data Mining and Knowledge Discovery Handbook. Springer, New York (2005)

Sobecki, J.: Rekomendacja Interfejsu Użytkownika w Adaptacyjnych Webowych Systemach Informacyjnych. Of. Wyd. Politechniki Wrocławskiej, Wrocław (2009). (in Polish)

Kulikowski, J.L.: Pattern recognition based on ambiguous indications of experts. In: Kurzyński, M. (ed.) KomPuterowe Systemy Rozpoznawania KOSYR 2001, pp. 15–22. Wyd. Politechniki Wrocławskiej, Wroclaw (2001)

Rutkowski, L.: Metody i Techniki Sztucznej Inteligencji, Sect. 4.7, pp. 90–94. Wyd. Naukowe PWN, Warszawa (2006). (in Polish)

Abdoullayev, A.: Reality, Universal Ontology, and Knowledge Systems, Toward the Intelligent World. IGI Publishing, Hershey (2008)

Kulikowski, J.L.: Description of irregular composite objects by hyper-relations. In: Proceedings of the International Conference on ICCVG 2004, Computer Vision and Graphics, Warsaw, Poland, September 2004, Springer CIaV vol. 32, Dordrecht, pp. 141–146 (2006)

Prediction of Power Load Demand Using Modified Dynamic Weighted Majority Method

Radoslav Nemec[✉], Viera Rozinajová, and Marek Lóderer

Faculty of Informatics and Information Technologies, Slovak University
of Technology, Ilkovičová 2, 841 04 Bratislava, Slovakia
{xnemecr, viera.rozinajova, marek_loderer}@stuba.sk

Abstract. The paper deals with the prediction of electricity demand, using data from smart meters obtained in defined time steps. We propose the modification of ensemble learning method called Dynamic Weighted Majority (DWM). The data are represented by data streams. According to our experiments, the proposed solution offers favorable alternative to current solutions. We also focus on the comparison of proposed ensemble method with single predictions used in the model.

Keywords: Prediction · Data streams · Ensemble learning · Dynamic weighted majority · Smart meter

1 Introduction

Predicting future values is a topic that fascinates mankind for decades. Man is a curious creature and his interest in knowing the future directly or indirectly affects his life. The known and popular applications of prediction are the weather forecasting prediction of growth or depreciation of finance. However, an increasingly popular area of prediction is power load demand, which is the main goal of this paper. Due to non-stationarity and special characteristic of dynamicity, the prediction of electricity loads is a difficult problem. Moreover, predicting electricity loads is very important for suppliers as well as for its customers. If it is precise, it could help companies to create major decisions about its purchase. Using the correct prediction could also give customers an opportunity to schedule their energy consumption, which may affect the saving of finance.

In this paper we will focus on different approaches of machine learning and data mining algorithms. The main part will describe the proposed approach, which is based on ensemble learning method called Dynamic Weighted Majority [12]. This method was modified to fit and also solve our prediction problem. Ensemble learning approaches are nowadays becoming very popular and currently they are subject to many research works in various cases of study. Their relevance and achieved results are one of the main motivating factors for their usage in our work.

This paper is organized as follows: Sect. 2 contains a summary of the related work. In Sect. 3 we describe our proposed approach (the Modified Dynamic Weighted Majority method and Dynamic Weighted Majority method with decomposition and

© Springer International Publishing AG 2017
J. Świątek and J.M. Tomczak (eds.), *Advances in Systems Science*, Advances in Intelligent Systems and Computing 539, DOI 10.1007/978-3-319-48944-5_4

parameter optimization). Description of used datasets, experimental evaluation and results are presented in Sect. 4 and the conclusion is in Sect. 5.

2 Related Work

The prediction is based on the values measured in the past, which are used to predict the future values. To achieve the precise prediction, it is very important to adjust the prediction to the problem, which is to be solved. The load prediction, also known as load forecasting, could be in general divided into three categories: short-term, medium-term and long-term forecasts [4, 20]. The type of prediction is one of the facts, which had to be considered before implementing the solution. The second important fact, which should be considered, is related to the type of learning we want to use. There are two known types of frequently used learning algorithms for the prediction. The first one is called offline learning. In this case, the learning methods are basically focused on datasets which are static. The arrival of any other data after training is not expected [14]. The electricity load data are formed as data streams and they arrive in specified time intervals meaning that the size of this dataset grows continuously. This fact led us to analyze and use the second type of learning method that is called online learning. Methods for this type of learning could provide i.e. lifelong learning. This means that the models can obtain new information that can consequently be used to adapt to new changes in data stream [14].

Over the last decades various online and offline models for improving electricity load prediction accuracy have been developed. For example regression models [25], methods designed for time series, ARMA and ARIMA models [16] or Holt-Winters exponential smoothing [21]. In recent decade, intelligent forecasting models like Support Vector Regression, Artificial Neural Networks or Expert Systems have been developed and used [8, 20].

Till now, there is no single model which is capable to provide best forecasting results for every kind of data. Big effort has been spend to overcome this issue.

One of the most popular methodologies used to solve this issue is called Ensemble Learning [13]. The main idea of Ensemble Learning is that proper combination of predictions of different base models can create more accurate result in comparison to the result provided by the best individual model.

Recently, an interesting example of Ensemble Learning was proposed. It used a cuckoo search algorithm to find optimal weights to combine four forecasting models based on different types of neural networks [24]. Each neural network forecasts electric load demand based on historical load data. The forecasting results of the combined models were significantly improved compared to the results of the individual models.

In work [18] a Pattern Forecasting Ensemble Model (PFEM) for electricity demand time series is proposed. It is based on the previous PSF algorithm, but uses a combination of five separate clustering models. Published results indicate that proposed approach gives more accurate results compared to all the other five individual models.

Another approach is proposed in work [7], where an ensemble of online regression and option trees is introduced. Mendes-Moreira et al. provide an exhausting research on topic of ensemble approaches for Ensemble [13].

Other successful implementation of Ensemble Learning approach is Dynamic Weighted Majority (DWM). It was introduced by Kolter and Maloof in 2003 [12]. The method was presented as a new ensemble method for tracking concept drifts in data streams. It was mainly designed to solve the classification problems. In our work, we adjust the DWM method to solve regression problems e.g. prediction of electricity loads. During the analysis of available solutions we also found a modification of DWM that is called Additive Expert Ensembles (AddExp) [11]. This modified method became also one of the methods which are included in our proposed solution.

3 Proposed Method

We propose the Modified Dynamic Weighted Majority method for time series prediction. This approach was chosen for its ability to adapt to changes in the distribution of a target variable in time series data. It has a potential to obtain more accurate prediction results than a single base prediction method.

3.1 Modified Dynamic Weighted Majority Method

Authors of the original DWM describe the main functionality of the method by four mechanisms: (1) models of the ensemble are trained based on their performance, (2) each model of ensemble is weighted, (3) models are removed from ensemble based on their performance, and (4) new models are added to the ensemble based on global performance i.e. the performance of the whole ensemble. As mentioned in previous chapter, DWM method was until now mainly used for solving classification problems and we could not find any described modification of this method for the prediction of electricity loads. So we decided to modify it to fit to our problem [12]. Following pseudo-code represents the simplified modified version of DWM method, which we have used in our implementation.

```
1. Choose_Experts()                                    //choose, set initial experts
2. e_m ← Train_Expert(train_1)
3. pred_m ← Predict(e_m)
4. for i ← 2, ..., n                                    //loop over examples
5.      for j ← 1, ..., m                               //loop over experts
6.              error ← Mape(pred_j, real_{i-1})        //evaluate prediction
7.              if (error ≥ γ)                          //modify expert weight
8.                  w_j ← Decrease_Expert_Weight(w_j, β)
9.              else
10.                 w_j ← Increase_Expert_Weight(w_j, β)
11.             end;
12.     end;
13.     Normalize_Experts_Weights()
14.     Λ ← Integration(pred, w)                        //compute the global prediction
15.     if(i mod p = 0)
16.             r ← Remove_Experts(e, θ)                //remove "weak" experts
17.             for k ← 1, ..., r                       //replace removed experts
18.             e_k ← Create_New_Expert
19.             end;
20.     end;
21.     for j ← 1, ..., m                               //train experts on next example
22.             e_j ← Train_Expert(train_i)
23.             pred_j ← Predict(e_j)
24.     end;
25.     output Λ
26. end;
```

The pseudo-code of DWM method use the following parameters:

- β – factor for increasing/decreasing model weight
- p – number of iterations between the models removal and creation
- Λ, λ – global prediction of ensemble and local prediction of model
- w_j – actual weight of model j
- win – predefined length of sliding window (e.g. number of days for training)
- $train_i$ – training data chunk of size win for iteration i
- $real_{i-1}$ – data chunk of real values from previous iteration i
- $pred_m$ – predicted values from model m
- θ – threshold for removing experts according to their actual weights
- γ – parameter, which represents a threshold for an acceptation of expert prediction.

The DWM method begins with initialization of prediction models (lines 1–3). During the initialization, the method sets the initial weights, trains chosen models on the first data chunk and obtains the prediction. After the initialization phase, the method continues in two loop cycles. The outer loop goes through the data chunks of incoming data stream and the inner loop passes through all models of the ensemble. For the evaluation of local prediction (line 6), we used the mean absolute percentage error (MAPE), which represents a frequently used prediction accuracy measure. MAPE is defined by Eq. (1).

$$MAPE = \frac{1}{n}\sum_{t=1}^{n} \frac{|\hat{y}_t - y_t|}{y_t} \times 100 \tag{1}$$

where y_t is a real consumption, $\hat{y}_t$ is a predicted load and n is a length of the time series.

Subsequently, the method evaluates the performance of current model by increasing or decreasing its actual weight (lines 7–11). The process of increasement of model weight is defined by Eq. (2).

$$w_j^{new} = w_j * (1 + \beta) \tag{2}$$

The process of decreasement of model weight is defined by Eq. (3).

$$w_j^{new} = w_j * \beta \tag{3}$$

Then the method normalizes the models weights to an interval <0, 1> by sub-procedure 'Normalize_Experts_Weights'. Subsequently the global prediction of the whole ensemble is computed by Eq. (4).

$$\hat{f}_{global} = \frac{\sum_{i=1}^{n}(w_i * \hat{f}_i)}{\sum_{i=1}^{n}(w_i)} \tag{4}$$

where n is the number of prediction models in ensemble, w_i is the assigned weight of *ith* model and $\hat{f}_i$ is the prediction of *ith* model.

If the predefined number of iterations was reached or the prediction error of the ensemble is higher than the specified threshold (parameter θ), the method proceeds into the removal phase, where the "weak" models are removed and replaced by new created models (line 18). Subprocedure 'Create_New_Expert' is used to choose new prediction model from available set of base learners that is not presented in ensemble yet. Subsequently, all models are trained on the following data chunk by subprocedure 'Train_Expert' (line 22).

As the base learners of our DWM method, we chose six prediction models that are frequently used for prediction of electricity consumption:

1. Resilient Backpropagation Neural Network (NN) is an artificial neural network with learning heuristic for supervised learning which performs a local adaptation of the weight-updates according to the behavior of the error function [1]. Attributes to neural network are load and lag of 3 days and training window of size 3 days. NN used tanh activation function and sum of squares error as an error function. NN had 3 hidden layers. The number of neurons in each hidden layer and learning rate was determined by Particle Swarm Optimization algorithm (PSO).

2. Recursive Partitioning and Regression Trees (RT) is a method using 2 stages [2]. The first stage the model is created by splitting the data into subgroups using best split variables. The second stage consists of cross-validation used to trim the created tree. The attributes for RT are load and time vector. See Eqs. (5), (6) and (7).

$$\{Day_i\}_{i=1}^{d} \tag{5}$$

$$Day_i = 1, 2, \ldots, period \tag{6}$$

$$|Day_i| = period \tag{7}$$

where *period* is the number of daily electric load measurements per end-user and d is the number of days in training window. The maximum depth of final tree was set to value 20.

3. Support Vector Regression (SVR) tries to find regression function that can best approximate the actual output vector with an error tolerance [25]. SVR maps the input data in to a higher dimensional feature space, in which the training data may exhibit linearity. For this purpose various kernels are used. Attributes for SVR were modeled as binary (dummy) variables representing the sequence numbers in regression model. Variable equals 1 in the case when they point to the *ith* value of the period, where $i = (1, 2, \ldots, period)$.

4. Seasonal decomposition of time series by loess (STL): is a method [6] that decomposes a seasonal time series into three components: trend, seasonal and irregular. The resulting three components are forecasted separately by ARIMA (STL + ARIMA). ARIMA is one of the most frequently used forecasting methods [8]. It consists of three parts: autoregressive (AR), moving average (MA) and the differencing processes (I).

5. Exponential smoothing (STL + EXP) forecasts the future values of a time series as a weighted average of past values [15]. The weights decay exponentially with time as the observations get older.

6. Holt-Winters prediction method (HW) is used when the data shows not only the trend, but the seasonality as well [23]. The additional formula and smoothing parameter are introduced to handle the seasonality. The additive version of HW is used.

All three methods STL + ARIMA, STL + ETS and HW receive only the load time series.

The main differences between the original and modified version of DWM are:

- Integration function – the original version uses the result of the best model as global prediction. In modified version, the global prediction is calculated as a weighted average of predictions of all models in ensemble that allows solving regression problems.
- New threshold parameter γ for an evaluation of expert prediction based on prediction error.

- The weights of models are updated at each iteration, unlike the original version, where the weights were updated periodically after p iterations.
- Modified version of DWM starts with m prediction models. The original version starts with one model.

3.2   Dynamic Weighted Majority with Decomposition

In addition to the modified DWM described in previous section, we have proposed further modification of this method. The main difference lies in decomposition of the time series. Each part of incoming data stream is represented as a time series that can be further categorized to three types of time series patterns: trend, seasonal and cyclic [3]. Our method is based on the prediction of each of these patterns using different prediction model. Obtained results can be then combined to one final prediction. Therefore, from six models, which were used as prediction models in modified DWM, variations of size three without repetitions were created. As the result we have created 120 different variations where each one represents one predictive model. We suppose that the prediction of each part with different model could lead to higher probability to create a better model, whose prediction error will be lower than error of previous six models. Results of this proposed method (named as DWM+) can be found in evaluation chapter.

In this case, we have used the method of additive expert ensembles with its core functionality based on classical DWM. This method was proposed in 2005 by the same authors as the original DWM method [14]. The main difference lies in modified training process of models that are currently included in ensemble. However, in original DWM method, each predictive model learns i.e. updates its learned information every time when the new data chunk of data stream arrives. Here comes the main question: Why are we updating the learned information, if the last prediction error was low?

In this modified version, we re-train only those models, whose prediction error from the last iteration is higher than defined threshold i.e. an acceptable error rate was achieved. We have used other models, which had prediction error under the threshold, in iteration with previously learned information.

As the name of method AddExp already suggests, another feature of this method is based on adding new models to the ensemble. In our proposed method, we add a new model to ensemble every time when the prediction of the whole ensemble i.e. global prediction exceeds the defined threshold. New added model will be trained for the first time on new data chunk and this prediction can help the whole ensemble to decrease the global prediction error.

3.3   Technical Issues

After the implementation and initial testing of this method, we encountered the problem of the rapid increase in the number of models over the longer term prediction (e.g. prediction for the one-year period). This problem significantly affects the time and space complexity of this method. To solve this problem, we have suggested a pruning method which is based on following two parameters: maximal number of models in ensemble and pruning threshold. The pruning method is applied every time when the size of ensemble reaches the defined maximal value. After that, all models, whose prediction error from the last iteration was higher than pruning threshold, were removed from the ensemble. Removed models were then replaced by new models that were chosen by proposed heuristic. This heuristic relies on choosing the best predictive models from the last iteration of method i.e. iteration where the last models have been removed. The main idea of this heuristic is to add models with better results in comparison with the results of removed models. So, if the number of better models is lower than number of removed models, it is not necessary to preserve the original number of models. Pruning will be applied again when the size of ensemble reaches the maximum number of models.

During implementation and initial testing of proposed method and predictive models, we used recommended configuration parameters, described in the studied papers [10, 11, 15, 19]. The electricity load forecasting represents a special type of prediction problem therefore we decided to optimize parameters of DWM method and its modified version AddExp, to fit our electricity loads dataset exactly.

To perform the optimization task, one of the Biologically Inspired Algorithms (BIA), called Particle Swarm Optimization (PSO), was used. The main reason of choosing the BIAs is their excellent ability to optimize various problems. This approach can lead to solving the optimization problem in a different way in comparison to classical optimization methods [5].

3.4   Optimization of Parameters by Particle Swarm Optimization

PSO algorithm represents a popular biological inspired method, which is frequently used for optimization problems. More information about this algorithm and used implementation library could be found in [9].

Before the optimization of DWM parameters took places, we decided to optimize parameters of neural network predictive model [17]. The main reason for the optimization of this model was its high time complexity that was needed in order to obtain the prediction. Optimized parameters include: number of hidden neurons and hidden layers, interval for learning rate of network and threshold for stopping criteria.

Figure 1 shows a comparison of an optimized version of model to a neural net model with default parameters. The comparison is based on the average execution time.

The time interval, which was chosen for the optimization task of this model, represents one month period of predictions. During this period, PSO algorithm was optimizing the obtained results in terms of execution time and prediction error i.e. MAPE. The x axis in Fig. 1 represents selected end-users from our dataset. The y axis shows an average time of execution for described time period. As we can see, the optimization by PSO algorithm helped the neural network to decrease the execution time for some end-users more than fifteen times. It is important to mention that this time reduction also helped to improve the whole ensemble of predictive models. Then we continued with optimization of DWM method and its modified version.

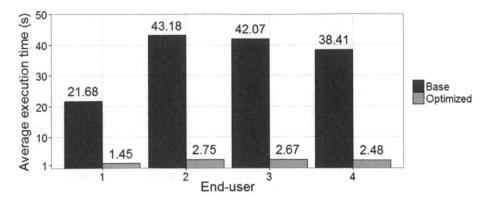

Fig. 1. Average execution time of prediction for 4 end-users using neural network optimization

The Fig. 2 shows the obtained results from optimization of proposed DWM methods for the same period as previously optimized neural network model. We have optimized following parameters: decrease/increase factor for model weights, prediction acceptance threshold and parameter representing initial ensemble size. In Fig. 2 the x axis represents the tested methods and the y axis shows the mean absolute percentage error of prediction. Except the time reduction, the optimization brought also a moderate decrease of prediction error. The results indicate that the prediction error of the optimized version was reduced in some cases by more than one percent in comparison to original method.

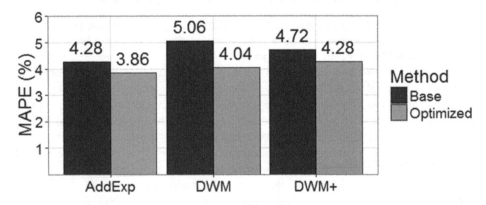

Fig. 2. MAPE of AddExp and DWM methods optimized by PSO

4   Evaluation

In this chapter we focus on comparison of errors between the predictions of three modifications of original DWM method and six base predictive models. Before we describe the designed experiments and results, we focus on electricity loads dataset used in our study.

4.1   Data Description

Dataset used in this paper is represented by electricity load records of Slovak companies from 1.7.2013 to 16.02.2015. These records were obtained by smart meters that send information about the actual electricity consumption every 15 min. Our dataset consists of ca. 490mil records from 21 502 different end-users.

Records in a modified version of the dataset contain the following attributes: date, time and electricity load.

The whole dataset was transformed to a stream of chunks where each part represented one shift of sliding window for 96 records i.e. one day load records.

Figures 3 and 4 represent the comparisons of all predictive models that were used in this study. For the purpose of testing we have chosen two different time periods from the data set. Figure 3 shows results from one month prediction and Fig. 4 represents the results obtained from prediction of one year period. For better evaluation of the results, we also provide a numeric comparison, which can be seen in Table 1.

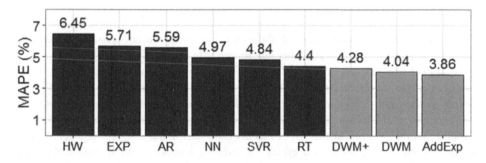

Fig. 3. Comparison of all tested predictive models (month prediction)

As we see, prediction errors obtained from one year period are higher than prediction errors obtained from shorter period i.e. one month. This fact was caused mainly due to the presence of concept drifts that occurred in this one year period. However, in both tested time intervals, all three modified versions of original DWM reached the lowest prediction errors in comparison to other predictive models. This result represents a main achievement of this evaluation, which was based on the assumption that ensemble methods could obtain more precise prediction than single models. The modified method AddExp, which represents an extended version with a decomposition of time series, reached the lowest prediction error.

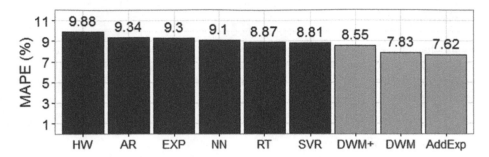

Fig. 4. Comparison of all tested predictive models (year prediction)

Table 1. Obtained results of Mean Absolute Percentage Error (MAPE)

Name of predictive model	MAPE (%) month	MAPE (%) year
Additive Expert Ensembles (AddExp)	3,86	7,62
ARIMA (AR)	5,59	9,34
Dynamic Weighted Majority (DWM)	4,04	7,83
Dynamic Weighted Majority with decomp. (DWM +)	4,28	8,55
Exponential Smoothing (EXP)	5,71	9,30
Holt-Winters (HW)	6,45	9,88
Neural Network (NN)	4,97	9,10
Regression Tree (RT)	4,40	8,87
Support Vector Regression (SVR)	4,84	8,81

Based on further investigation of obtained predictions, we also provide a study on prediction error progress on previous one month interval.

In Fig. 5 we can see a more detailed view of prediction error progress of six models. Graphs on the left side represent the implemented methods based on DWM. Graph on the right side represent the error of three models, whose error development compared to previous methods looks more irregular. This fact can also be seen in Table 2 that represents a histogram of models used during the one year prediction. The methods with lowest prediction errors are preferred mostly and therefore they are frequently used.

In Table 2, we can see three models that were characterized by their irregular error development in Fig. 5. They were also less used in original DWM, where all parts of time series are predicted by same model. On the other hand, we can notice that these same three models were the most common in AddExp method. This fact proves the following statement: Predictive models ARIMA, Holt-Winter and model of exponential smoothing achieve better prediction results, if they predict only a part of time series than a whole data stream chunk. This statement could lead to reverse fact that the predictive models like Regression tree and Support Vector Regression model can be more frequently used in AddExp model in case they are a part of ensemble, where they

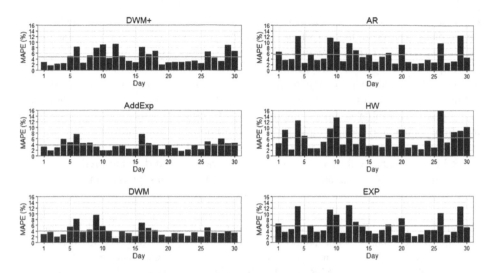

Fig. 5. Prediction error progress (month prediction)

Table 2. Histogram of models usage

Model name	AddExp decomposition			AddExp	DWM
	Trend	Season	Cyclic	Sum	Sum
AR	655	481	386	**1522**	**72**
EXP	568	460	458	**1486**	**93**
HW	450	636	616	**1702**	**60**
NN	55	82	25	**162**	**102**
RT	470	447	508	**1425**	**104**
SVR	359	451	564	**1374**	**106**

could predict all three parts of time series. Consequently, this fact can lead to a reduction of prediction error of AddExp method.

5 Conclusion

In this paper, we proposed three modifications of original Dynamic Weighted Majority method that were applied as a solution to prediction problem of electricity load records. We also focused on decomposition of time series, optimization of parameters by PSO algorithm, or modifications based on extension of original DWM method with the aim of further prediction accuracy improvement.

The results of proposed solutions were compared to six predictive models. Tested models were compared by their prediction error that was represented by Mean Absolute Percentage Error metric. The prediction was performed for two different time intervals from dataset. In all tested cases, which were applied, the proposed methods achieved a lower prediction error than other prediction models. The best results i.e. the lowest prediction error or most regular error development, were reached by method AddExp. We believe that further improvements of this method could be reached by applying described modification of models variations in time series decomposition. Additional modifications could be a subject of future work of our study. The accuracy of proposed ensemble could be improved by developing a mechanism to keep the threshold values and models input settings continuously up-to-date.

Our modified ensemble learning methods i.e. DWM, DWM + and AddExp method proved their suitability for solving the electricity power load demand predictions. The proposed method is applicable generally for the time series prediction problems in various domains.

Acknowledgements. This work was partially supported by the Research and Development Operational Programme for the project "International Centre of Excellence for Research of Intelligent and Secure Information-Communication Technologies and Systems", ITMS 26240120039, co-funded by the ERDF and the Scientific Grant Agency of The Slovak Republic, grant No. VG 1/0752/14 and VG 1/0646/15. The authors would also like to thank for financial contribution from the STU Grant scheme for Support of Young Researchers.

References

1. Anastasiadis, A.D., Magoulas, G.D., Vrahatis, M.N.: New globally convergent training scheme based on the resilient propagation algorithm. Neurocomputing **64**, 253–270 (2005)
2. Archer, K.J.: rpartOrdinal: an R package for deriving a classification tree for predicting an ordinal response. J. Stat. Softw. **34**(7), 1–17 (2010)
3. Athanasopoulos, G., Hyndman, R.J.: Forecasting: principles and practice. OTexts, Melbourne (2013)
4. Aung, Z., Toukhy, M., Williams, J.: Towards Accurate Electricity Load Forecasting in Smart Grids, vol. 8, pp. 51–57 (2012)
5. Binitha, S., Siva Sathya, S.: A survey of bio inspired optimization algorithms. Int. J. Soft Comput. Eng. **2**(2), 137–151 (2012)
6. Cleveland, R.B., Cleveland, W.S., McRae, J.E., Terpenning, I.: STL: a seasonal-trend decomposition procedure based on loess. J. Official Stat. **6**(1), 3–73 (1990)
7. Ikonomovska, E., Gama, J., Džeroski, S.: Learning model trees from evolving data streams. Data Min. Knowl. Disc. **23**(1), 128–168 (2011)
8. Hong, W.-C.: Modeling for Energy Demand Forecasting. In: Hong, W.-C. (ed.) Intelligent Energy Demand Forecasting. Lecture Notes in Energy, vol. 10, pp. 21–40. Springer, Heidelberg (2013)
9. Kennedy, J., Eberhart, R.: Particle swarm optimization. In: Proceedings of IEEE International Conference on Neural Networks, Washington, DC (1995)
10. Maqsood, I., Khan, M.R.: Dynamic weighted majority: an ensemble method for drifting concepts. JMLR **8**, 2755–2790 (2007)
11. Maqsood, I., Khan, M.R.: Using additive expert ensembles to cope with concept drift. In: Proceedings of the Twenty Second ACM International Conference on Machine Learning (ICML 2005), Bonn, Germany (2005)

12. Maqsood, I., Khan, M.R.: Dynamic weighted majority: a new ensemble method for tracking concept drift. In: Proceedings of the Third International IEEE Conference on Data Mining (ICDM 2003), Los Alamitos (2003)

13. Mendes-Moreira, J., Soares, C., Jorge, A.M., Freire de Sousa, J.: Ensemble approaches for regression: a survey. ACM Comput. Surv. **45**(1), 1–40 (2012)

14. Narasimhamurthy, A., Kuncheva, L.I.: A framework for generating data to simulate changing environments. In: Proceedings of the 25th IASTED International Multi-Conference on Artificial Intelligence and Applications (AIA 2007), Innsbruck, Austria (2007)

15. Neupane, B.: Ensemble Learning-based Electricity Price, Masdar Institute of Science and Technology (2013)

16. Pappas, S.Sp., Ekonomou, L., Karampelas, P., Karamousantas, D.C., Katsikas, S.K., Chatzarakis, G.E., Skafidas, P.D.: Electricity demand load forecasting of the Hellenic power system using an ARMA model. In: Electric Power Systems Research, vol. 80, pp. 256–264 (2010)

17. Riedmiller, M., Braun, H.: Rprop - a fast adaptive learning algorithm. In: Proceedings of the International Symposium on Computer and Information Science VII, Antalya, Turkey (1992)

18. Shen, W., Babushkin, V., Aung, Z., Woon, W.L.: An ensemble model for day-ahead electricity demand time series forecasting. In: Proceedings of the 4th International Conference on Future Energy Systems - e-Energy 2013. pp. 51–62. ACM, New York (2013)

19. Sidhu, P., Bhatia, M.P.S.: Empirical support for concept drifting approaches: results based on new performance metrics. IJISA **7**(6), 1–20 (2015)

20. Singh, A., Khatoon, S.: An overview of electricity demand forecasting techniques. Netw. Complex Syst. **3**(3), 38–48 (2013)

21. Taylor, J.W., McSharry, P.E.: Short-term load forecasting methods: an evaluation based on european data. IEEE Trans. Power Syst. **22**(4), 2213–2219 (2007)

22. Therneau, T., et al.: rpart: recursive partitioning and regression trees. R package version 4.1–10, (2014). http://cran.r-project.org/package=rpart

23. Winters, P.R.: Forecasting sales by exponentially weighted moving averages. Manage. Sci. **6**(3), 324–342 (1960)

24. Xiao, L., Wang, J., Hou, R., Wu, J.: A combined model based on data pre-analysis and weight coefficients optimization for electrical load forecasting. Energy **82**, 1–26 (2015)

25. Zhang, F., et al.: A conjunction method of wavelet transform-particle swarm optimization-support vector machine for streamflow forecasting. J. Appl. Math. **2014**, 1–10 (2014)

Estimating Cluster Population

Laxmi Gewali[✉], K.C. Sanjeev, and Henry Selvaraj

University of Nevada, Las Vegas, USA
{laxmi.gewali,henry.selvaraj}@unlv.edu

Abstract. Partitioning a given set of points into clusters is a well-known problem in pattern recognition, data mining, and knowledge discovery. One of the widely used methods for identifying clusters in Euclidean space is the K-mean algorithm. In using K-mean clustering algorithm it is necessary to know the value of k (the number of clusters) in advance. We present an efficient algorithm for a good estimation of k for points distributed in two dimensions. The techniques we propose is based on bucketing method in which points are examined on the buckets formed by carefully constructed orthogonal grid embedded on input points. We also present experimental results on the performances of bucketing method and K-mean algorithm.

Keywords: Clustering · Data partitioning · Bucketing method

1 Introduction

Clustering is a technique of identifying 'closely related points' from a collection of large number of given data points. Closely related points in terms of some distance metric are grouped together as a cluster. In most input data there could be several blocks of clusters. Cluster analysis is extensively used in many fields that include statistics, medicine, social sciences and humanities [1, 6]. In fact, any study that uses collection of data can make productive use of cluster analysis.

Most of the early research on cluster analysis is done by considering point distribution in Euclidean space, where Euclidean metric is used to measure distance between points. In this setting, distance between a pair of points in the same cluster is distinctly smaller than the distance between a pair formed by taking one point from the cluster and the other from outside the cluster. After the advent of computer science, researchers considered the problem of developing efficient algorithms for extracting clusters [3, 4]. K-Mean algorithm and its variations are example of practical algorithms for identifying clusters in Euclidean space. In recent years, there is surge in research interest for identifying clusters in big-data. In the normal data we can assume that all the data is available in the main memory and algorithms are developed by considering standard RAM model. In big-data, not all the data can be stored in RAM. The challenge is to develop cluster identification algorithms when data is available in external memory and the cloud. In some applications, Euclidean metric cannot be used to measure distance between points. For example, the data points could be visitors to Las Vegas entertainment sites and we may be interested to identify cluster of visitors who frequent casino sites and are coming from Hong Kong. Straightforward use of Euclidean metric

© Springer International Publishing AG 2017
J. Świątek and J.M. Tomczak (eds.), *Advances in Systems Science*, Advances in Intelligent Systems and Computing 539, DOI 10.1007/978-3-319-48944-5_5

may not be applicable for such data to extract clusters. We need to come up with appropriate metric other than the Euclidean. In statistics, a widely used technique for cluster analysis is the method of principal component analysis (pca). In this approach orthogonal transformation is performed to obtain linearly uncorrelated data from possibly correlated ones [5].

In this paper we address the issues of estimating the number of clusters for points distributed in Euclidean space. In Sect. 2, we present a brief review of the prominent existing methods for extracting clusters. In Sect. 3, we present the main contribution of the paper. We present an efficient algorithm for estimating the number of clusters and the location of their centers. In Sect. 4, we present preliminary results of the experimental investigation of the proposed algorithm.

2 Review of Existing Approaches

Clustering algorithms have been reported in Engineering and Statistics literature for almost last one hundred years [6, 7]. Most of the clustering algorithms are developed by using some variations of the following two general methods. In *hierarchical scheme*, each of the point p_i in the input data is considered as a cluster. Each cluster is associated with its centroid point which is taken as the arithmetic mean of the coordinates of the points in the cluster. Two clusters are picked to combine by formulating some metric. One simple way of combining clusters is to pick a pair of clusters whose centroids are closest. Another way to combine clusters is to consider the smallest distance between nodes from one cluster to the other. When a new cluster is formed by combining two smaller clusters, the corresponding centroid is also computed. The process of combining two clusters is continued until all points are grouped into one cluster. In some sense the hierarchical clustering scheme works by following the spirit similar as the construction of minimum spanning tree by using Kruskals' algorithm [2]. We can illustrate this strategy by an example shown in Fig. 1, where the top part shows the recursive process of cluster combination and the bottom part shows the implied tree.

In the *point assignment* strategy, clustering algorithms are developed by making an initial estimate of the number of clusters and approximate locations of their centers. The K-mean algorithm described next is an example of this strategy. The K-Mean Algorithm was first formally introduced by Stuart Lloyd [7] in connection with its application to pulse code modulation at Bell Lab. This algorithm is perhaps the most widely referred clustering algorithm for almost 35 years. The algorithm works for points distributed in Euclidean space. The algorithm assumes the number of clusters as a part of the input. The location of the initial k points is also specified by the user of the algorithm. The algorithm grows clusters by adding carefully selected nodes to the partially constructed clusters.

Initially, the k clusters have one node each. The locations of the initial single member in the clusters are taken as their centroids. The algorithm progresses through a series of steps to grow clusters by adding one node at a time. The nodes outside the clusters are unprocessed nodes. The algorithm examines an unprocessed node p_i as the next candidate point. The candidate point p_i is added to the cluster whose center is closest to p_i. This process of "adding a candidate point" is continued until all nodes are

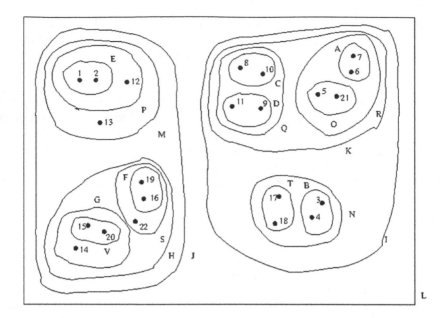

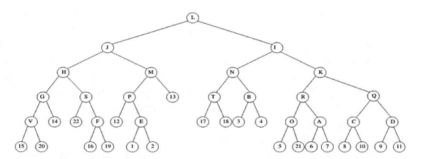

Fig. 1. Illustration of hierarchical clustering

processed. When all points are processed, one pass of the "clusters construction" is completed. After the completion of a pass the centroids are recomputed. The updated centroid of a cluster C_i is the centroid of all points included in it. A new pass of computation starts with respect to the newly updated centroids. In each pass the estimation of centroids and their memberships are updated. The initiation of the next pass stops when cluster members do not change or the change in the location of centroids is below a certain predetermined threshold value.

3 Estimation of Cluster Population

3.1 Preliminaries

As mentioned in the previous section, one of the most popular methods for constructing clusters from a given set of points distributed in Euclidean space is the K-mean algorithm [7]. This algorithm assumes that the number of clusters k *(cluster population)* is known in advance. If the value of k is not given as a part of the input then we need to estimate it 'somehow'. One straightforward technique would be to repeat the execution of the algorithm for several values of k and evaluate the quality of resulting solutions. The value of k that corresponds to the best value of cluster quality is the desired answer. A brute-force method is to try all values of $k = 2,3,4,\dots n$. A faster method based on the binary search technique has been suggested [6], for searching the value of k. Obviously the binary search technique is only effective where the quality of cluster as a function of k is a monotone function. An exhaustive searching approach has several demerits: (i) executing the clustering algorithm repeatedly is time consuming, (ii) measuring the quality of a candidate solution is not precise, and (iii) locating the cluster center for a given value of k is itself a difficult problem. We present next an innovative approach for estimating the value of k and the locations (co-ordinates) of cluster centers for points distributed in the Euclidean plane.

3.2 Adaptive Bucketing

Without loss of generality we can assume that the input point-sites $S = p_0, p_1, \dots p_{N-1}$ are inside a rectangular box R of height $= h$ and width $= w$. The box R can be divided into n by m orthogonal buckets. The value of bucket size m can be pre-determined by examining the distribution of the nearest neighbor distance distribution for n input points. An example of partitioning the bounding box R into orthogonal buckets is shown in Fig. 2 (top part).

A straight-forward approach for counting the points in each bucket is to check for point inclusion in each bucket. The bucket that returns 'true' for point inclusion is the bucket containing the point. Since the buckets are disjoint, only one bucket will return true for inclusion for a given point. To implement this approach we maintain a count array *cnt[ ]* whose entries are initialized to zeros. An array *bx[ ]* holds the coordinates of the top left corner of buckets. If the inclusion test for point $p_i(x_i, y_i)$ against bucket *bx[j]* returns true then *cnt[bx[j]]* is increased by 1. When this check is repeated for all points, point counts for all buckets are complete. A formal sketch of the algorithm based on this approach is listed as Straightforward Count Algorithm (Algorithm 1). It is easily seen that he time complexity of Algorithm 1 is $O(Nnm)$. If $n*m$ is comparable to N then the time complexity becomes $O(N^2)$ which is rather high.

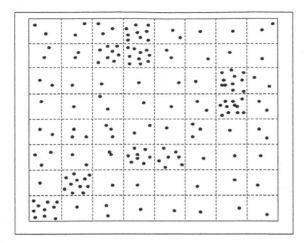

Fig. 2. Illustrating grid embedding and index mapping

Algorithm 1. Straightforward Count Algorithm

Input: (i) *p[N]*; // Input points in 2D
　　　(ii) int *n, m*; // Number of bucket rows and columns
　　　(iii) int *bx[n,m]*; //To hold top left corner of buckets
　　　(iv) int *kv, kh*; // length and width of each bucket
Output: *cnt[]*; // Array to hold count of bucket
Step 1: read *p[N], n, m ,kv,kh*
Step 2: for (int *i* = 0; *i* < *n*m*; *i*++)
　　　　　cnt[i] = 0;
Step 3: for (int *i* = 0; *i* <*N*; *i*++) *f*
　　　　　for (int *j* = 0; *j* < *n*m; j*++) *f*
　　　　　　if (inside(*bx[j], p[i]*))　*cnt[bx[j]]*++;
Step 4: Output *cnt[]*;

Mapping Count Approach: This approach is used to directly map p_i to the bucket b [j] where it falls. Since the size of buckets are the same and rectangular, the index of the bucket where point p_i falls can be computed in term of the row number, column number, width, and height of the bucket. It is given that the outer rectangle R bounding the input points is partitioned into n columns and m rows of buckets, each of size k_v*k_h. Here k_v is the vertical extent of the bucket and k_h its horizontal width. For a given point $p_i(x_i, y_i)$, its row number r_n is given by $rn = y_i/k_v + 1$ and column number $c_n = x_i / k_h + 1$. We can index buckets left to right and top to bottom as 1, 2,......, $n*m$ as shown in Fig. 2 (bottom part). Then the bucket index corresponding to point $p_i(x_i, y_i)$ is $(r_n - 1) * n + c_n$. As an example, point $p_1(55, 25)$ is mapped bucket $(3-1)*5 + 4 = 14$.

Based on this mapping, the following is a faster algorithm (Algorithm 2) called Mapping Count Algorithm. The time complexity of Algorithm 2 is $O(N + nm)$. This time complexity is optimal in the sense that it takes $O(N)$ time to read the points and nm is at most N

> **Algorithm 2.** Mapping Count Algorithm
> **Input:** (i) $p[N]$; // Input points in 2D
>   (ii) int $n;m$; // Number of rows and columns
>   (iii) int $bx[n;m]$; //To hold top left of buckets
>   (iv) int k_v,k_h; // length and width of each bucket
> **Step 1:** read $p[N]$, n, m, k_v, k_h // Read input
> **Step 2:** for(int $i = 0$; $i < n*m$; i++)
> $cnt[i] = 0$;
> **Step 3:** for(int $i = 0$; $i < N$; i++)
> $r_n = yi/kv + 1$;
> $c_n = xi/kh + 1$;
> $j = (r_n - 1) * n + c_n$
> $cnt[bx[j]]$++;
> **Step 4:** Output $cnt[]$

Remark 1 (*Number of buckets*): The very purpose of using buckets fails if there are too many buckets. For making the bucketing approach valid we do not want to have many empty buckets. At the same time to identify the boundaries of clusters we should have enough buckets. A good upper bound for the number of rows and columns in bucket partitioning is $N^{1/2}$. In some applications, the number of rows and columns is much smaller than $N^{1/2}$, and in some cases it is even constant.

3.3   Aggregating Bucket Clusters

After identifying buckets containing a high concentration of points, it is now necessary to aggregate buckets together belonging to the same cluster. We can clarify this with the following example in Fig. 3.

In this example there are two clusters $C1$ and $C2$. Cluster $C1$ has 10 buckets $b1$, $b2$, $b3$, $b4$, $b5$, $b6$, $b7$, $b8$, $b11$, $b12$ and cluster $C2$ has five buckets $b9$, $b10$, $b13$, $b14$, $b15$. Suppose the starting bucket is $b6$. The algorithm proceeds by initializing a queue Qb by

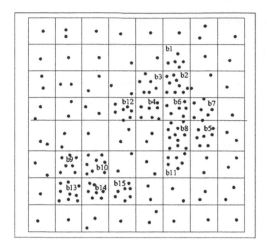

Fig. 3. Aggregating 'H'-buckets

inserting the starting bucket $b6$ to Qb. The algorithm then repeats the following generic task until all buckets of the cluster are aggregated.

Generic Task: Pick the bucket b_j from the front of the queue Qb and inset into queue all four connected neighbors of b_i that are marked H. Bucket b_j is pushed onto stack S_b and b_j is marked processed.

In our running example, bucket $b6$ is removed from the front of the queue Q_b and its 'H' marked neighbors that have not been processed yet ($b2$, $b4$, $b8$ and $b7$) are inserted into queued onto queue Q_b. Bucket $b6$ is marked processed. Next bucket $b2$ is removed from the queue and its unprocessed neighbors that are marked 'H' ($b1$ and $b3$) are inserted into queue onto the queue.

Bucket $b2$ is marked processed. These operations on stack and queue are repeated until the queue is empty. When the queue is empty all the buckets of the cluster in the context are present in the stack. A formal sketch of the algorithm which we refer to as Bucket Clustering Algorithm is listed as Algorithm 3.

Algorithm 3. Bucket Clustering Algorithm
    Input: (i) An array $b[]$ of size $m * n$ representing
                the top left co-ordinates of buckets
        (ii) A given starting bucket index q that
                belongs to current cluster
    Output: A stack containing the buckets
                representing the cluster counting $b[q]$
    Step 1: $Q = b[q]$; // Initialize queue Q
    // Initialize stack Sb to be empty
    Step 2: while (Q is not empty)
          a. $Px = Q.delete()$;
          b. Let Rc be set of unprocessed
              4-neighbors of Px
          c. Insert points in Rc into Q
          d. Push Px into stack Sb
          e. Mark points in Rc 'processed'
    Step 3: Output Sb

3.4 Nudging

A straightforward application of the bucketing technique aggregates high count buckets (H-buckets) to extract a cluster. We refer to the clusters constructed in this way as *coarse clusters* and their boundaries as *coarse boundaries*. Some points in L-clusters adjacent to coarse boundaries are not included in the cluster even if they are very close to the fence of a H-bucket. Of course, points in L-buckets adjacent to a coarse boundary should not be included in the cluster if such points are farther away from the boundary and appear disconnected to the cluster.

In Fig. 4 (top part), the cluster at the center is formed by aggregating 8 buckets [4, 4], [5,4], [4,5], [5,5], [6,5], [3,6], [4,6] and [5,6]. However, boundary points in low count buckets [4, 7], [5, 7], [6, 7] and [6, 6] should be included in the cluster. When such boundary points are included in the cluster we get better estimation of the cluster as shown in Fig. 4 (bottom part). Now we describe a formal way of identifying points near the coarse boundary that can be included in the cluster. Our approach is to nudge coarse boundaries to capture proximity points in the corresponding cluster. Consider a H-bucket adjacent to a coarse boundary as shown in Fig. 5. If a L-bucket shares an edge with a H-bucket, then we can inspect points inside a rectangle of size l × l/4 (strip rectangle) as shown in Fig. 5 to possibly include in the cluster, where l is the side length of the bucket. This is called strip nudging. If a L-bucket is adjacent to a corner of a H-bucket then we should inspect points inside an arc of radius l/4 and angle $3\pi/4$, as shown in the lower left of Fig. 5. This technique is called arc nudging.

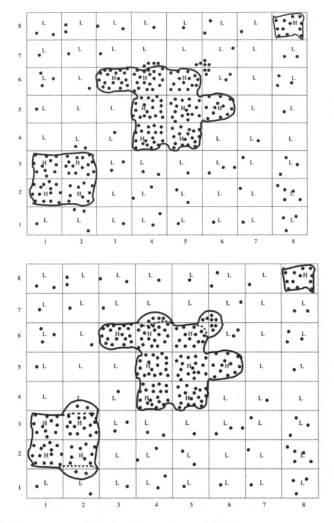

Fig. 4. Aggregation and nudging

4 Results and Discussion

We generated different examples with varying number of clusters and initial representative points to test the performance of the bucketing algorithm. To measure the
quality of the solution obtained using the bucket clustering algorithm, we used the *sum
of squared error (SSE)* as our objective function [3, 9].

   We first calculate the squared error of each point to its closest centroid and computed the total sum of the squared errors for the clusters. A small SSE means the
generated clusters better represent the points in the cluster. We calculated SSE for
clusters generated using both the standard K-means algorithm and bucketing algorithm.
The results are tabulated in Table 1.

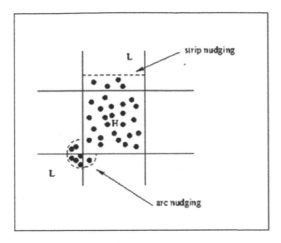

Fig. 5. Strip nudging and arc nudging

Table 1. Experimental Results

Method	Avg. SSE	Total clusters	Threshold points
K-mean (k = 4) Set-1	901703.25	4	N/A
K-mean (k = 4) Set-2	907126.25	4	N/A
Bucketing	24114140	1	4
Bucketing and Nudging	24566045	1	4
Bucketing	1279606	4	5
Bucketing and Nudging	1595448	4	5
Bucketing	599789	4	9
Bucketing and Nudging	881676	4	9
Bucketing	373290	3	13
Bucketing and Nudging	592832	3	13
Bucketing	122360	2	15
Bucketing and Nudging	282545	2	15

From the experimental results it is clear that when the threshold points are carefully selected, the clusters obtained using the bucketing algorithm, in most cases, have either less or almost equal SSE compared to the standard K-means algorithm. Due to the wrong selection of threshold points, in some cases, the SSE obtained from the bucketing algorithm is higher than the standard K-means. Overall, the bucketing algorithm provides almost the same or better SSE compared to original K-means. In addition, bucketing algorithm removes the necessity of providing the number of clusters at the beginning.

References

1. Berg, M., Krevald, M., Overmars, M., Schwarzkopf, O.: Computational Geometry: Algorithms and Applications, 2nd edn. Springer, Heidelberg (2000)
2. Cormen, T.H., Lieserson, C.E., Ronald, L., Rivest, R.L., Stein, C.: Introduction to Algorithms, 3rd edn. MIT Press, Cambridge (2009)
3. Guha, S., Rastogi, R., Shim, K.: CURE: An efficient clustering algorithm for large databases. Inf. Syst. **26**(2), 35–58 (2009). MIT Press
4. Indyk, P., Motwani, R.: Approximate nearest neighbors: towards removing the curse of dimensionality. In: Proceedings of STOC, pp. 604–13 (1998)
5. Jolliffe, I.T.: Principal Component Analysis, 2nd edn. Springer, New York (2002)
6. Leskovec, J., Ullman, J.D., Rajaraman, A.: Mining of Massive Datasets. Cambridge University Press, New York (2014)
7. Lloyd, S.P.: Least square quantization in PCM. IEEE Trans. Inf. Theory **28**(2), 129–137 (1982)
8. O'Rourke, J.: Computational Geometry in C, 2nd edn. Cambridge University Press, Cambridge (1998)
9. Tan, P., Steinbach, M., Kumar, V.: Introduction to Data Mining. Pearson Addison Wesley, Boston (2005)

Evaluation of Particle Swarm Optimisation for Medical Image Segmentation

Mohammad Hashem Ryalat[1]([⊠]), Daniel Emmens[2], Mark Hulse[4],
Duncan Bell[4], Zainab Al-Rahamneh[3], Stephen Laycock[1], and Mark Fisher[1]

[1] University of East Anglia, Norwich Research Park, Norwich NR4 7TJ, UK
{M.Ryalat,S.Laycock,Mark.Fisher}@uea.ac.uk
[2] Department of Clinical Oncology, Ipswich Hospital NHS Trust, Ipswich, UK
Daniel.Emmens@ipswichhospital.nhs.uk
[3] Department of Computer Information Systems,
AlBalqa Applied University, Salt, Jordan
Zainab@bau.edu.jo
[4] Faculty of Health and Science, University Campus Suffolk, Ipswich, UK
m.hulse@ucs.ac.uk, Duncan@fisherideas.co.uk

Abstract. Otsu's criteria is a popular image segmentation approach that selects a threshold to maximise the inter-class variance of the distribution of intensity levels in the image. The algorithm finds the optimum threshold by performing an exhaustive search, but this is time-consuming, particularly for medical images employing 16-bit quantisation. This paper investigates particle swarm optimisation (PSO), Darwinian PSO and Fractional Order Darwinian PSO to speed up the algorithm. We evaluate the algorithms in medical imaging applications concerned with volume reconstruction, with a particular focus on addressing artefacts due to immobilisation masks, commonly worn by patients undergoing radiotherapy treatment for head-and-neck cancer. We find that the Fractional-Order Darwinian PSO algorithm outperforms other PSO algorithms in terms of accuracy, stability and speed which makes it the favourite choice when the accuracy and time-of-execution are a concern.

Keywords: Particle swarm optimisation · Medical image segmentation · Volume reconstruction

1 Introduction

Segmentation aims to simplify image representation into something that is more meaningful and easier to analyse and it is a common step in medical image processing pipelines since diagnosis often relies on visually enhancing medically relevant structures within the body. Segmentation of medical images can be performed manually, automatically (computerized) or using a combination of both (computer assisted). Manual segmentation is time-consuming compared to automatic segmentation and the results may be prone to observer variability whereas using computer-aided segmentation techniques significantly improves

© Springer International Publishing AG 2017
J. Świątek and J.M. Tomczak (eds.), *Advances in Systems Science*, Advances in Intelligent Systems and Computing 539, DOI 10.1007/978-3-319-48944-5_6

the precision of segmentation outcomes. There is no standard classification of segmentation algorithms applied to medical images. However, a review of algorithms for medical image segmentation [1] categorised them according to the primary methodologies employed, for example, thresholding, clustering techniques, deformable models, etc. Otsu's method is a thresholding technique that automatically clusters image pixels [2,3]. The basic assumption of Otsu's approach is that the image contains two classes of pixels (i.e. foreground pixels and background pixels). The algorithm computes the optimum threshold separating the two classes so that their inter-class variance (i.e. between-class variance) is maximal. An extension of Otsu's method, which classifies pixels as belonging to one of many classes using multi-level thresholding, is presented in [4].

Particle swarm optimization (PSO) is a population based stochastic optimization algorithm inspired by social behaviour of bird flocking or fish schooling in search of food [5]. The algorithm adopts a genetic search strategy and is initialized with a population of random solutions. The basic drawback of the PSO algorithm (as with other optimisation algorithms) is the possibility of becoming trapped in a local maxima/minima. Darwinian particle swarm optimization (DPSO) [6] extends PSO by adding a natural selection mechanism (i.e. survival of the fittest) to improve its ability to escape from local maxima/minima. In DPSO, many parallel PSO algorithms, each forming a swarm, operate in the same search space. Fractional-order Darwinian PSO (FODPSO) [7] represents a further extension that introduces concepts from fractional calculus to control the rate of convergence.

Exhaustive search for optimal thresholds requires the evaluation of $(n + 1)(D - n + 2)^n$ combinations of thresholds [8] where n represents the number of thresholds and D represents the absolute difference between the $max\_intensity$ pixel value and the $min\_intensity$ pixel value in the image. The computational complexity motivates a search for more efficient methods that operate under Otsu's criterion. A few studies have applied PSO, DPSO and FODPSO algorithms for image segmentation. For example Ghamisi et al. [9] study remote sensing applications involving hyperspectral images comprising many data channels, and [10] is one of a number of studies applied PSO to segment medical images. This paper evaluates and compares PSO, DPSO and FODPSO algorithms in terms of speed, accuracy and stability for segmentation of medial relevant structures in CT and MRI. We also extend FODPSO, to create a segmentation tool that automatically removes artifacts produced by immobilisation masks, that are fixations routinely worn by patients undergoing radiotherapy treatment for tumours affecting the head and neck. The experiments demonstrate improvements in the speed and accuracy of PSO-based algorithms when applied for medical imaging.

The remaining sections of this paper is organised as follows. Section 2 gives an overview on the PSO, DPSO and FODPSO algorithms. Section 3 presents the datasets, experiments, applications and results of the work. Finally, Sect. 4 draws conclusions.

2 Overview of PSO-Based Algorithms

2.1 Particle Swarm Optimisation (PSO)

Particle swarm optimisation (PSO) algorithms are inspired by the behaviour of swarms of biological organisms that can be observed to move in a synchronized way. Algorithms typically model populations initialized with a random solutions called particles, each distinguished by its own position and velocity. The following equations describe how velocity V_{id} and position X_{id} are updated at each iteration k.

$$V_{id}(k) = wv_{id}(k-1) + c_1 r_{1id}(k)(Xpbest_{id} - X_{id})$$
$$+ c_2 r_{2id}(k)(Xgbest_d - X_{id}) \tag{1}$$
$$X_{id}(k) = X_{id}(k-1) + V_{id}(k) \tag{2}$$

where w represents the inertia weight, r_1 and r_2 are random numbers with a uniform distribution in the range [0,1], and c_1 and c_2 assign weights to the local and global best solutions respectively. Each particle has a kind of memory which stores the position where it had the lowest cost ($Xpbest_{id}$), and the position of the best particle in the population ($Xgbest_d$).

The PSO algorithm starts by initialising swarm parameters such as population-size, number-of-iterations, c_1(cognitive weight), c_2(Social weight), w(Inertial factor), and V_{max} and V_{min} to set the limits of velocities. Then the algorithm iterates through all particles to calculate the fitness function (inter-class variance between pixels intensities). The aim is to search for the threshold(s) that globally optimise the fitness value until the stopping criteria is satisfied. The algorithm terminates when the incremental improvement in fitness falls below a certain threshold (or the number of iterations reach the maximum allowed).

2.2 Darwinian Particle Swarm Optimisation (DPSO)

A general problem with optimization algorithms is that they have difficulty in finding global optima and as such specific techniques tend to work well on one problem but may fail on another. Darwinian Particle Swarm Optimisation [6] is an approach that addresses the problem by using natural selection to search areas within the solution space that have been under explored. DPSO deploys multiple independent PSO swarms, organised as a group in a way that is intended to simulate natural selection. The algorithm starts by setting initial values in a collection of parameters. Those parameters include number of swarms, maximum and minimum possible number of swarms, maximum and minimum possible population size in addition to parameters of the basic PSO. DPSO spawns (reproduces) a new particle if it finds a new global maxima/minima, and a particle is removed if the swarm has been unsuccessful in improveing the fitness criteria in a fixed number of steps.

2.3   Fractional Order Darwinian Particle Swarm Optimisation (FODPSO)

The basic idea behind the FODPSO algorithm is that it utilises concepts from fractional calculus in order to control the convergence rate of the DPSO algorithm. Fractional calculus is a generalization of ordinary differential calculus [11] and there are numerous uses of fractional calculus in physics, mechanics, chemistry, and computational mathematics. Systems that employ fractional processes are characterised by residual memory and their fractional order is understood as a measure of their memory strength [12,13].

Equation 3 presents the Grunwald-Letnikov description based on the concept of fractional differential of a general signal $x(t)$:

$$D^\alpha[x(t)] = \lim_{h \to 0} \left[\frac{1}{h^\alpha} \sum_{k=0}^{+\infty} \frac{(-1)^k \Gamma(\alpha+1)x(t-kh)}{\Gamma(k+1)\Gamma(\alpha-k+1)} \right] \tag{3}$$

where Γ is the gamma function and α is the fractional coefficient such that $\alpha \in$ C. It is worthy to notice that while an integer-order derivative is evaluated as an finite series, the fractional-order derivative is evaluated as an infinite number of terms. Consequently, integer-order derivative behaves like a local operator, while fractional-order derivative behaves like a structure that has a memory of all past events [7].

3   Experimental Work

To evaluate PSO, DPSO and FODPSO algorithms for medical image segmentation we performed experiments focusing on three applications: segmentation of brain in MRI image slices; volume reconstruction from a stack of CT image slices; removal of CT artefacts due to immobilisation masks used in radiotherapy treatment of cancer tumours located in the head and neck.

3.1   Segmentation of Brain in MRI Images

The information provided by MRI images about the soft tissue anatomy has significantly improved the study of neuropathology [14] and provides medical decision support in numerous areas such as cancer research, heart disease and brain trauma [15,16]. In the context of brain MRI, segmentation means to divide up the image/volume into a number of regions, each of which is homogeneous in some sense. This involves assigning a label to each point in the MRI image/volume to identify a tissue class, such as grey matter, white matter, cerebral spinal fluid and air. In our study we are not concerned with evaluating the accuracy of such class labels, instead, we evaluate PSO, DPSO and FODPSO algorithms, in terms of CPU execution time, fitness criteria and stability.

The experiments we present in this study were performed on an Intel®. Core™ i7-3770 CPU @ 3.40 GHz 3.40 GHz / 16.0 GB (RAM)/ Windows 7 Enterprise 64-bit operating system. The algorithms were applied to three MRI image

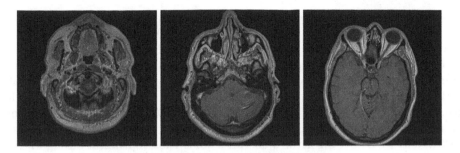

Fig. 1. Three MRI images from three patients.

slices from datasets representing three different randomly selected anonymised patients (Fig. 1).

In order to present the difference that is gained by applying the FODPSO algorithm, we compared it with PSO, DPSO and exhaustive search methods in terms of speed and accuracy. The PSO, DPSO and FODPSO need, as any optimization algorithm, to define search space, candidate solutions and global optima. In the case of images, the pixel intensities of the image will form the search space, the possible threshold values (i.e. $[min - intensity, max - intensity]$) will form the candidate solutions, and maximising the inter-class variance will be our global optimisation criteria. Table 1 presents the initial values of parameters used in the algorithm.

Table 1. Initial parameters of the PSO, DPSO and FODPSO algorithms

Parameter	PSO	DPSO	FODPSO
Population size	180	35	35
Number of iterations	25	25	25
Cognitive weight	0.8	1.1	1.1
Social weight	0.8	0.9	0.9
Inertial factor	1.2	1.2	1.2
Vmax	3	3	3
Vmin	−3	−3	−3
Number of swarms	N/A	5	5
Max number of swarms	N/A	7	7
Min number of swarms	N/A	3	3
Max population size	N/A	50	50
Min population size	N/A	20	20
Stagnancy	N/A	8	8
Fractional coefficient	N/A	N/A	0.8

In order to evaluate the accuracy of the results generated by each one of the three optimisation algorithms (i.e. PSO, DPSO and FODPSO), we measured the fitness (i.e. inter-class variance) for each algorithm and compared the outcomes with the brute-force (i.e. exhaustive search) method. Since the brute-force method explores all combinations of threshold values, then it will certainly find the optimal threshold(s) and the optimal fitness value. Table 2 presents the average fitness values generated by the PSO, DPSO and FODPSO algorithms against the fitness value generated by the brute-force method. We run PSO, DPSO and FODPSO algorithms 50 times since those optimisation algorithms are stochastic and random and we calculate the average fitness values for 1, 2, 3 and 4 thresholds.

Table 2. Average fitness values of Brute-Force, PSO, DPSO and FODPSO algorithms for different number of thresholds over three MRI images.

Image No.	T.holds	Brute-Force	PSO	DPSO	FODPSO
Image1	1	1985.55	1984.78	1985.55	1985.55
	2	2218.33	2216.98	2218.33	2218.33
	3	2273.58	2272.91	2273.38	2273.49
	4	2300.49	2297.44	2299.21	2299.27
Image2	1	2187.61	2187.43	2187.61	2187.61
	2	2426.88	2425.53	2426.88	2426.88
	3	2505.68	2504.62	2505.59	2505.59
	4	2536.87	2535.89	2536.76	2536.79
Image3	1	2306.73	2306.64	2306.73	2306.73
	2	2547.22	2546.12	2547.22	2547.22
	3	2618.67	2617.97	2618.63	2618.65
	4	2645.43	2643.89	2645.17	2645.31

From Table 2 we see that the three optimisation algorithms achieved fitness values which are either the same or very close to that generated by the brute-force method. In general, the fitness values returned by FODPSO and DPSO were better than that of the PSO algorithm. FODPSO generates either the same or a slightly better fitness value compared to the DPSO algorithm. Although the fitness values of the three algorithms were close to the one of the brute-force method, it is noticable that as the number of thresholds increase, the differences between fitness values of the three optimisation algorithms compared with the brute-force method marginally increases. Fortunately, the trend is the same for the three test images whichtends to confirm the robustness of the procedure. The average optimal threshold(s) that were generated by PSO, DPSO and FODPSO against the ones generated by the brute-force method are presented in Table 3. The readings of this table supports the fact that PSO algorithm is prone to give a local maxima.

Table 3. Average thresholds of PSO, DPSO and FODPSO over three MRI images against the thresholds coming from Brute-Force.

No.	T.holds	Brute-Force	PSO	DPSO	FODPSO
Img1	1	{582}	{575}	{582}	{582}
	2	{418,1077}	{407,1066}	{418,1077}	{418,1077}
	3	{341,835,1242}	{330,824,1231}	{343,837,1244}	{342,837,1244}
	4	{275,681,1022,1374}	{256,655,996,1346}	{270,672,1015,1363}	{270,672,1012,1358}
Img2	1	{473}	{468}	{473}	{473}
	2	{373,946}	{365,937}	{373,946}	{373,946}
	3	{241,622,1020}	{231,613,1011}	{240,621,1018}	{240,621,1018}
	4	{224,581,887,1186}	{213,569,876,1175}	{220,578,885,1184}	{220,577,884,1184}
Img3	1	{427}	{423}	{427}	{427}
	2	{381,983}	{374,976}	{381,983}	{381,983}
	3	{221,579,1006}	{212,571,998}	{220,578,1004}	{219,578,1004}
	4	{206,541,793,1098}	{196,534,788,1095}	{203,543,799,1106}	{203,541,797,1105}

Table 4. Average execution time (in sec) of the Brute-Force, PSO, DPSO and FODPSO over three MRI images.

No.	T.holds	Brute-Force	PSO	DPSO	FODPSO
Img1	1	0.1168	0.2792	0.1225	0.1118
	2	1.466	0.4376	0.3732	0.3501
	3	158.8388	0.5575	0.4575	0.4373
	4	14623	0.6840	0.5490	0.5170
Img2	1	0.1386	0.2673	0.1291	0.1155
	2	1.47438	0.4337	0.3526	0.3428
	3	157.7319	0.5680	0.4628	0.4425
	4	12821	0.6846	0.5426	0.5255
Img3	1	0.1369	0.2611	0.1148	0.1089
	2	1.4568	0.4328	0.3432	0.3317
	3	157.1496	0.5703	0.4570	0.4327
	4	12765	0.7088	0.5564	0.5224

In order to evaluate the speed of each algorithm, we measured the average CPU processing time that PSO, DPSO, FODPSO and brute-force methods need to produce results. Table 4 presents the average execution time in seconds for each method. The table confirms that FODPSO is always slightly faster than the DPSO algorithm and DPSO is significantlyfaster than the PSO algorithm.

Comparing the brute-force method against the FODPSO algorithm in terms of speed in Table 4, we can conclude that the speed of the brute-force search is similar to, but less than the speed of FODPSO when the number of thresholds equals one. But as the number of thresholds increases, the difference between the speed of the brute-force and the speed of the other three optimisation algorithms becomes significant. This makes FODPSO an attractive choice for automatically labelling tissues in MRI and CT. The fractional coefficient used in FODPSO allows the convergence rate of the algorithm to be controlled and this explains why FODPSO outperforms the DPSO algorithm. Previous researchers [17] report results that indicate the time required by the brute-force method is less than that for the PSO algorithm when the number of thresholds equals one. However, their experiments are performed on images quantised at 8-bits (i.e. gray values lie in the range [0, 255]). The range of values, in the case of medical images adopting the dicom standard), is [−1000, 3000] which explains why the brute-force method (i.e. exhaustive search) takes considerably more time (Table 4).

Table 5. Standard deviation of fitness for PSO, DPSO and FODPSO after running each algorithm 50 times over three MRI images.

No.	T.hold	PSO	DPSO	FODPSO
Img1	1	0.8134	0.0259	0.0213
	2	0.9821	0.0329	0.0255
	3	1.1711	0.6257	0.0341
	4	1.2513	0.4251	0.1002
Img2	1	0.3207	0.0053	0.0012
	2	0.0893	0.0082	0.0013
	3	0.9625	0.0823	0.0149
	4	1.0261	0.0928	0.0356
Img3	1	0.0251	0	0
	2	0.0011	0	0
	3	0.0966	0.0622	0.0112
	4	0.3227	0.1925	0.0409

Since PSO-based algorithms are stochastic and random we use the standard deviation σ of the fitness value, and stability generated by each run as an evaluation metric of stability by finding the deviation of the fitness value generated in each run. Table 5 shows that FODPSO produces the most stable results when compared to the PSO and DPSO and that σ increases as the number of thresholds increase in most cases. Typical results of segmentation using FODPSO in three MRI images using different number of thresholds is shown in Fig. 2.

3.2 Volume Reconstruction

Reconstructing volumes from CT images takes considerable time. A fundamental reason for this is due to the fact that there is normally a large number of 2D slices generated by the CT scanner. That number depends on the resolution of the CT-scanner, spacing-between-slices and the dimensions of the scanned object. Before the process of volume reconstruction starts, the stack of 2D slices generated from the CT-scanner must be segmented. We then use the marching

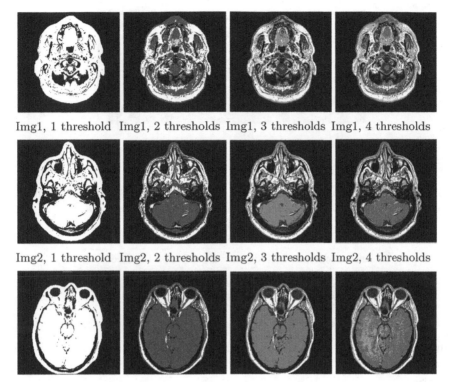

Img1, 1 threshold Img1, 2 thresholds Img1, 3 thresholds Img1, 4 thresholds

Img2, 1 threshold Img2, 2 thresholds Img2, 3 thresholds Img2, 4 thresholds

Img3, 1 threshold Img3, 2 thresholds Img3, 3 thresholds Img3, 4 thresholds

Fig. 2. Results of segmentation when applying FODPSO over the three MRI images using different number of thresholds

cubes algorithm [18] to construct the volume from the stack of segmented images. The objects shown in Fig. 3 were used in this part of the study. The first object (a) is the scaled head of a 19th century Cantonese chess piece, the original being delicately carved from ivory [19]. The second (b) is a nested cubes and the third (c) a plastic hemisphere. All the objects were manufactured by a 3D printer using (a) data acquired by a GE Discovery CT590 RT CT scanne r(helical pitch = 0.562:1, collimation 16×0.625 mm (10 mm)).

Table 6 shows a comparison, in terms of execution time, for volume reconstruction. The table gives the time for constructing the mesh, in seconds, for the three objects (i.e. the Cantonese head, nested cubes and the hemisphere). Table 6 shows clearly that the FODPSO algorithm is the fastest algorithm and PSO the slowest. It is also obvious from that table the speed of DPSO is very close to the speed of the brute-force method. It is worth noting that in order to construct a volume from the CT images, we chose the number of thresholds to be equal to one for the 2D segmentation process. Consequently the difference in CPU execution time between the brute-force method and the FODPSO algorithm is not significant. We recommend using the FODPSO algorithm as a

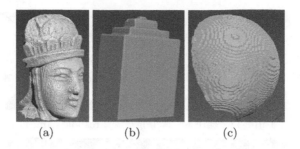

(a)   (b)   (c)

Fig. 3. Three objects used in experiments described in Sect. 3.2

Table 6. Comparison of brute-force, PSO, DPSO and FODPSO algorithms in terms of execution time (in sec) when they are used to construct 3D models.

Object name	Brute-force	PSO	DPSO	FODPSO
Cantonese head	24.193	49.257	24.815	21.669
Nested cubes	8.538	15.355	8.187	6.814
Hemisphere	17.031	32.913	16.896	14.766

segmentation method to construct volumes when the total number of slices is large since this will make a direct impact on the total time for the construction process.

3.3 Automatic Editing of CT Images

Immobilisation masks are routinely used for patients undergoing radiotherapy treatment of tumours affecting the head and neck and CT images of those patients form the basis of their radiotherapy treatment plans. Manually editing the data to remove artefacts due to the mask is time consuming and error prone, particularly so in areas where the mask contacts the skin.

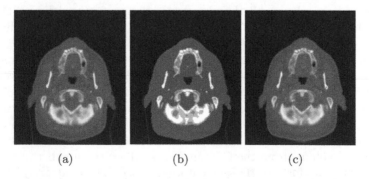

(a)   (b)   (c)

Fig. 4. (a) The input CT image with a mask. (b) After using FODPSO algorithm to segment it. (c) Output image (mask removed)

We find that setting the number of thresholds to be equal to five tends to lead to better results as this number classifies all or most of the pixels belonging to the mask as one class. We applied our procedure over a group of image (slices) for the same subject and our experiments produced 90 % correctly segmented results (e.g. see Fig. 4).

4   Conclusion

This paper evaluated PSO, DPSO and FODPSO algorithms in the field of medical image segmentation. When using PSO-based algorithms, some parameters need to be initialised; we chose these empirically and then performed experiments comparing PSO, DPSO, FODPSO and brute-force methods. The objective function used in this study is Otsu criterion (i.e. inter-class variance). Our data sets were acquired using CT and MRI and comprised both human and manufactured (3D printed) objects. We demonstrated that the algorithm was of practical use by employing FODPSO to automatically remove features arising from immobilisation masks from CT slices. Future research will investigate the feasibility of applying the FODPSO algorithm in medical image segmentation using other objective functions.

References

1. Ma, Z., et al.: A review of algorithms for medical image segmentation and their applications to the female pelvic cavity. Comput. Meth. Biomech. Biomed. Eng. **13**(2), 235–246 (2010)
2. Otsu, N.: A threshold selection method from gray-level histograms. Automatica **11**(285–296), 23–27 (1975)
3. Sezgin, M., et al.: Survey over image thresholding techniques and quantitative performance evaluation. J. Electron. Imaging **13**(1), 146–168 (2004)
4. Liao, P.-S., Chen, T.-S., Chung, P.-C., et al.: A fast algorithm for multilevel thresholding. J. Inf. Sci. Eng. **17**(5), 713–727 (2001)
5. Eberhart, R.C., Kennedy, J., et al.: A new optimizer using particle swarm theory. In: Proceedings of the Sixth International Symposium on Micro Machine and Human Science, vol. 1, pp. 39–43. New York (1995)
6. Tillett, J., Rao, T., Sahin, F., Rao, R.: Darwinian particle swarm optimization. In: Prasad, B. (ed.) Proceedings of the 2nd Indian International Conference on Artificial Intelligence (IICAI 2005), pp. 1474–1487 (2005)
7. Couceiro, M.S., Rocha, R.P., Fonseca Ferreira, N.M., Tenreiro Machado, J.A.: Introducing the fractional-order darwinian PSO. Signal Image Video Process. **6**(3), 343–350 (2012)
8. Kulkarni, R.V., Venayagamoorthy, G.K.: Bio-inspired algorithms for autonomous deployment, localization of sensor nodes. IEEE Trans. Syst. Man Cybernetics Part C (Applications, Reviews) **40**(6), 663–675 (2010)
9. Ghamisi, P., Couceiro, M.S., Martins, F.M.L., Benediktsson, J.A.: Multilevel image segmentation based on fractional-order darwinian particle swarm optimization. IEEE Trans. Geosci. Remote Sens. **52**(5), 2382–2394 (2014)

10. Ait-Aoudia, S., Guerrout, E.-H., Mahiou, R.: Medical image segmentation using particle swarm optimization. In: 2014 18th International Conference on Information Visualisation, pp. 287–291. IEEE (2014)
11. Petráš, I.: Fractional derivatives, fractional integrals, and fractional differential equations in Matlab. Eng. Educ. Res. MATLAB, InTech, kap 10, 239–264 (2011)
12. Marinov, T., Ramirez, N., Santamaria, F.: Fractional integration toolbox. Fractional Calc. Appl. Anal. 16(3), 670–681 (2013)
13. Oldham, K.B., Spanier, J.: The fractional calculus: theory and applications of differentiation and integration to arbitrary order. Academic, New York (1974). Republished by Mineola, 2006
14. Fletcher-Heath, L.M., Hall, L.O., Goldgof, D.B., Reed Murtagh, F.: Automatic segmentation of non-enhancing brain tumors in magnetic resonance images. Artif. Intell. Med. 21(1), 43–63 (2001)
15. Gorunescu, F.: Data mining techniques in computer-aided diagnosis: non-invasive cancer detection. PWASET 25, 427–430 (2007)
16. Hema Rajini, N., Bhavani, R.: Classification of MRI brain images using k-nearest neighbor and artificial neural network. In: 2011 International Conference on Recent Trends in Information Technology (ICRTIT), pp. 563–568. IEEE (2011)
17. Wei, C., Kangling, F.: Multilevel thresholding algorithm based on particle swarm optimization for image segmentation. In: 2008 27th Chinese Control Conference, pp. 348–351. IEEE (2008)
18. Lorensen, W.E., Cline, H.E., Marching cubes: A high resolution 3D surface construction algorithm. In: ACM SIGGRAPH Computer Graphics, vol. 21, pp. 163–169. ACM (1987)
19. Laycock, S.D., Bell, G.D., Mortimore, D.B., Greco, M.K., Corps, N., Finkle, I.: Combining x-ray micro-ct technology, 3D printing for the digital preservation, study of a 19th century cantonese chess piece with intricate internal structure. J. Comput. Cult. Heritage (JOCCH) 5(4), 13 (2012)

Automated Processing of Micro-CT Scans Using Descriptor-Based Registration of 3D Images

Jakub Kamiński$^{(\boxtimes)}$, Bartłomiej Trzewiczek, Sebastian Wroński, and Jacek Tarasiuk

Faculty of Physics and Applied Computer Science, AGH University of Science and Technology, al. Mickiewicza 30, 30-059 Krakow, Poland
`kaminski@fis.agh.edu.pl`

Abstract. Under the study, 6 bovine femur heads and overall number of 70 dissected cuboid specimens of cancellous bone were intended for micro-CT. Descriptor-based approach was used for 3D images registration from different resolutions using translation and rotation invariant local geometric descriptors based on 3D Laplace filter and nearest neighbours identification using 6-dimensional scalar vectors. Presented approach is simple and effective and can be processed using macros for ImageJ tool. Obtained accuracy of registration with error lower than 1 pixel allows for further analysis of bone mechanical properties, enabling precise determination of orientation for anisotropy and therefore the study of behaviour of the bone under load.

Keywords: Micro-ct · Image processing · Volumes registration · Bones

1 Introduction

The problem of registration of 3D volumetric images is well known in medical imaging and usually it is accomplished through sample independent fiduciary markers [1]. Because the amount of fiduciary markers available for registration is generally low, error analysis is the most widely studied rather than efficiency of matching of multiple markers with only partial overlap [2]. As opposed to medical imaging, in the robotics and automation studies there is interest in localization of large amounts of objects that are usually extracted from images and compared with data from databases to determine their orientation and type [3]. The use of local descriptors instead of complete images for matching is applied in many fields including computer vision and image registration [4], as well as robotics and autonomous systems [5]. Therefore, approach that combines the idea of using fiduciary markers, local descriptors and applying global optimization for whole problem might potentially be used to register 3D complex volumetric structures efficiently. Such a descriptor-based approach was originally used for data from multi-view technique Selective Plane Illumination Microscope (SPIM) and can register an arbitrary number of partially overlapping point clouds [6]. Although it

© Springer International Publishing AG 2017
J. Świątek and J.M. Tomczak (eds.), *Advances in Systems Science*, Advances in Intelligent Systems and Computing 539, DOI 10.1007/978-3-319-48944-5_7

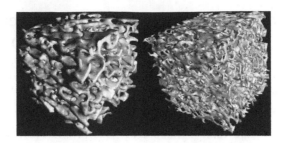

Fig. 1. 3D reconstructions of cuboid cancellous bone specimens from different sites of bovine femur obtained using micro-CT imaging (mesoscopic level).

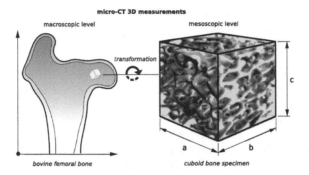

Fig. 2. Visualization of the input data registration consisting of 3D images from micro-CT measurements of bovine femoral bone using two different resolutions for 3D scans.

was defined for SPIM, due to its efficiency the method meets the requirements for the registration of 3D measurements from MicroComputed Tomography (micro-CT) - high-resolution non-destructive imaging technique based on X-rays [7].

2   Problem Statement

3D reconstructions of cuboid cancellous bone specimens from different sites of bovine femur obtained using high-resolution micro-CT imaging are illustrated in Fig. 1. As can be seen there, even within the same macroscopic bone, such volumes of interest are uniquely characterized by a complex network of connections with specific pore sizes and thickness of the trabeculae. Therefore the problem of digital registration of such complex structures, measured in different resolutions and orientations as illustrated in Fig. 2, requires the application of efficient computerized techniques and extraction of local image features. Developed automated methodology that uses such an approach is illustrated in Fig. 3.

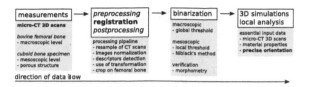

Fig. 3. Scheme of the automated methodology that uses descriptor-based registration.

3 Materials and Methods

The Difference of Gaussian (DoG) function provides a close approximation to the Laplacian of Gaussian (LoG) and produce the most stable image features compared to a range of other possible solutions, such as the Hessian, gradient or Harris corner function [4]. DoG function can be used to define local descriptor of a bead for the image I approximating 3D Laplace filter $\nabla^2 I$ and detecting all beads with sufficient accuracy while effectively suppressing high frequency noise. The local descriptor of a bead is defined by the locations of its 3 nearest neighbours in 3D image space ordered by their distance to the bead. All local minima in a $3 \times 3 \times 3$ neighbourhood in $\nabla^2 I$ represent intensity maxima - sub-pixel location is then estimated by fitting a 3D quadratic function to this neighbourhood. The DoG detector identifies beads even if they are close to each other.

Based on the above assumptions Descriptor-Based Registration (DBR) plugin for open source ImageJ tool [9] was originally developed for 2D or 3D images by Stephan Preibisch [6]. DBR uses translation and rotation invariant local geometric descriptors based on 3D Laplace filter and nearest neighbours identification using 6-dimensional scalar vectors. To efficiently extract the nearest neighbours in image space kd-tree implementation of the WEKA framework is used and DoG function is fixed with a standard deviation of 1.4 pixels and 1.8 pixels respectively. To eliminate false correspondence of extracted nearest neighbours, learning technique Random Sample Consensus (RANSAC) is used on the affine transformation model to estimate parameters of a model by random sampling of observed data [8]. Result of RANSAC is then followed by robust regression. Originally plugin performs all the steps of the registration pipeline: bead segmentation, correspondence analysis of bead-descriptors, outlier removal (RANSAC and global regression), global optimization and fusion.

In order to perform the automated registration of two 3D images from different micro-CT measurements, plugin modification were necessary. Therefore DBR was enriched with the option to adjust the input 3D images by the histogram-based normalization of the volumes. The implemented algorithm selects the cut-off level of the histogram to a minimum following a global maximum. Thereby data from lower-resolution micro-CT measurements can be fitted to data from higher-resolution scans allowing for precise registration.

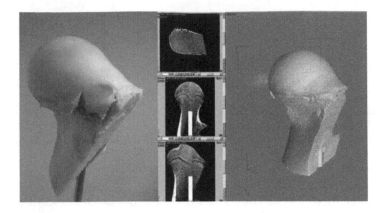

Fig. 4. Part of the femur head before micro-CT measurement (on the left), cross-sections of the measurement in three perpendicular directions (in the center), 3D reconstruction of the measured bone (on the right).

4   Results and Discussion

In the presented work 6 bovine femur heads and overall number of 70 dissected cuboid specimens of cancellous bone were intended for micro-CT. The experimental part of the research was conducted in Laboratory of Nano and Micro-tomography (AGH University of Science and Technology, Cracow). All micro-CT measurements were performed using the X-ray nanotomograph Nanotom S (General Electric) with supply voltage of 100 kV and current of 160 µA and isotropic voxel (high-resolution 6.5 µm for 10 mm cuboid bone specimens and low-resolution 45.6 µm for femur heads). 3D reconstruction of the structures using Feldkamp algorithm was performed using datosX software (ver. 2.1) dedicated to GE Nanotom S. Pre-processing of the data was performed in the VG Studio Max (ver. 2.1) as illustrated in Fig. 4.

Descriptor-based approach was used for 3D volumes registration of bones from different resolutions, namely automated matching of high-resolution cuboid bone specimens (to register) with low resolution femur heads (reference) after resampling to fit the resolution of 4 times resampled low-resolution data. The number of descriptors was set to value of 3, redundancy for descriptor matching to 1, significance required for a descriptor match to 1 and 3D rigid transformation model using maxima only type of detection. The result of such a registration is illustrated in Fig. 5. Figure 6 shows a percentage value of the descriptors match in function of cut-off level for one of the statistically representative volume of cancellous bone - automatically selected cut-off level value was marked using vertical dashed line. Similar curves were obtained for all of the cuboid specimens and in most cases it results in registration with a lower error and higher number of matched descriptors – several percentage points greater value in average.

Morphometrical analysis of the high-resolution structures after binarization using automatically determined global threshold was performed using ImageJ

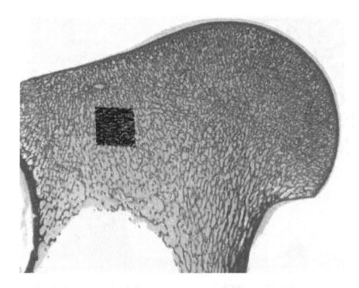

Fig. 5. The graphical result of descriptor-based registration of cuboid cancellous bone sample and femur head (overlapping stacks).

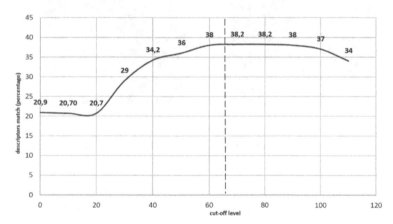

Fig. 6. Percentage value of the descriptors match in function of cut-off level for the histogram for lower-resolution bone measurements (automatically selected cut-off level value was marked using vertical dashed line).

enriched with the plug-in BoneJ, where direct three-dimensional methods are implemented [10]. Similar analysis was performed for low-resolution structures after inverse transformation obtained through registration, appropriate cropping and binarization using Auto Local Threshold plugin for ImageJ and modified Niblack's method [11]. From the definition the threshold value for each pixel is computed with a window of radius r around it based on the local mean m, standard deviation s, multiplier k and offset c as follows: $T = m + k \cdot s - c$.

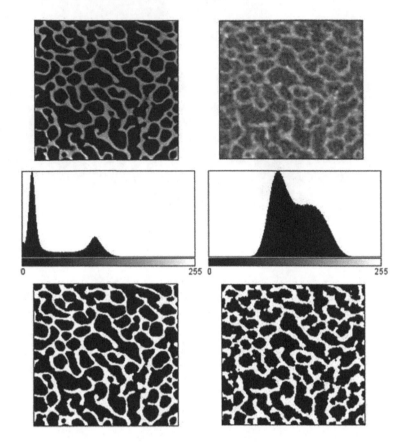

Fig. 7. The comparison between high-resolution and low-resolution gray-scale images after registration (on the top) and binarization (on the bottom) with the information about initial histograms (in the center).

After trial and error method with morphometrical verification, k was fixed to value of 0.43 and r to 17 pixels with c equals to 0 as in original algorithm. The comparison of stacks after registration and processing is illustrated in Fig. 7.

Regarding the effectiveness of the method of registration for different specimens, the mismatch problem due to a lack of finding an adequate number of descriptors or error value greater than 1 pixel was eliminated through bigger value of virtual cut for the femur heads measurements (reference). Therefore it is important to keep information about physical cut sites which can be useful in such cases. For all investigated cases it was possible to set a constant threshold parameter for a given femur head allowing for automation. Additionally implemented cut-off level calculation for reference 3D image based on histogram allowed to increase the number of matched descriptors and achieve error lower than 1 pixel for all specimens. Obtained accuracy of registration allows for further analysis of the anisotropy and mechanical properties with experimental validation based on in-situ compression tests and thereby a study of behaviour of the bone under load.

5   Conclusion

Automated processing of micro-CT scans was successfully applied to register 3D images of cancellous bone porous tissue from different resolutions. Descriptor-based registration approach used in the study allows for integration with developed automated methodology for verification of experimental mechanical properties of bone, being valuable tool for in vitro and in vivo studies in the future.

Acknowledgments. J. Kamiński. acknowledges benefit from Ph.D. scholarship by Marian Smoluchowski Cracow Scientific Consortium – KNOW.

References

1. Wiles, A., Likholyot, A., Frantz, D., Peters, T.: A statistical model for point-based target registration error with anisotropic fiducial localizer error. IEEE Trans. Med. imaging **27**, 378–390 (2008)
2. Fitzpatrick, J.M., West, J.B.: The distribution of target registration error in rigid-body point-based registration. IEEE Trans. Med. Imaging **20**, 917–927 (2001)
3. Canny, J.: A computational approach to edge detection. IEEE Trans. Pattern Anal. Mach. Intell. **8**(6), 679–698 (1986)
4. Lowe, D.G.: Distinctive image features from scale-invariant keypoints. Int. J. Comput. Vis. **60**(2), 91–110 (2004)
5. Kuipers, B., Byun, Y.T.: A robot exploration and mapping strategy based on a semantic hierarchy of spatial representations. J. Robot. Auton. Syst. **8**, 47–63 (1991)
6. Preibisch, S., Saalfeld, S., Schindelin, J., Tomancak, P.: Software for bead-based registration of selective plane illumination microscopy data. Nat. Meth. **7**(6), 418–419 (2010)
7. Stock, S.R.: MicroComputed Tomography: Methodology and Applications. CRC Press, Boca Raton (2008)
8. Fischler, M.A., Bolles, R.C.: Random sample consensus. Commun. ACM **24**(6), 381–395 (1981)
9. Abramoff, M.D., Magalhaes, P.J., Ram, S.J.: Image processing with ImageJ. Biophotonics Int. **11**(7), 36–42 (2004)
10. Doube, M., et al.: BoneJ: free and extensible bone image analysis in ImageJ. Bone **47**, 1076–1079 (2010)
11. Niblack, W.: An introduction to Digital Image Processing. Prentice-Hall, Englewood Cliffs (1986)

Gender Recognition Based on Speaker's Voice Analysis

Joanna Świebocka-Więk[✉]

Faculty of Physics and Applied Computer Science,
AGH University of Science and Technology, 30-059 Cracow, Poland
jsw@agh.edu.pl

Abstract. Methods for automatic gender recognition based on voice allow the person identification only by analysis of this person speech recording. Human speech, beyond verbal communication, gives many information about the speaking person, such as identification personal characteristics: gender, age, and even emotional state. This paper proposes an approach for personal gender recognition, based on their speech. The efficiency of the method was tested using speech synthesizer. It was demonstrated that the method is effective in the all of test cases.

Keywords: Voice recognition · Gender detection · Acoustic signal analysis · Speaker classification · Human identification · Biometrics

1 Introduction

Due to its characteristics, voice is regarded as one of the most common biometric features [1]. The human voice is unique in many aspects: both objective: linguistic (word choice), prosodic (accents, intonation), articulation (position of the tongue and palate) and subjective such as depth (the way of ejecting the air from the lungs) nasality (the amount of air that escapes through the nose), the height (the frequency of opening and closing of the vocal cords. On the basis of these and many other information many voice-based methods of automatic gender recognition have been developed. Mainly these approaches are based on objective features, which guarantee that obtained results will be independent from examiner's personal experience.

It is also worth to mention there are many potential applications of Automatic Speech Recognition (ASR) systems related to the biometrics, personal identification and verification [2]. The most common are: voice access control, banking services (phone identity confirmation) and criminal cases (evidence in the form of tape recordings or telephone conversations) [1]. For these reasons, in last few years we experienced very strong development of methods, associated with voice identification and automation of the procedure. Among the many approaches the methods of recognition the person speaking by phone, discussed in [3,4], like PPR (Parallel Phone Recognizer), Bayesian networks, mixed Gaussian model based on MFCC coefficients (Mel-Frequency Cepstral Coefficients) should be distinguished.

© Springer International Publishing AG 2017
J. Świątek and J.M. Tomczak (eds.), *Advances in Systems Science*, Advances in Intelligent Systems and Computing 539, DOI 10.1007/978-3-319-48944-5_8

2 Algorithm Description

The aim of the project was to create a program that recognizes gender of the speaker with the disposal of only a sample of the sound. The procedure was realized in the Matlab environment, due to its enormous capabilities and built-in database of subroutines [10]. Among other functions, a Fourier transform was applied, to analyze the frequency of signals [9], the command plot for the presentation of results and wavread command to load samples stored in *.wav format. The basic requirement of the algorithm is to provide recognition and reasonable voice analysis. For this purpose the program was equipped with a small database containing voice samples of four people; two women and two men. Registered polish words was chosen not to be phonetically similar; these are words "*część*", "*kot*", "*żóty*" (polish words meaning in order: part, cat and yellow). All samples contained in the reference database were recorded using a simple, standard microphone, a sampling frequency was 16 kHz with a resolution 16 B.

After a preliminary analysis of the collected samples, the background noise was identified (especially the noise of 50 Hz frequency which was removed each time before launching the program and further processing.

In Fig. 1 the waveforms sample signal ("kot") was presented before (left graph) and after (right graph) filtration. Filtering was performed automatically by writing *GetFreq* function, defining frequencies contained in the signal and identifying background noise. After that, by using another function created for the project, the redundant frequencies, detected GetFreq, were removed.

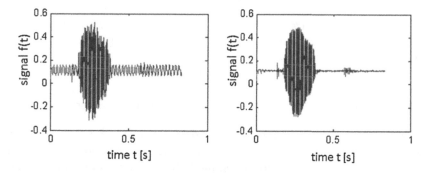

Fig. 1. Waveform for the male voice sample while pronouncing the word "kot" (polish word meaning cat) before (left graph) and after (right graph) noise removing filtration

After noise removing form the samples, analysis was proceeded. It was based on a few simple regularities observed during the samples frequency analysis.

In Fig. 2 the Fourier transform for all samples was shown (four persons, three samples for each person). The frequencies which were identified in men's voice were marked as blue, while frequencies which were identified in women's voice were marked as red. Points on the graph are the tops of the peaks obtained from the Fast Fourier Transform (FFT).

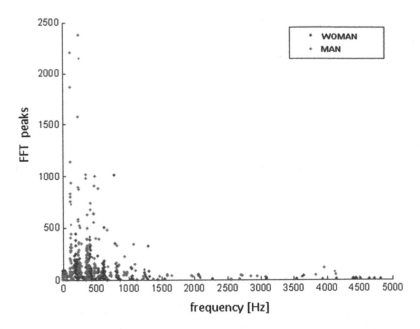

Fig. 2. Frequency spectrum with marked samples of the male (blue points) and female (red points) voice (Color figure online)

The following conclusions were drawn:

- male voice is characterized by high powers transmitted at low frequencies
- frequency spectrum for female's voice is more stretched (frequency comes to 5000 Hz);
- it can be clearly seen that certain frequency bands are reserved just for men, while others - for women,
- in case of women, the frequency of maximum power is greater

Taking above into consideration it was decided to use five separate criteria for the diagnosis and voice classification. If prediction was not explicit, a new decision was taken by voting and taking into account the overwhelming number of results. In the case of unreliability of some of the criteria which made decision impossible to take, program stated that the gender was not possible to recognize. The following criteria were chosen:

1. **Analysis of frequency and power**. The frequency of tested sample was marked in the graph presented in Fig. 2, after that program was checking how many points was situated in its closest environment (defined circles of a given radius around the tested sample). On this basis decision was made; if the predominant number of points belonged to men - the program determined that the voice belongs to a man, and vice versa. In case no points were found inside the circle, the radius was increased and the testing procedure

was applied once again. If the situation was repeated several times, the test was recognized as unreliable and it was interrupted. Initial radius was 10, increment was equal to 1 and number of tries was set up as 50 ($r = 10$, $i = 1$, $N = 50$).

2. **Analysis of the frequency bands.** The test is fairly similar to the previous one, except that now power is not analyzed. Surround of the tested samples was defined as vertical bands with certain width. The principles of the test evaluation were the same as in the case of frequency and power analysis; if the test for the first time gave negative results (none samples in the sample surrounding), the radius was increased.

3. **Analysis of the medium frequency.** The average frequency of samples of male and female in the database was set and compared with the average frequency of the test sample. The result of the comparison is a divergence (distance) that determines sex.

4. **Analysis of the frequency of maximum power (I).** Next procedure was to check the distance between the frequency at which the maximum power of the tested sample is transmitted and the frequency of the maximum transmitted power for both men and women.

5. **Analysis of the frequency of maximum power(II).** The test is fairly similar to the previous step, although also power not only distance between frequencies was taking into account. The distance between the tops of the peak maximum was measured.

The defined criteria, despite its simplicity give reasonable results. In all tested cases, the program coped with the task, even when a man deliberately used a high-pitched voice. The test samples were divergent patterns in content from the database. Another interesting test is an artificial male voice synthesis (reading program). All five proposed indicators pointed that the voice as belonging to a man.

The results presented by the procedure can be seen partially in the Fig. 3, partly in Matlab command window. In Fig. 3 red points presents women voice frequency, blue points presents men's voice and green are correlated with test sample. Means of mass was marked by circles, and the maximum power included in the squares. In the Matlab window, the sequence of tests can bee seen, and the results. Based on them, program decides whether it is dealing with a male or a female voice.

Following steps and analysis results for example word *"kot"* was presented below (M-men, W-women, R-surrounding radius):

1. Background noise filtration
   - Following frequency was detected in the background: F = 49.6709 Hz
   - Power transmitted on different frequencies P = 58.1778 Hz (background noise has been removed from the samples
2. Frequency analysis (Fifty most significant (with the highest power) frequencies from samples vector were selected.

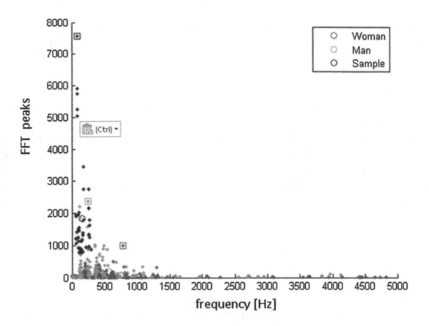

Fig. 3. The results of the algorithm. Summary of the position of the test samples (green points) in the frequency spectrum and samples characteristic of the male (blue points) and female (red points) voice (Color figure online)

3. Sample analysis
   - **TEST 1**: Analysis of frequency and power. Compatibility: W = 0 M = 2 R = 20
   - **TEST 2**: Analysis of the frequency bands. Compatibility: K = 79 M = 102 R(vertical band) = 3
   - **TEST 3**: Analysis of the medium frequency. Divergence (distance): K = 597 M = 411.
   - **TEST 4**: Distances frequency on whom is maximum power. Divergence (distance): K = 697 M = 167.
   - **TEST 5**: Distances between power maxima. Divergence (distance): K = 6612 M = 5203.
4. Ending the test
   - Number of tests not reliable: 0.
   - Number of results further features female: 0
   - Number of results closer masculine characteristics: 5
   - Procedure response: **It is probably a man.**

3   Summary and Conclusions

In this paper a gender recognition procedure was presented. The algorithm gave satisfactory and promising results. Implemented tests allowed to conclude that the method is effective in gender classification. In addition to testing for samples with known gender affiliation, further tests with un-typical voice (a high-pitched male voice) and speech synthesizer were applied. In both cases gender has been recognized properly.

Without a doubt, the algorithm requires further development and specialization. It is planned to perform the algorithm's evaluation on a large database of records to confirm its effectiveness. In further work is is considered to use ELSDSR *(Language Speech Database for Speaker Recognition)* base, containing recordings of voice messages of 23 people ranging in age from 24 to 63 years, including 13 men and 10 women [7]. This will further extend the analysis of the gender of the person speaking on the analysis of their age, as both of these features are often examined in correlation [5, 6]. Introducing the additional features might increase the algorithm efficiency in recognition and identification for the purpose of biometrics. Surely, proposed solution has relatively low complexity, and thus it is limited. It is difficult to define clear and unique criteria for distinguishing male and female voices; accomplish this would be a subject of more complex study and developing showed procedure.

Acknowledgement. The work was financed (co-financed) by the Polish Ministry of Science and Higher Education (MNiSW).

References

1. Bahari M.H., Van Hamme H.: Speaker age estimation and gender detection based on supervised non-negative matrix factorization. In: Biometric Measurements and Systems for Security and Medical Applications (BIOMS), pp. 1–6 (2011)
2. Metze, F., Ajmera, J., Englert, R., et al.: Comparison of four approaches to age and gender recognition for telephone applications. In: Acoustics, Speech and Signal Processing (2007)
3. Bocklet, T., Maier, A., Bauer, J.G.: Age and gender recognition for telephone applications based on GMM supervectors and support vector machines. In: Acoustics Speech and Signal Processing, pp. 1605–1608 (2008)
4. Chaudhari, S.J., Kagalkar, R.M.: A methodology for efficient gender dependent speaker age and emotion identification system. Int. J. Adv. Res. Comput. Commun. Eng. 4(7), 1–6 (2015)
5. Hubeika, V.: Estimation of gender and age from recorded speech. In: Proceedings of the ACM Student Research competition (2006)
6. Burkhardt, F., Feld, M., Muller C: Automatic speaker age and gender recognition in the car for tailoring dialog and mobile services. In: Proceedings of the Interspeech (2010)
7. Feng, L.: Speaker Recognition. Technical University of Denmark, DTU, Informatics and Mathematical Modelling (2004)
8. Jitendra A.: Effect of age and gender on LP smoothed spectral envelope. In: Proceedings of the Speaker Odyssey. IEEE (2006)
9. Zielinski, T.: Cyfrowe przetwarzanie sygnalow. Od teorii do zastosowan, Wydawnictwa Komunikacji i Lacznosci WKL (2013)
10. http://www.mathworks.com

Topic Modeling Based on Frequent Sequences Graphs

Piotr Ozdzynski and Danuta Zakrzewska(✉)

Institute of Information Technology, Lodz University of Technology,
ul. Wolczanska 215, 90-924 Lodz, Poland
dzakrz@ics.p.lodz.pl

Abstract. Huge amount of documents in the digital libraries requires automatic and efficient techniques for their management. Topic modeling is considered as one of the most effective method of automatic document categorization. In the paper, contrarily to using "bag of words", phrase based topic modeling is considered. We propose a methodology, which consists in building frequent sequences graph and finding significant word sequences. Graph structure makes possible selecting sequences of words which are characteristics for different topics. The methodology is evaluated on experiments performed on real document collections. The results are compared with the ones received by using LDA algorithm.

1 Introduction

Nowadays there have been arising huge collections of documents, which effective exploring and browsing have become challenging tasks. Big amount of data available in repositories such as digitized libraries resulted in growing needs of effective methods of information management. Automatic document categorization seems to be one of the crucial job in this area. It consists in assigning to a document one or more predefined classes, to which the document may belong. As one of the recently developed approaches to document classification there should be mentioned topic modeling, where each document can be labeled with topic names. Topic models are based on the idea that documents are mixtures of topics, where a topic is a probability distribution over words [1]. Topic modeling consists in mining the collections through the underlying and constantly reappearing topics. Such approach can be used for searching similar documents, what plays important role in information retrieval tasks.

Most topic modeling algorithms consider "bag of words" text representations and use only single words to depict topics. As human interpretation of a text is rather based on the recognition of the meaning of phrases than on the separated words, current topic modeling methods use phrases instead of words to build a model. In the paper a topic modeling approach, based on frequent sequences graph is considered. The proposed method aims at finding significant word sequences. The basic assumption of the considered approach is an ability to find the most informative sequences and omit meaningless phrases from an analyzed collection of text documents. These sequences together with documents

© Springer International Publishing AG 2017
J. Świątek and J.M. Tomczak (eds.), *Advances in Systems Science*, Advances in Intelligent
Systems and Computing 539, DOI 10.1007/978-3-319-48944-5_9

in which they occurred can be further used to find document topics. Phrases and connections between them are analyzed by building graph structures. Using graphs makes possible to select sequences of words which are characteristic for different topics. Weights assigned to graph edges indicate number of sequence occurrences in documents.

The remainder of the paper is organized as follows. In the next section the topic modeling approaches are described. Then the proposed methodology including techniques of finding frequent sequences and relations between them is depicted. Next experiments carried out on real document collections are discussed. Finally, some concluding remarks and future research are presented.

2   Topic Modeling Methods

Topic modeling problem consists in automatic discovering topics from a collection of documents. The central computational task for topic modeling concerns using observed documents to discover the hidden topic structure, such as per-document topic distributions, and per-document per-word topic assignments. Such approach can be regarded as "reversing" of the generative process. Papadimitriou et al. [2] defined topic models as probability distributions on terms. They considered a set of documents as a combination of terms from the selected universe. However they assumed that documents are not represented by terms but by the underlying (latent, hidden) concepts referred by the terms. They proposed using techniques from linear algebra to capture hidden document structure. Accordingly they introduced the information retrieval method, known as Latent Semantic Indexing. Blei et al. considered Latent Dirichlet Allocation (LDA) method [3,4] based on a generative probabilistic model of a corpus. The documents are represented as random mixtures over latent topics, where each topic is characterized by a distribution over words. Formally a topic is defined to be a distribution over fixed vocabulary. There is assumed that topics are specified before any data has been generated. Then, for each document in the collection, the process is divided into two stages: (1) the distribution over topics is randomly chosen; (2a) the topic is randomly chosen; (2b) words from the corresponding distribution over the vocabulary are being selected.

This statistical model assumes that documents are represented by multiple topics. Each document demonstrates topics with different proportion (step 1); each word in each document is drawn from one of the topics (step 2b), with selected topics chosen from the per-document distribution over topics (step 2a).

In [5] there is considered another strategy for topic modeling. The authors introduced the framework named KERT. The technique is based on finding phrases instead of single words. Topical keywords are found out by LDA method. Then frequent sequences are generated using an efficient pattern mining algorithm FP-growth [6]. Candidate topical keyphrases are selected from the ones containing topical keywords. Then the phrase qualities are evaluated using a characteristic function, and ranked accordingly. Top ranked phrases are selected as a representation of the topic.

The similar approach is used in a method called Scalable Topical Phrase Mining from Text Corpora [7]. The technique is also based on the frequent sequences, however in this case frequent sequences are generated before applying the modified LDA method. The last technique is used for finding topical keyphrases. This modification is called PhraseLDA.

Wang et al. [8] considered semantic information in semi-structured contexts conveyed by hashtags. They constructed different kinds of hashtag graphs based on statistical information of hashtag occurrence in a crowdsourcing manner. Based on these hashtag graphs, they proposed a framework of Hashtag Graph-based Topic Model (HGTM). The method was applied to Twitter microblogs topic modeling [8].

3   Methodology

3.1   Method Overview

A collection of text documents will be represented as a graph, where documents and selected frequent sequences of words form its nodes. The edges connect sequences and the documents containing them. Weights assigned to edges are calculated as the reciprocal of the number of edges linking the node sequence and documents. Therefore, the sum of the weights of all edges of node sequence is equal to 1.

As a word sequence there will be considered an ordered list of consecutive words. Sequences $A = (a_1, a_2, a_3, ..., a_k)$ and $B = (b_1, b_2, b_3, ..., b_m)$ are equal if they are of the same length and in the both of the sequences the same words are at the same positions. The length of the sequence is calculated as the number of its words.

For the given threshold N a sequence will be considered as frequent if it appears more than N times in the whole document set. N is also called a sequence support. A frequent sequence is *closed* when there does not exist other frequent sequences including this sequence.

In the first step, the analyzed document collection is pre-processed by converting words to lower cases, removing stop-words, punctuation marks and numbers and finally applying stemming procedure. Then the graph with nodes consisting of documents and frequent sequences and respective edges is built.

In the next step the edges between sequence nodes are created. Each new edge connects directly a pair of sequence nodes. The weight of this edge is calculated on the basis of existing paths between two nodes. Such a path consists of two edges from one sequence node to the document node and from the document to the second sequence node. The amount of these paths is equal to the number of documents such that each of them contains both phrases. An example structure of connections between joined sequences is presented in Fig. 1.

The weight of the new edge W_E is a function of the number of paths and the edge weights, and is calculated by multiplying the number of paths and the lower value of weights of edges between the document and the sequence nodes:

$$W_E = c * min\left\{w_{e_1}, w_{e_2}\right\}. \tag{1}$$

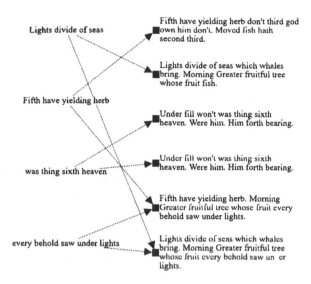

Fig. 1. A sample structure of connections between joined sequences.

where w_{e_1} and w_{e_2} are respectively weights of the sequence nodes outgoing edges and c is the number of pairs of edges, which are to connect. Thus connections between sequences, which frequently occur in the same document are preferred. If one sequence occurs more frequently than the other, then the smaller weight value is used to calculate the weight of an edge between sequences.

In further analysis the graph consisting only of sequence nodes and direct edges between these sequences is considered. Document nodes as well as edges to them are ignored. In the obtained graph there are distinguished the groups of nodes connected by edges of significantly higher weight values. Sequences, which represent these nodes are used to build up the topics for a set of documents in which they occur. For each group of phrases there can be assigned a set of documents containing at least one phrase from the group. Collections of documents related to phrase sets do not need to be separable, as one document may cover more than one topic. The architecture of the proposed system is presented in Fig. 2.

3.2 Building Frequent N-grams

The sequence consisting of n words will be referred to the name of the n-gram. In particular, the bigram and trigram will mean the sequence of the two and three words. In further considerations the word sequence will be used interchangeably with the word phrase. In the algorithm, each new frequent sequence of length $n + 1$ is built on the basis of the existing sequence of length n and information concerning bigrams location. This technique is derived from the observation that if a sequence of length $n + 1$ is frequent, all subsequences of the sequence are also frequent. Such approach is used in the algorithm apriori [9], which is

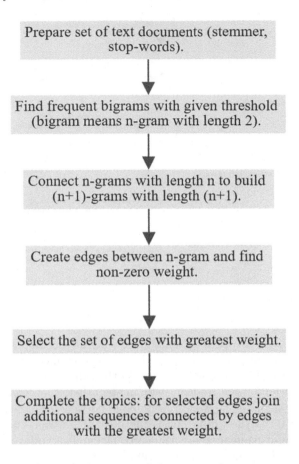

Fig. 2. Architecture of the proposed system.

a reference for many other algorithms for searching frequent patterns. Thus, we assume that frequent sequences of length $n + 1$ consist of frequent sequences of length n. A data structure, which stores all pairs of consecutive words as well as additional information about their positions is built. The occurrence of each pair is associated with a specific document and positions of pairs being an offset from the beginning of the document. As all the bigrams are indexed the structure is called the inverted bigram index.

To indicate only frequent sequences support threshold should be an input parameter. Further steps of the algorithm is performed only for these n-grams for which the number of occurrences is greater or equal to the threshold. Thus it is practical to sort bigram keys in a descending order. The starting set of frequent sequences is the set of bigrams thus $n = 2$. For each sequence of length n (denoted by $Q_i(n)$) a list of candidate sequences of length $n + 1$ ($Q_i(n+1)$) is created. All n-grams whose first $n - 1$ words are the same as the last $n - 1$ words of the starting n-gram are searched. Since this operation is repeated many times

Table 1. Pairs of phrases connected by edges with the largest weight

Lp.	Phrase 1 (occurrences)	Phrase 2 (occurrences)	Edge weight
1	year old (1488)	old (2892)	0.8690
2	potenti (341)	somatosensori evok potenti (51)	0.7875
3	depend (1079)	insulin (1226)	0.6559
4	neck (559)	head (591)	0.6192
5	valve (1207)	mitral (665)	0.4969
6	otiti media (144)	middl ear (137)	0.3030
7	bone marrow transplant (183)	graft versu host diseas (73)	0.2909
8	southern blot analysi (53)	t cell receptor (76)	0.2553
9	comput tomographi ct (137)	magnet reson imag mri (111)	0.1971
10	type iv collagen (50)	epidermolysi bullosa (52)	0.1428

it is reasonable to hold a map of n-grams in a memory. The list of potential sequences of the length greater by one is formed by joining two n-grams with the same subsequence of the length of $n - 1$. The new sequence has to be more frequent than the specified support threshold. Therefore it is necessary to count the number of times that the sequence occurs in the text. There is no need to search all the set of documents. It is enough to compare an array of positions of the sequence $Q_i(n)$ and the position of the n-gram that expands this sequence. If a candidate sequence $Q_i(n + 1)$ occurs in the text, the position of the ending is greater by one than the index of the starting n-gram. The example of such relationship is illustrated in Fig. 3.

After completion of the cycle of the algorithm for the next n all the data is stored in the structure similar to the one presented in Fig. 3. This graph structure allows to analyze the links between documents and sequences as well as the links between sequences of different lengths. Only the longest sequences are considered. The list is reduced to the closed frequent sequences, which are the ones not contained in longer sequences. Such filtration is linear to the length of all frequent sequences.

Sequences of type $\{a, b_1..b_k\}$ and $\{b_1..b_k, p\}$ are special cases. These kind of sequences are replaced with a substitute sequence $\{b_1..b_k\}$. Hence, single words which were not taken into account at the initial stage may appear in the result set. The selected sequences are used as the document representation, which will be applied in the next step of documents grouping.

3.3 Finding Significant Edges

Creating a graph of linked n-grams is equivalent to finding the coefficients in the square matrix of the size equal to the number of the n-grams. Each matrix element represent the weight of the edge between the two sequences. As weights

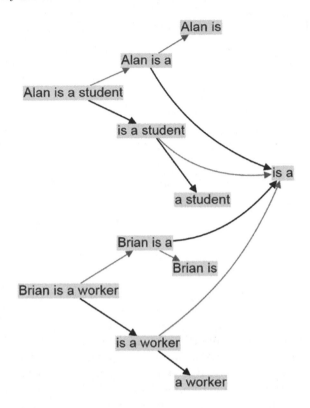

Fig. 3. A structure of connections between joined n-grams.

are not specified for loops, only elements below the main diagonal need to be calculated.

The complexity of these calculations depending on the number of sequences is expressed as $O(n^2)$. It can be reduced taking into account the fact that if two sequences do not occur together in a document, the weight will be equal to 0. It means that there is a lack of connection between the two nodes, so the edge is not created.

At the beginning, each document node needs to have assigned a list of n-grams, which it contains. This operation is made only once. List of document indexes containing the sequence is read from each frequent sequence. Each document which is on the list has added the reference to processed n-gram. Using the above relationship a list of n-grams connected via document is created for each n-gram and respective weights of edges are calculated.

According to (1), to determine weight values of created edges the number of paths should be calculated. Each n-gram is linked with a list of positions in documents. This list can be reduced to a list of indexes of documents. Since the list is created sequentially while reading the subsequent documents, their indexes are put in ascending order. For two sequences we search for common

Table 2. Selected groups of phrases

Lp.	Proposed method	LDA (Mallet 2.0.7)
1	**gene**, epstein barr viru, **receptor**, lymphoma, acut myeloid **leukemia**, t, t **cell receptor**, hepat, chemotherapi, t **cell** lymphoma, hemolyt anemia, achiev, late, southern blot **analysi**, symptomat	**cell**, human, alpha, growth, **gene**, express, beta, il, normal, **receptor**, cd, factor, **leukemia**, protein, dna, **analysi**, marrow
2	continu, **mg** kg, microgram, kg, infus, partial, achiev, **dai**, min, **dose**, median, m2, phase, nausea vomit	treatment, therapi, treat, **mg**, **dose**, group, studi, drug, **dai**, effect, receiv, placebo, respons, week, trial, oral
3	**plasma**, dose, depend, **concentr**, **glucos**, patient **insulin** depend **diabet**, **control subject**, **diabet**, **insulin**, beta, non, growth	level, **plasma**, serum, **insulin**, normal, **subject**, **concentr**, **glucos**, **diabet**, increas, **control**, elev, high, cholesterol, significantli, low
4.	neck, **chemotherapi**, free, head neck **cancer**, head, injuri, **primari**, **local**, brain, squamou **cell carcinoma**, advanc, trauma, femor	tumor, **cancer**, **carcinoma**, **cell**, **primari**, stage, malign, case, breast, **local**, grade, lymphoma, **chemotherapi**, radiat, metastat, bladder, tissu, small
5	**hiv infect**, **hiv**, lymphocyt, **human**, **human immunodefici viru hiv infect**, **human immunodefici viru**, **infect**, **posit**, **relat**, **human immunodefici viru type hiv**, **aid**	**infect**, **hiv**, **viru**, **human**, **aid**, **immunodefici**, **type**, **relat**, acquir, diseas, **posit**, viral, htlv, dna, case, hpv, clinic

pairs of indexes. For this purpose a binary search algorithm is used alternately. From the first list of indexes of documents the first one is selected and searched in the second list. Then next element is searched in the other list. The second list is shortened respectively no matter if a common pair is found or not. Assuming that the two lists of length k and l contain m common values, the computational complexity is equal to $m*(\log(k)+\log(l))$. As the number of pairs is calculated for each non-zero-edge, performance of the algorithm has a significant impact on the overall performance of the whole system.

3.4   Topic Modeling

All previously created edges have non-zero weights. The higher the weight, the more times phrases connected by the edge occur simultaneously in the document. Therefore, the edges are sorted in descending order taking into account weight values. The given number N of edges is selected from the sorted list. They comprise the set E of edges, which meet the given condition. There are considered all sequences connected by edges of the greatest weights. For these edges the maximum weight of the edge between all pairs of vertexes from the set E (v_{max})

is calculated. The lower the weight of that edge, the less frequently sequences from this edge occur together with sequences from the set E. Such edge is the preferred one to constitute the nucleus of a new topic. On the other hand, it is also important that the weight value of that edge is as big as possible. Therefore, the edge, for which the ratio of the weight of edges and v_{max} is the maximum, is included to the result set. Iteration is repeated until the set E reaches the expected cardinality N. The resulting collection contains pairs of sequences that occur in the document at the same time.

Each edge from the set E is a collection of two sequence nodes. This collection is enlarged by joining nodes connected to the both starting sequences by edges of the greatest weights. The group of sequences that determines the topic of the documents in which these sequences occur is created.

The lower number of edges is chosen the more consistent the topic is. On the other hand, not all the documents from the analyzed set will be linked to the topic. Therefore, the amount of attached adjacent vertexes should be selected in the way which will assure, that topics cover the greatest number of documents.

4   Experiment Results and Discussion

The proposed method was evaluated by experiments done on two document collections: the Ohsumed one (OM) [12] and 20Newsgroups corpus (NG) [13]. The first document collection contains medical abstracts from Medical Subject Headings categories of the year 1991 [12]. The 20 Newsgroups data set is a collection of newsgroup documents, partitioned across 20 different newsgroups. Topics were found for both of the document sets by using the presented method. In both of the cases collections of frequent sequences was divided into groups of related phrases, which represented topics related to documents. A threshold for a frequent sequence was set to 50.

For the first set (OM) frequent phrases were found in 21641 documents of the number of 23166 analyzed. This number covers over 93 % of all documents. Finding the key phrases of the remaining 7 % of the documents failed. In the case of NG set, 20417 documents have been examined, frequent phrases were found in 18377 of documents(90 %). Exemplary pairs of phrases connected by edges of the highest weight values, for OM set, are presented in Table 1.

Obtained results were compared to the ones got by LDA method implementation taken from Mallet 2.0.7 framework [10]. Table 2 shows the results obtained for the considered document collections. Selected phrases for both of the methods are presented in two columns. Fonts for phrases which appear in the both of the columns are bold. The number of presented phrases is limited to 16. Qualitative analysis of the results of both of the methods showed that in many topics keywords presented in both of the columns of Table 2 are repeated. However many differences between the words can be observed. Although the relevance of the selected words and key phrases is a subjective opinion based on knowledge of the document characteristics.

For quantitative analysis search engine based similarity scoring is used. The two considered methods for evaluating quality of topics are proposed by Newman et al. [14]. Generated topics are treated as queries. Two scores obtained from search results have been chosen. TITLES counts all occurrences of words from a topic that exists in top-100 search results. A LOGHITS score is defined as a log number of hits for a query. The second method of evaluation uses the Palmetto quality measuring tool for topics. All measures are obtained using the C_A parameter method [15]. C_A is based on a context window, a pairwise comparison of the top words and an indirect confirmation measure that uses normalized pointwise mutual information (NPMI) and the cosinus similarity. Both values are computed for 20 topics generated using LDA and the proposed method. The average values and standard deviations (σ) are presented for both of the sets in Table 3. LOGHITS score values are very similar for both of the obtained topic sets in the case of the Ohsumed document set. For the 20Newsgroups document collection LOGHITS value calculated by the proposed method is lower than in the case of LDA technique. Average number of words found in search results for topics indicated by the proposed method is greater than for the ones found by LDA. One can conclude that generated topics are more coherent than the ones obtained by using LDA. Results got from the Palmetto tool are very similar for both of the methods and both of the document sets.

Table 3. TITLES and LOGHITS results

	TITLES (σ)	LOGHITS (σ)	C_A (σ)
Proposed method (OM)	1554 (1731)	4.718 (1.844)	0.210 (0.108)
LDA (OM)	971 (344)	4.699 (1.919)	0.221 (0.053)
Proposed method (NG)	1319 (2317)	3.022 (2.587)	0.200 (0.110)
LDA (NG)	1187 (348)	5.465 (1.584)	0.183 (0.053)

As the advantage of the proposed method there should be mentioned appearance of complete phrases in the results. For example, let us consider row number 4 in OM collection. LDA method has found words *'carcionma'* and *'cell'*. The proposed technique shows that these words are parts of frequent sequence *'squamous cell carcinoma'*. Other examples can be noticed in row 5 where the significant part of words found by LDA forms common phrases: *'human immunodeficienci virus hiv infection'* and *'human immunodeficienci virus type hiv'*.

5 Concluding Remarks and Future Research

In the paper the topic modeling approach based on frequent sequences graph is considered. Building of graph structure enables selecting word sequences concerning different topics. Weights assigned to graph edges are connected with number of sequence occurrences and indicate their importance in the text. Experiments done on real documents have shown the big potential of the proposed method. Its performance has been compared with well known LDA method. As the main advantage of the proposed technique one should mention indicating longer phrases than in the case of LDA, what makes topic models more complete.

Future research will consist in further development of the proposed method. As one of the most important amendments there should be considered improvement of the filtration algorithm, which will enable avoiding the redundancy problem. In the current state the algorithm is adopted to non separable closed sequences. However as the result of their combination we may obtain two kind of sequences: not closed and containing them the closed ones. Solving this problem will definitely improve the performance of the algorithm.

Computing of edge weights is also worth investigations. Currently less frequent phrases appearing together are promoted. Considered improvement will consist in taking into account how close selected pairs of phrases appear in the document. Further amelioration will concern reduction of amount of edges for weights calculation taking into account their features. For example, for phrases which do not occur together, weights are always equal to zero. What is more, some of phrases are omitted in further computing. Including such premises into the algorithm will significantly diminish its computational complexity. Finally, in the proposed solution the resulting phrases are limited to directly connected nodes. It seems that taking into account nodes linked indirectly, but of sufficiently high edge weights will also improve the method performance.

References

1. Steyvers, M., Griffiths, T.: Probabilistic topic models. In: Landauer, T., McNamara, D., Dennis, S., Kintsch, W. (eds.) Latent Semantic Analysis: A Road to Meaning, pp. 1–15. Laurence Erlbaum, Hillsdale (2007)
2. Papadimitriou, C., Raghavan, P., Tamaki, H., Vempala, S.: Latent semantic indexing: a probabilistic analysis (1998)
3. Blei, D., Ng, A., Jordan, M.: Latent Dirichlet allocation. J. Mach. Learn. Res. 3, 993–1022 (2003)
4. Blei, D.: Probabilistic topic models. Commun. ACM 55(4), 77–84 (2012)
5. Danilevsky, M., Wang, C., Desai, N., Ren, X., Guo, J., Han, J.: Automatic construction and ranking of topical keyphrases on collections of short documents. In: SDM 2014 (2014)
6. Han, J., Pei, J., Yin, Y., Mao, R.: Mining frequent patterns without candidate generation: a frequent-pattern tree approach. Data Min. Knowl. Discov. 8(1), 53–87 (2004)
7. El-Kishky, A., Song, Y., Wang, C., Voss, C., Han, J.: Scalable topical phrase mining from text corpora. Proc. VLDB Endow. 8(3), 305–316 (2014)
8. Wang, Y., Liu, J., Huang, Y., Feng, X.: Using hashtag graph-based topic model to connect semantically-related words without co-occurrence in microblogs. IEEE Trans. Knowl. Data Eng. 13(9), 1–14 (2014)

9. Agrawal, R., Srikant, R.: Fast algorithms for mining association rules in large databases. In: Proceedings of the 20th International Conference on Very Large Data Bases, VLDB 1994, pp. 487–499. Morgan Kaufmann Publishers Inc., San Francisco, CA, USA (1994)
10. Machine learning for language toolkit. http://mallet.cs.umass.edu/
11. Hamming, R.W.: Error detecting and error correcting codes. Bell Syst. Tech. J. **29**(2), 147–160 (1950)
12. ftp://medir.ohsu.edu/pub/ohsumed
13. http://www.ai.mit.edu/people/jrennie/20Newsgroups/
14. Newman, D., Lau, J.H., Grieser, K., Baldwin, T.: Automatic evaluation of topic coherence. In: Human Language Technologies: The 2010 Annual Conference of the North American Chapter of the ACL, pp. 100–108 (2010)
15. Röder, M., Both, A., Hinnenburg, A.: Exploring the space of topic coherence measures. In: Proceedings of the Eighth ACM International Conference on Web Search and Data Mining, pp. 399–408 (2015)

Gaussian Process Regression with Categorical Inputs for Predicting the Blood Glucose Level

Jakub M. Tomczak[✉]

Department of Computer Science, Faculty of Computer Science and Management,
Wrocław University of Science and Technology, Wrocław, Poland
jakub.tomczak@pwr.edu.pl

Abstract. In diabetes treatment, the blood glucose level is key quantity for evaluating patient's condition. Typically, measurements of the blood glucose level are recorded by patients and they are annotated by symbolic quantities, such as, date, timestamp, measurement code (insulin dose, food intake, exercises). In clinical practice, predicting the blood glucose level for different conditions is an important task and plays crucial role in personalized treatment. This paper describes a predictive model for the blood glucose level based on Gaussian processes. The covariance function is proposed to deal with categorical inputs. The usefulness of the presented model is demonstrated on real-life datasets concerning 10 patients. The results obtained in the experiment reveal that the proposed model has small predictive error measured by the Mean Absolute Error criterion even for small training samples.

Keywords: Gaussian process · Categorical data · Diabetes · Nonparametric regression

1 Introduction

Diabetes is reported to be one of the most dangerous chronic disease that afflicted around 171 million people in the world in the year 2000 [26]. The increasing number of diabetics entails growing total costs of a treatment, *i.e.*, pharmacological treatment, hospitalization, laboratory test, medical visits, and constant patient health monitoring. Therefore, there is a need to propose personalized therapy to lower the costs and make the disease bearable for patients [22]. In diabetes treatment, the blood glucose level is crucial quantity for evaluating patient's condition. Understanding the influence of different factors on the glucose level and possibility to predict its values for new measurements would give opportunity to design therapy-effective decision-support systems.

First mathematical models for diabetes aimed at understanding the biochemical processes governing the blood glucose level. The mechanistic models were proposed to explain dependencies between the glucose level and insuline and food ingestion [9]. However, such approach fails in several apsects. First, it is troublesome to propose correct relations using dynamical systems. Second, in practice it

© Springer International Publishing AG 2017
J. Świątek and J.M. Tomczak (eds.), *Advances in Systems Science*, Advances in Intelligent Systems and Computing 539, DOI 10.1007/978-3-319-48944-5_10

is almost impossible to force patients to record all meals represented by ingested calories. Additionally, patients have tendency to forget to report all important information which results in unreliable models. Third, sometimes parameters of the models do not represent any physical quantities and hence the justification of applying the mechanistic models is put into question. Fourth, there are many external factors which affect the blood glucose level, *e.g.*, lifestyle, which cannot be included in the mechanistic models. Fifth, except numeric values of the blood glucose level, we can have access to a non-numeric (*symbolic*) description of the measurement, *e.g.*, day of a week, period of day, measurement's code representing insuline dose, or food ingestion. Typically, such information cannot be included in the mechanistic models.

All these issues cause a demand to formulate new models that allow to predict the blood glucose level for symbolic data. In other words, we need to propose a model for a relation between non-numeric inputs (symbolic variables representing measurements) and numeric output (the blood glucose level). The problem of symbolic variables is an important issue in the modern modelling and different types of symbolic data are distinguished. There are *categorical (nominal) variables* that take values in a finite set with no ordering, and *ordinal variables* that take values in a finite set with an ordering between the values but no metric notion is appropriate [1]. Moreover, there are *structural variables* that take values in sets of mathematical structures, *e.g.*, graphs [4].

The theoretical inquires about symbolic data for classification or regression models are forced by many practical applications, *e.g.*, nominal data in credit scoring [27] and medicine [22], structural data in biology [10], biochemistry [21], and chemistry [25]. Therefore, practice requires developing new models to cope with symbolic data [4]. There are methods for clustering, see [14], dimensionality reduction, see [18], mixture models, see [16], classifiers, *e.g.*, logistic regression [5], and regression models, *e.g.*, CART [6].

In this paper, we cope with the regression problem with categorical inputs, in which mechanistic models fail completely. Moreover, because of the specificity of the domain, we would like to apply a non-parametric model in order to avoid proposing explicit parametrization of the model. In machine learning, one of the most successful non-parametric regression model is *Gaussian process regression* [19]. It has been applied to numerous applications, *e.g.*, biosystems [3], discovering biomarkers in microarray gene expression data [7], chemical plants [17], non-linear system identification [24], predicting Quality-of-Service in Web service systems [23]. Additionally, Gaussian process regression allows to find a relation between any kind of inputs and output because similarity between objects is expressed by a kernel function. Hence, the core of the approach is to define appropriate kernel function for symbolic inputs [12,20].

The contribution of the paper is the following. First, the Gaussian process regression as the predictive model for the blood glucose level is outlined. Second, the covariance function for categorical inputs is presented. Third, the mean function for categorical inputs is proposed. Fourth, the learning of hyperparameters is outlined. Fifth, the experiment with real-life data is conducted.

2   Methodology

Let us consider a dataset $\mathcal{D}$ of N measurements of patient's blood glucose level. Each observation is represented by measurement's description, denoted as a vector of D categorical (nominal) variables $\mathbf{x} \in \mathcal{X}$,[1] and measured blood glucose level, $y \in \mathbb{R}_+$. The variables $\mathbf{x}$ will be called *inputs* and y – *output*. Further, we write $\mathbf{X}$ to denote a matrix of training inputs, and $\mathbf{y}$ – a vector of training outputs.

2.1   Gaussian Process Regression Model

In the regression problem it is assumed that there exists a mapping between inputs and output, denoted by $f(\mathbf{x})$, with an additive Gaussian noise

$$y = f(\mathbf{x}) + \varepsilon, \tag{1}$$

where ε is a zero mean Gaussian random variable with variance σ^2, that is, $\varepsilon \sim \mathcal{N}(\cdot|0, \sigma^2)$. If the mapping f is parameterized by $\mathbf{w} \in \mathbb{R}^D$, there exist a set of features ϕ transforming the original input space to a new space, and the mapping is linear with respect to parameters, that is, $f(\mathbf{x}, \mathbf{w}) = \mathbf{w}^\top \phi(\mathbf{x})$, then such model is known as *linear regression model* [5].

The linear regression models have limiations because they require explicit form of the features and the number of parameters. Therefore, it would be beneficial to assume that the mapping f is unknown and try to induce it from data. In the probabilistic (Bayesian) framework it is accomplished by treating the mapping f as a latent variable that results in obtaining a flexible non-parametric regression model. In fact, this is the idea standing behind the *Gaussian processes* [19]. The final regression model is the following:

$$
\begin{aligned}
y &= f(\mathbf{x}) + \varepsilon \\
f &\sim \mathcal{GP}(\cdot|\mu(\mathbf{x}), k(\mathbf{x}, \mathbf{x}')) \\
\varepsilon &\sim \mathcal{N}(\cdot|0, \sigma^2)
\end{aligned} \tag{2}
$$

where $\mathcal{GP}$ denotes the Gaussian process, $\mu(\mathbf{x})$ is the mean function, $k(\mathbf{x}, \mathbf{x}')$ is the covariance function.

There are two components in the model to be determined: the covariance function and the mean function for categorical inputs. Both issues are crucial to allow calculating similarities between measurements' descriptions. In order to solve these problems we need to propose proper kernel function for the covariance function and a domain-specific expression for the mean function.

---

[1] Further, in the experiment, we will consider only three inputs ($D = 3$) which are typical in the diabetes treatment, namely, day of a week, period of a day, and a measurement code. However, the presented idea is given in a general case for any number of inputs.

Covariance Function for Categorical Inputs. The covariance function of two function values corresponding to the inputs $\mathbf{x}$ and $\mathbf{x}'$ is a kernel function[2] [5,19], denoted by $k(\mathbf{x}, \mathbf{x}')$. Our goal is to propose a proper kernel function for the categorical inputs in the context of diabetes. Here, we restrict ourselves to nominal variables, however, in general, there are many possible kernels for other types of symbolic data like strings, trees, and graphs [12]. We propose to apply the following kernel function for nominal inputs:

Proposition *(Covariance Function for Categorical Inputs). Let $\mathbf{x}$ be a vector of categorical inputs, $x_d \in \mathcal{X}_d$, $\mathrm{card}\{\mathcal{X}_d\} < \infty$, for all $d = 1 \ldots D$ and there is no ordering between the values, and $\delta_d(\mathbf{x}, \mathbf{x}')$ be the Kronecker's delta,*

$$\delta_d(\mathbf{x}, \mathbf{x}') = \begin{cases} 1, \; if x_d = x'_d, \\ 0, \; otherwise. \end{cases} \tag{3}$$

Then the following function

$$k(\mathbf{x}, \mathbf{x}') = \prod_{d=1}^{D} \delta_d(\mathbf{x}, \mathbf{x}') \tag{4}$$

is a valid kernel function.

Proof. First, let us prove that the Kronecker's delta is a kernel function. From the definition of kernel function we need to show that for any set $\{x_n\}_{n=1}^{N}$

$$\sum_{i=1}^{N} \sum_{j=1}^{N} x_{d,i} \, \delta_d(\mathbf{x}_i, \mathbf{x}_j) \, x_{d,j} \geq 0.$$

The Kronecker's delta returns 1 if the two values are equal and 0 otherwise, thus we get a sum of squares of those objects which have equal values. The sum of squares for any objects is nonnegative that yields the Kronecker's delta is a valid kernel function.

Second, we need to prove that the product of Kronecker's deltas is also a valid kernel function. We use the fact that product of any valid kernels is also a kernel [5,20]. Hence, we get that the proposed kernel function for categorical inputs (4) is a valid kernel function. □

Our proposition of the kernel function could be presented in a simpler form as the Kronecker's delta for whole vectors $\mathbf{x}$ and $\mathbf{x}'$. However, we present the kernel in the given form (4) because of two reasons. First, we aim at distinguishing our proposition to the one proposed in [8] which is a sum of Kronecker's deltas (and kernels for continuous variables). Second, it is easier to interpret the product of the Kronecker's deltas for each input as a partition of the input space into single conjunctions of values.

---

[2] Kernel function is a symmetric function and the Gram matrix whose elements are given by $k(\mathbf{x}_n, \mathbf{x}_m)$ is positive semidefinite for any set $\{\mathbf{x}_n\}_{n=1}^{N}$ [20].

Mean Function. It is common practice to use Gaussian processes with a zero mean function [19]. However, explicit modelling of the mean function allows us to incorporate additional information about the considered phenomenon. In the case of diabetes and the categorical inputs we can take advantage on the character of the input space which is finite and propose different mean values for different combinations of values of selected or all inputs. In other words, the mean function can be parameterized as follows: for each combination of inputs' values a fixed nonnegative real number is assigned, namely

$$\mu(\mathbf{x}) = \mu_{\mathbf{x}}, \tag{5}$$

where $\mu_{\mathbf{x}} \in \mathbb{R}_+$ is a fixed value for given inputs $\mathbf{x}$.

The form of the proposed mean function requires to calculate as many values of mean as the cardinality of the input space which grows exponentially. However, we can limit the number of mean values by considering only selected inputs, *e.g.*, the code of the measurement, and do not include others in the calculations, *e.g.*, day of a week.

2.2 Prediction

Let us take a test measurement $\mathbf{x}_t$ for which we want to predict an output y_t. The similarities between the new observation $\mathbf{x}_t$ and the training examples $\mathbf{X}$ are defined by the kernel function $k(\mathbf{x}_t, \mathbf{x}_n)$ as in Eq. (4). We write $\mathbf{k}_t$ to denote the vector of covariances between the test point and the N training examples, and $\mathbf{K}$ is the kernel matrix for $\mathbf{X}$. According to the regression model in Eq. (2) we get the predictive distribution for given $\mathbf{x}_t$ with the following mean [19]:

$$\mu_t = \mu(\mathbf{x}_t) + \mathbf{k}_t^\top (\mathbf{K} + \sigma^2 \mathbf{I})^{-1} (\mathbf{y} - \boldsymbol{\mu}), \tag{6}$$

where $\boldsymbol{\mu}$ is a vector of means for $\mathbf{X}$, and variance [19]:

$$\text{var}(\mathbf{x}_t) = k(\mathbf{x}_t, \mathbf{x}_t) - \mathbf{k}_t^\top (\mathbf{K} + \sigma^2 \mathbf{I})^{-1} \mathbf{k}_t. \tag{7}$$

The predictive distribution is Gaussian distribution and thus the most probable value is chosen as the prediction, *i.e.*, $y_t = \mu_t$. Additionally, we can provide the uncertainty of the prediction which is equivalent to the variance $\text{var}(\mathbf{x}_t)$.

2.3 Learning

In practical applications, we need to determine a covariance function, a mean function and values of free parameters (*hyperparameters*), *e.g.*, the noise variance σ^2. While selection of the covariance function and mean function may be accomplished basing on the considered domain, setting the hyperparameters requires application of techniques of *model selection* [19]. We refer to the determination of values of the hyperparameters as *learning*.

Learning Mean Function. The mean function is parameterized by $\mu_{\mathbf{x}}$ for all possible combinations of values of selected inputs. Therefore, we propose to calculate mean values of outputs for all possible $\mathbf{x}$ as follows:

$$
\mu_{\mathbf{x}} = \begin{cases} \dfrac{\sum\limits_{\mathbf{x}_n \in \mathbf{X}} \mathbb{1}\{\mathbf{x}_n = \mathbf{x}\} y_n}{\sum\limits_{\mathbf{x}_n \in \mathbf{X}} \mathbb{1}\{\mathbf{x}_n = \mathbf{x}\}}, & \text{if } \sum\limits_{\mathbf{x}_n \in \mathbf{X}} \mathbb{1}\{\mathbf{x}_n = \mathbf{x}\} > 0, \\ 0, & \text{otherwise,} \end{cases} \tag{8}
$$

where $\mathbb{1}\{\cdot\}$ is the indicator factor.

Rationale behind the formula in Eq. (8) is that the prediction is made as a mean value of the same situations in the past. Such approach represents an assumption that patient's life is repeatable and her customs are essentially the same during a week. This is a manner how to incorporate *context* of daily routines into the model. On the other hand, the correlations among past observations are introduced by the covariance function.

Learning Covariance Function. In the literature, there are several approaches to model selection, *e.g.*, cross-validation, approximate methods like Laplace's Approximation, Variational Bayes, Expactation Propagation [19]. However, in this work we use the procedure based on the maximization of the marginal likelihood [5,19]. Once we have determined mean values using (8), we deal with one parameter only, *i.e.*, the variance of the noise σ^2. Let us denote the difference between outputs and means by $\bar{\mathbf{y}} = \mathbf{y} - \boldsymbol{\mu}$. Then the objective function is the log likelihood function in the following form:

$$
\ln p(\mathbf{y}|\mathbf{X}, \sigma^2) = -\frac{1}{2}\bar{\mathbf{y}}^\top (\mathbf{K} + \sigma^2 \mathbf{I})^{-1}\bar{\mathbf{y}} - \ln|\mathbf{K} + \sigma 2\mathbf{I}| - \frac{N}{2}\ln 2\pi, \tag{9}
$$

where $|\cdot|$ denotes the determinant of a matrix. Next, we need to calculate derivative of (9) w.r.t. σ^2 which leads to the following equation:

$$
\frac{\partial}{\partial \sigma^2} \ln p(\mathbf{y}|\mathbf{X}, \sigma^2) = -\frac{1}{2}\bar{\mathbf{y}}^\top (\mathbf{K} + \sigma^2 \mathbf{I})^{-1} \frac{\partial(\mathbf{K} + \sigma^2 \mathbf{I})}{\partial \sigma^2}(\mathbf{K} + \sigma^2 \mathbf{I})^{-1}\bar{\mathbf{y}}+
$$
$$
- \frac{1}{2}\mathrm{tr}\!\left((\mathbf{K} + \sigma^2 \mathbf{I})^{-1} \frac{\partial(\mathbf{K} + \sigma^2 \mathbf{I})}{\partial \sigma^2}\right), \tag{10}
$$

where $\mathrm{tr}(\cdot)$ denotes the trace of a matrix. Notice that $\frac{\partial(\mathbf{K}+\sigma^2 \mathbf{I})}{\partial \sigma^2} = \mathbf{I}$ which yields

$$
\frac{\partial}{\partial \sigma^2} \ln p(\mathbf{y}|\mathbf{X}, \sigma^2) = -\frac{1}{2}\bar{\mathbf{y}}^\top (\mathbf{K} + \sigma^2 \mathbf{I})^{-1}(\mathbf{K} + \sigma^2 \mathbf{I})^{-1}\bar{\mathbf{y}}+
$$
$$
- \frac{1}{2}\mathrm{tr}((\mathbf{K} + \sigma^2 \mathbf{I})^{-1}). \tag{11}
$$

Solving the optimization problem with the objective function given by (11) and without constraints requires inverting $(\mathbf{K} + \sigma^2 \mathbf{I})$. The computational cost is proportional to $O(N^3)$ and hence gradient-based optimization techniques can be successfully applied for reasonably small N.[3]

---

[3] By *reasonably small* we mean up to $N = 1000$.

3 Experiment

To evaluate the Gaussian process regression for predicting the blood glucose level a real-life datasets are used [11]. The original data covers 70 patients. Diabetes patient records were obtained from two sources: an automatic electronic recording device and paper records. The automatic device had an internal clock to time stamp events, whereas the paper records only provided "logical time" slots (breakfast, lunch, dinner, bedtime). Each patient's medical history corresponds to a period from 20 to 149 days of measurements, depending on a patient.

3.1 Data Description

Original diabetes files consist of four information per record: (i) date, (ii) time, (iii) code (categorical), (iv) blood glucose level (numeric). The code describes the measurement, *e.g.*, regular insulin dose, pre-lunch glucose measurement, typical meal ingestion, typical exercise activity, and others (details can be found in [11]).

The original records were transformed into the following inputs: x_1 – day of a week, $\mathcal{X}_1$ consists of the following values: Monday, Tuesday, Wednesday, Thursday, Friday, Saturday, Sunday, x_2 – part of a day, $\mathcal{X}_2$ consists of the following values: from 4:00 until 10:00, from 10:00 until 16:00, from 16:00 until 22:00, and from 22:00 until 4:00, x_3 – measurement code, $\mathcal{X}_3$ consists of 20 values, *e.g.*, insulin dose, measurement before breakfest.

In the experiment only 10 out of 70 patient records were used from which the smallest number of examples was 926 (116 days), and the biggest number was 1327 (149 days). The rest of records consist of too small number of observations to conduct statistically reliable experiments.

3.2 Experiment Details

Evaluation Metric. In order to evaluate the performance of the Gaussian process regression and compare it with other models we use the Mean Absolute Error (MAE). It is reported that this evalution metric is less sensitive to outliers in comparison to other metrics, *e.g.*, mean square error and root mean square error, and thus is preferred in forecast accuracy assessement [15].

Predictive Models. In the experiment, the prediction of the blood glucose level is made according to the following models:

1. *Mean Prediction* (MPred) is a model which always returns $y_{MP} = \dfrac{1}{N} \sum\limits_{n=1}^{N} y_n$
   as a prediction. This model is a baseline for comparing models in the experiment.
2. *Classification and Decision Tree* (CART) is a regression model which can be used for symbolic inputs. In this approach the input space is recursively partitioned due to given criterion which results in a tree-structured model. At each leaf a mean value of objects covered by decision criteria at each node up to the root. For details see [6].

3. *Gaussian process regression* (GP) model with the covariance function in the form (4), and the mean function (5) calculated only for the input x_3, that is, the measurement code.[4] This assumption results in 20 mean values to be determined. The prediction for new object is made according to (6).

Experiment Setting. The considered data in the experiment consists of 10 datasets representing different patients' records. Each dataset formulates time series. We decided to fix test set to contain last 300 examples in the time series (the most recent examples). However, the training set consists of varying number of observations equal 100, 200, 300, 400, 500, and 600. This aspect allows us to analyze the sensitivity of the considered models to different number of observations. Additionally, we use 100 examples before training set as a validation set to determine the noise parameter σ^2.

Implementation Details. The experiment was carried out in MATLAB environment. The GP regression model with the proposed covariance function and mean function were implemented in MATLAB. For CART model the built-in MATLAB implementation was used. In order to determine the hyperparameter σ^2 MATLAB optimization function was used with the objective function given by (11) which was calculated basing on data included in the validation set.

3.3   Results and Discussion

The results of the prediction of the blood glucose level averaged over 10 patients are presented as boxplots in Figs. 1 and 2. In the Fig. 1 the predictive models are compared for different size of the training set. In the Fig. 2 detailed performance on all predictive models are presented as a function of varying number of the training examples.

CART and Gaussian process regression performed significantly better than the baseline. CART and GP obtained mean MAE at the level of 23–27 mg/dl (see Fig. 2(a)), and 21.5-23 mg/dl (see Fig. 2(b)), respectively, whilst MPed – 77–78 mg/dl (see Fig. 2(c)). GP model behaved more stable for varying size of the training set than CART and it was enough to have $N = 300$ of training examples to obtain the best predictive accuracy (see Fig. 1). Moreover, for $N = 100$ and $N = 200$ the GP model performs only slightly worst (about 1.5 mg/dl) than for N greater than 300 (see Fig. 2(a)).

There are two main conclusions following from the experiment. First, both CART and GP models achieved good prediction accuracy and thus could be succesfully used in real-life applications. Second, the GP regression obtained better results than CART for smaller training samples. This issue is especially important from the practical point of view because of smaller memory requirements and lower computational costs. These aspects are crucial in modern eHealth systems [2], *e.g.*, as mobile services [13].

---

[4] We have omitted the day of a week and the part of a day because of two reasons. First, we wanted to have less parameters of the mean function. Second, in the preliminary experiments, including also x_1 and x_2 resulted in no significant change in the performance of the GP.

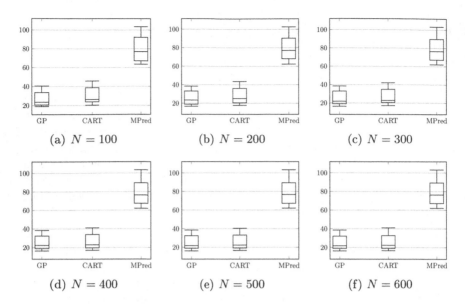

Fig. 1. Boxplots representing comparison between methods using MAE evaluation metric for changing number of training examples N. GP stands for Gaussian process regression, CART – Classification and Regression Tree, MPred – prediction with mean value basing on training examples.

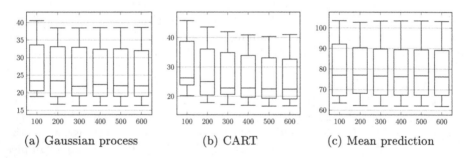

Fig. 2. Boxplots for each method with changing number of training examples (x-axis) and MAE evaluation metric (y-axis).

4   Conclusion

In this paper, we have presented the model based on Gaussian processes for the prediction of the blood glucose level. Considering the specificity of the problem, *i.e.*, the symbolic character of inputs, the covariance function and the mean function have been proposed. The learning of the hyperparameter, *i.e.*, mean values and the variance of the noise, has been presented using maximization of the marginal likelihood. At the end, the experiment with 10 real-life datasets has been conducted. The results indicate high predictive accuracy of the proposed approach (see Figs. 1 and 2). Moreover, our model can be easily implemented in mobile eHealth systems and this would be a focus of our future work.

Acknowledgements. The research is partially supported by the grant co-financed by the Ministry of Science and Higher Education in Poland.

References

1. Agresti, A.: An Introduction to Categorical Data Analysis. Wiley-Interscience, New York (2007)
2. Alemdar, H., Ersoy, C.: Wireless sensor networks for healthcare: a survey. Comput. Netw. **54**(15), 2688–2710 (2010)
3. Ažman, K., Kocijan, J.: Application of Gaussian processes for black-box modelling of biosystems. ISA Trans. **46**, 443–457 (2007)
4. Billard, L., Diday, E.: From the statistics of data to the statistics of knowledge: symbolic data analysis. J. Am. Stat. Assoc. **98**(462), 470–487 (2003)
5. Bishop, C.: Pattern Recognition and Machine Learning. Elsevier, Amsterdam (2006)
6. Breiman, L., Friedman, J., Olshen, R., Stone, C., Steinberg, D., Colla, P.: CART: Classification and Regression Trees. Wadsworth, Belmont (1983)
7. Chu, W., Ghahramani, Z., Falciani, F., Wild, D.: Biomarker discovery in microarray gene expression data with Gaussian processes. Bioinforma **21**(16), 3385–3393 (2005)
8. Daemen. A., De Moor, B.: Development of a kernel function for clinical data. In: Annual International Conference of the IEEE Engineering in Medicine and Biology Society (EMBC 2009), pp. 5913–5917. IEEE (2009)
9. De Gaetano, A., Arino, O.: Mathematical modelling of the intravenous glucose tolerance test. J. Math. Biol. **40**, 136–168 (2000)
10. Fischer, I., Meinl, T.: Graph based molecular data mining - an overview. In: IEEE International Conference on Systems, Man and Cybernetics, vol. 5, pp. 4578–4582. IEEE (2004)
11. Frank, A., Asuncion, A.: UCI machine learning repository (2010). http://archive. ics.uci.edu/ml
12. Gärtner, T.: A survey of kernels for structured data. ACM SIGKDD Explor. Newsl. **5**(1), 49–58 (2003)
13. Grzech, A., Juszczyszyn, K., Swiatek, P., Mazurek, C. Sochan, A.: Applications of the future internet engineering project. In: International Conference on Software Engineering, Artificial Intelligence, Networking and Parallel & Distributed Computing (SNPD), pp. 635–642. IEEE (2012)
14. Huang, Z.: Extensions to the k-means algorithm for clustering large data sets with categorical values. Data Min. Knowl. Discov. **2**(3), 283–304 (1998)
15. Hyndman, R., Koehler, A.: Another look at measures of forecast accuracy. Int. J. Forecast **22**(4), 679–688 (2006)
16. Iannario, M.: Preliminary estimators for a mixture model of ordinal data. Adv. Data Anal. Classif. **6**, 163–184 (2012)
17. Likar, B., Kocijan, J.: Predictive control of a gas-liquid separation plant based on a Gaussian process model. Comput. Chem. Eng. **31**, 142–152 (2007)
18. Makosso-Kallyth, S., Diday, E.: Adaptation of interval PCA to symbolic histogram variables. Adv. Data Anal. Classif. **6**, 1–13 (2012)

19. Rasmussen, C.E., Williams, C.K.I.: Gaussian Processes for Machine Learning. MIT Press, London (2006)
20. Shawe-Taylor, J., Cristianini, N.: Kernel Methods for Pattern Analysis. Cambridge University Press, Cambridge (2004)
21. Srinivasan, A., King, R.D.: Feature construction with inductive logic programming: a study of quantitative predictions of biological activity aided by structural attributes. Data Min. Knowl. Discov. 3(1), 37–57 (1999)
22. Tomczak, J., Gonczarek, A.: Decision rules extraction from data stream in the presence of changing context for diabetes treatment. Knowl. Inf. Syst. 34, 521–546 (2013)
23. Tomczak, J., Świątek, J., Latawiec, K.: Gaussian process regression as a predictive model for quality-of-service in web service systems. arXiv preprint arXiv: 1207.6910 (2012)
24. Turner, R., Deisenroth, M.P., Rasmussen, C.E.: System identification in Gaussian process dynamical systems. In: Görür, D. (ed.) NIPS Workshop on Nonparametric Bayes. Whistler, Canada (2009)
25. Węglarz-Tomczak, E., Vassiliou, S., Mucha, A.: Discovery of potent and selective inhibitors of human aminopeptidases erap. 1 and erap. 2 by screening libraries of phosphorus-containing amino acid and dipeptide analogues. Bioorg. Med. Chem. Lett. 26(16), 4122–4126 (2016)
26. World Health Organization. Definition and diagnosis of diabetes mellitus and intermediate hyperglycemia. Report of a WHO/IDF Consultation (2006)
27. Zięba, M., Świątek, J.: Ensemble classifier for solving credit scoring problems. IFIP AICT 372, 59–66 (2012)

Automated Information Extraction and Classification of Matrix-Based Questionnaire Data

Damian Dudek[(✉)]

Department of Information Technology, The President Stanislaw Wojciechowski
Higher Vocational State School in Kalisz, ul. Nowy Swiat 4, 62-800 Kalisz, Poland
d.dudek@pwsz.kalisz.pl

Abstract. Survey research is an important part of teaching quality evaluation systems in higher education. Although universities more and more
often use software with electronic questionnaires for surveying, the results
are not always satisfactory due to low response rates or high workload of
data processing. In this paper an effective model of interviewing, based on
compact matrix questionnaires is presented. Innovative methods for automated extraction, transformation and classification of survey responses
are introduced. They were validated on real-life university questionnaire
data showing very good performance, especially for closed questions.

Keywords: Information extraction · Pattern recognition ·
Classification · Questionnaire data · Survey research · Teaching
evaluation in universities

1 Introduction

Survey research is an important evaluation tool used by universities [2,3,7,16,
17,28]. Questionnaire response can be valuable feedback showing students' perception of classes, strengths and weaknesses of teaching methods, and possible
anomalies, which require appropriate reaction. All that can be used for systematic improvements of the educational process [24,25]. This kind of research has
become particularly important for Polish universities since 2010, when the higher
education reform and the National Qualification Framework (NQF) were introduced, putting stress on learning outcomes [20,22]. Of course, more and more
universities use computer systems for surveying, as they are supposed to effectively support this process as well as further data analysing [5,11,14,18]. In fact
it is not always the case. One of the major problems of electronic questionnaires
is poor response rate – sometimes even worse than feedback of classical paper
ones [14,18]. For example, *ProAkademia* [1] and *USOS* [9] are Polish computer
systems for higher education, whose usage for surveying can be affected by the
low-response problem [4]. Due to their interviewing model, electronic questionnaires are prepared by university workers and then made available for students,

© Springer International Publishing AG 2017
J. Świątek and J.M. Tomczak (eds.), *Advances in Systems Science*, Advances in Intelligent
Systems and Computing 539, DOI 10.1007/978-3-319-48944-5_11

so that they can respond to them at convenient time. As one survey is typically meant for evaluation of a single class, a complete set of questionnaires for all courses run within a given period can be rather extensive (possibly spanned on many pages), and filling them out may be time-consuming for a student. Furthermore, in order to access such a survey, participants need to authenticate themselves using personal accounts and credentials, which does not guarantee sufficient anonymity [27]. These factors may discourage students, leading to low response rates and having negative impact on reliability of collected data. Another problem is limited efficiency of traditional questionnaire data processing methods based on spreadsheets.

In this paper an alternative, effective surveying model based on concise matrix questionnaires is described. Filling out surveys is partly supervised, while true anonymity is provided by shared access keys. An innovative computer system for automated questionnaire response processing is introduced. It uses original algorithms for data extraction and classification: A-QDX and A-QCC. Experiments based on real-life university questionnaire response data showed good performance of the proposed solutions.

The remainder of the paper is organized as follows. The proposed, effective model of surveying is introduced in the Sect. 2. In the next section practical implementation of the model is described. Then, in the Sect. 4 experimental results based on real-life data are presented.

2   The Effective Model of Surveying

2.1   Basic Assumptions

In order to counteract problems described in the introduction, the following assumptions of the effective surveying model were made.

- *Concise sets of questions.* It should be possible to fill out an electronic questionnaire within a short period (e.g. 15–20 min).
- *Uniform grading scales and their interpretation* – for all evaluated criteria. It enables computing valid summaries and rankings of survey responses.
- *Providing true anonymity of participants* – shared access keys, no logging of student computers' IP addresses or response date and time.
- *Partly supervised responding to questionnaires.* Students are assigned a defined time and place of surveying – possibly during classes at university. A supervisor lets a group of students enter a computer laboratory, delivers a shared access key and preferably leaves them alone for a given time.
- *Providing open questions* for free-form comments with size restrictions only.

2.2   Formal Model

We define *a matrix survey schema* as a set $Q_m \equiv Q_c \cup Q_f$, where $Q_c \equiv \{q_{c1}, q_{c2}, \ldots, q_{cn}\}$ for $n \in \mathbb{N}$ is a set of closed questions, graded by students in a discrete number scale $G \equiv \{g_1, g_2, \ldots, g_k\}$ for $g_i \in \mathbb{Q}$, $k \in \mathbb{N}$, and

$Q_f \equiv \{q_{f1}, q_{f2}, \dots, q_{fm}\}$ for $m \in \mathbb{N}$ is a set of open questions with response length limited to $l_{maxj} \in \mathbb{N}$, $j \in \{1, 2, \dots, m\}$. Let $C \equiv \{c_1, c_2, \dots, c_p\}$ for $p \in \mathbb{N}$ be a set of classes, each of which is described by a vector $c_i = (s, f, t)$, where $s \in S$ represents a course title, $f \in F$ – a format of classes (e.g. lecture, project), and $t \in T$ – university teacher's data. We define *a response* to a questionnaire of a schema Q_m as a set $r \equiv \{r_c, r_f\}$, consisting of a set of closed questions response vectors: $r_c \subseteq Q_c \times C \times G$, $r_c \equiv \{(q, c, g) : q \in Q_c, c \in C, g \in G\}$, and an open question response vector $r_f \equiv (a_{qf1}, a_{qf2}, \dots, a_{qfm})$, where a_{qfi} is an answer to a question $q_{fi} \in Q_f$. A set of all questionnaire responses r within a given surveying round is represented by a symbol R. A survey of a schema Q_m is called a *matrix* one, because a set r_c of closed question vectors for a single response can be represented by the following matrix M_{rc}.

$$M_{rc}(Q_c, C, G) = \begin{pmatrix} g_{11} & \cdots & g_{1n} \\ \vdots & \ddots & \vdots \\ g_{p1} & \cdots & g_{pn} \end{pmatrix}, \tag{1}$$

where $g_{ij} \in G$ is a grade awarded to a class $c_i \in C$ for $i \in \{1, 2, \dots, p\}$ in response to a closed question (criterion) $q_j \in Q_c$, where $j \in \{1, 2, \dots, n\}$.

Example. The Higher Vocational State School in Kalisz uses a survey schema $Q_{PWSZ} \equiv \{q_{c1}, q_{c2}, q_{c3}, q_{c4}, q_{c5}, q_{c6}, q_{c7}, q_{c8}\} \cup \{q_{f1}\}$ with the following closed questions concerning teaching performance [6,16]: {q_{c1} – *teacher's command of the course content*; q_{c2} – *effectiveness of teaching methods*; q_{c3} – *interestingness of classes*; q_{c4} – *alignment of course scope and final assessment*; q_{c5} – *objectivity of students' assessment*; q_{c6} – *respect and good manners in interaction with students*; q_{c7} – *teacher's punctuality and conscientiousness*; q_{c8} – *teacher's availability to students during office hours*}, which are graded in a numeric scale $G_{PWSZ} \equiv \{2, 3, 4, 5\}$ with the interpretation: 2 – *unsatisfactory*, 3 – *satisfactory*, 4 – *good*, 5 – *very good*. It is also possible to enter an optional, free-form comment (which can be associated with some classes or fully independent) q_{f1} with maximum length of $l_{max1} = 2000$ characters.

2.3 Processing of Survey Results

The major statistical summaries, which are computed as part of survey response data processing, are defined below, using the notation introduced before in the Sect. 2.2 *Formal Model*.

Report 1. The *summary of average grades* for particular criteria of the Q_c set for a given set of classes $C_s \subseteq C$ is defined by the following S_{sc} matrix. For instance, it can refer to all the classes that a group of students attended during one semester of their studies, or a subset of classes taught by a given faculty member within some period.

$$S_{sc}(Q_c, C_s, G, R) = \begin{pmatrix} |r_{c1}| & \bar{g}_{11} & \cdots & \bar{g}_{1n} & \bar{a}_1 \\ \vdots & \vdots & \ddots & \vdots & \vdots \\ |r_{ck}| & \bar{g}_{k1} & \cdots & \bar{g}_{kn} & \bar{a}_k \end{pmatrix}, \text{ where } C_s \equiv \{c_1, \ldots, c_k\} \subseteq C,$$

(2)

$$|r_{ci}| = |\{(q, c_i, g)\} \subseteq r_c|, \quad \bar{g}_{ij} = \frac{1}{|r_{ci}|} \sum g_{ij}, \quad \bar{a}_i = \frac{1}{n} \sum_{j=1}^{n} \bar{g}_{ij},$$

where: $g_{ij} \in G$ is a single grade awarded to a class $c_i \in C_s$, $i \in \{1, 2, \ldots, k\}$ in response to a closed question (criterion) $q_j \in Q_c$, $j \in \{1, 2, \ldots, n\}$, $|r_{ci}|$ is the number of responses for the class c_i, $r \equiv \{r_c, r_f\}$, $r \in R$, and $\bar{a}_i$ is the overall average grade given to the class c_i for all closed questions $q_j \in Q_c$.

Report 2. The *standard average* $\bar{a}_s(C_t, R)$ of all the grades awarded within a response set R to all the classes $C_t \subseteq C$ taught by a given faculty member $t \in T$, is defined by the following formula:

$$\bar{a}_s(C_t, R) = \frac{1}{k} \sum_{i=1}^{k} \bar{a}_i,$$

(3)

where: k – is the number of classes taught by a given academic staff member t, graded in r_c responses, and $\bar{a}_i$ – is the overall average grade of a single class $c_i \in C_t$, computed using the formula introduced in the definition of the Report 1. The standard average reflects grading behaviour of whole groups of students.

Report 3. The *weighted average* $\bar{a}_w(C_t, R)$ of all the grades awarded within the response set R to all the classes $C_t \subseteq C$ taught by a given faculty member $t \in T$, is defined by the following formula:

$$\bar{a}_w(C_t, R) = \frac{\sum\limits_{i=1}^{k} \bar{a}_i \cdot |r_{ci}|}{\sum\limits_{i=1}^{k} |r_{ci}|},$$

(4)

where: k – is the number of classes taught by a given academic staff member t, graded in r_c responses; $\bar{a}_i$ – is the overall average grade of a single class $c_i \in C_t$, computed using the formula introduced in the Report 1 definition; $|r_{ci}|$ is the number of responses for the class c_i. The weighted average reflects grades awarded by individual students.

Report 4. The *average score* $\bar{a}_R(C, R)$ of all the classes C graded within a response set R is defined as follows:

$$\bar{a}_R(C, R) = \frac{1}{n \cdot p} \sum_{i=1, j=1}^{i=p, j=n} g_{ij}, \text{ given } M_{rc}(Q_c, C, G) = \begin{pmatrix} g_{11} & \cdots & g_{1n} \\ \vdots & \ddots & \vdots \\ g_{p1} & \cdots & g_{pn} \end{pmatrix},$$

(5)

where the response matrix M_{rc} is defined as in the previous formula (1).

Computational Example. In a survey research the questionnaire schema $Q_{PWSZ} \equiv \{q_{c1}, q_{c2}, q_{c3}, q_{c4}, q_{c5}, q_{c6}, q_{c7}, q_{c8}\} \cup \{q_{f1}\}$ and the grading scale $G_{PWSZ} \equiv \{2, 3, 4, 5\}$ were used. Five classes $C \equiv \{c_1, c_2, c_3, c_4, c_5\}$ were scored: c_1 – *Databases – lecture* (t_1 – *dr Jan Nowak*); c_2 – *Databases – project* (t_1 – *dr Jan Nowak*); c_3 – *Web development – lecture* (t_2 – *prof. Maria Kowalska*); c_4 – *Web development – workshop* (t_1 – *dr Jan Nowak*); c_5 – *Project management – lecture* (t_2 – *prof. Maria Kowalska*). Four survey responses were received, represented by the following M_{rc} matrices, in which rows are grade vectors for particular classes and columns reflect the eight consecutive criteria (closed questions).

$$
\begin{array}{cccc}
M_{rc1} & M_{rc2} & M_{rc3} & M_{rc4}
\end{array}
$$

$$
\begin{pmatrix}
5 & 4 & 3 & 5 & 4 & 4 & 5 & 5 \\
4 & 4 & 5 & 2 & 5 & 5 & 3 & 3 \\
2 & 3 & 4 & 3 & 2 & 2 & 2 & 4 \\
4 & 2 & 5 & 2 & 4 & 5 & 2 & 3 \\
4 & 4 & 4 & 3 & 2 & 4 & 2 & 4
\end{pmatrix}
\begin{pmatrix}
4 & 5 & 3 & 4 & 5 & 4 & 5 & 3 \\
4 & 5 & 5 & 5 & 4 & 4 & 5 & 5 \\
2 & 3 & 2 & 4 & 4 & 2 & 2 & 3 \\
5 & 5 & 5 & 4 & 4 & 5 & 4 & 4 \\
4 & 4 & 3 & 3 & 4 & 5 & 4 & 5
\end{pmatrix}
\begin{pmatrix}
4 & 4 & 5 & 2 & 5 & 5 & 3 & 3 \\
5 & 4 & 3 & 5 & 4 & 4 & 5 & 5 \\
4 & 4 & 4 & 3 & 2 & 4 & 2 & 4 \\
2 & 3 & 4 & 3 & 2 & 2 & 2 & 4 \\
2 & 3 & 2 & 4 & 4 & 2 & 2 & 3
\end{pmatrix}
\begin{pmatrix}
4 & 5 & 5 & 5 & 4 & 4 & 5 & 5 \\
4 & 2 & 5 & 2 & 4 & 5 & 2 & 3 \\
5 & 5 & 5 & 4 & 4 & 5 & 4 & 4 \\
4 & 5 & 3 & 4 & 5 & 4 & 5 & 3 \\
4 & 4 & 3 & 3 & 4 & 5 & 4 & 5
\end{pmatrix}
$$

A summary S_{sc} of average grades for particular classes, computed due to the Report 1 definition is as follows. The response numbers are placed in the first column, while average grades of classes are highlighted in the last column.

$$
S_{sc}(Q_c, C_s, G, R) = \begin{pmatrix}
4 & 4.25 & 4.50 & 4.00 & 4.00 & 4.50 & 4.25 & 4.50 & 4.00 & \mathbf{4.25} \\
4 & 4.25 & 3.75 & 4.50 & 3.50 & 4.25 & 4.50 & 3.75 & 4.00 & \mathbf{4.06} \\
4 & 3.25 & 3.75 & 3.75 & 3.50 & 3.00 & 3.25 & 2.50 & 3.75 & \mathbf{3.34} \\
4 & 3.75 & 3.75 & 4.25 & 3.25 & 3.75 & 4.00 & 3.25 & 3.50 & \mathbf{3.69} \\
4 & 3.50 & 3.75 & 3.00 & 3.25 & 3.50 & 4.00 & 3.00 & 4.25 & \mathbf{3.53}
\end{pmatrix}
$$

The standard average (Report 2) and the weighted average (Report 3) of all the grades of the staff member t_1 (dr Jan Nowak), who taught classes $\{c_1, c_2, c_4\}$, is 4.00, and for the faculty member t_2 (prof. Maria Kowalska), who taught classes $\{c_3, c_5\}$, it is 3.44. The average score of all the graded classes (Report 4) is 3.78.

2.4   Classification Problems

The general purpose of classification methods is assigning class labels to test cases within analysed data. Thorough overviews of various classification methods – especially in context of machine learning – can be found in the works [10,12,19,21,29]. In this paper the following two classification problems concerning questionnaire data are considered. The solution of the first one makes it possible to import and process survey responses using a computer system, which is independent of the surveying platform itself. The other classification helps to assign free-form comments to related teachers. In the below definitions the same notation is used as in the Sect. 2.2 *Formal Model*.

Problem 1. Let a matrix survey schema Q_m and a response set R be given. For each response $r \in R$, $r \equiv \{r_c, r_f\}$ determine an unequivocal assignment of each numeric grade $g \in G$ within closed questions answers r_c to one closed question $q \in Q_c$ and one class $c = (s, f, t)$ of a course $s \in S$, taught in a format $f \in F$ by a staff member $t \in T$.

Problem 2. Let a matrix survey schema Q_m and a response set R be given. For each response $r \in R$, $r \equiv \{r_c, r_f\}$, $r_f \equiv (a_{qf1}, a_{qf2}, \ldots, a_{qfm})$, assign each non-empty answer a_{qfi} for a free-form question $q_{fi} \in Q_f$, $i \in \{1, 2, \ldots, m\}$, to zero, one or many faculty members $t \in T$.

3 Practical Deployment of the Surveying Model

3.1 The Surveying System

The matrix-based surveying model introduced in the Sect. 2 has been practically implemented using an open-source system *LimeSurvey* [23]. An example questionnaire is pictured in Fig. 1.

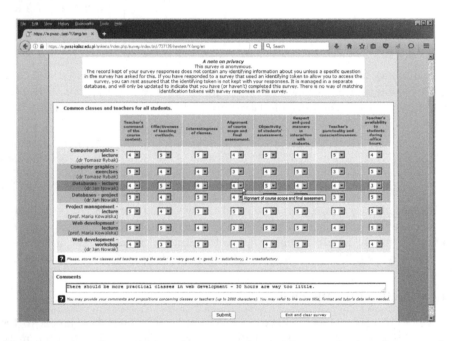

Fig. 1. An example matrix-based survey implemented using the *LimeSurvey* tool.

The *LimeSurvey* tool provides versatile survey management and export of results to CSV, XLS, DOC and PDF files. In order to protect our questionnaires from unauthorised access, the following methods and mechanisms were deployed.

- Access to each survey was restricted by a separate 10-character key (token), consisting of letters and digits, generated by the system and hard to guess.
- Students were delivered survey tokens and filled out the questionnaires only in computer laboratories at university (preferably during their classes).
- Each survey was accessible for a relatively short period, which was just sufficient for organised response.
- The client-server data transmission was encrypted using an SSL certificate.

3.2   The Questionnaire Analyzer

The proposed methods of questionnaire data processing (Sect. 2.3) were implemented in an original survey analyser using the PHP language [26] and the *Microsoft SQL Server 2014 Express Edition* (version 12.0) database system including the embedded *Transact-SQL* language [15]. The program automatically imports raw, semi-structural questionnaire responses, exported by the *LimeSurvey* system in the CSV (*comma separated values*) text format, transforms them into fully structural form and stores in tables of a relational database. The general schema of data structures, represented by a UML class diagram [13], is shown in Fig. 3. The analyser runs ETL (*extraction, transformation and loading*) processes, which resolve the classification Problem 1, defined in the Sect. 2.4. Then the application can be used for classifying free-form comments (the classification Problem 2) and generating reports defined in the Sect. 2.3. The general system architecture is shown in Fig. 2.

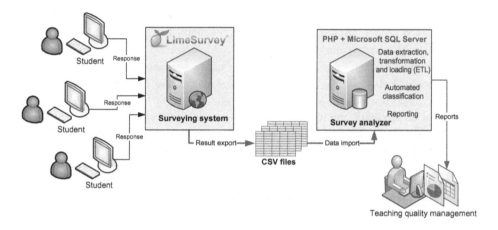

Fig. 2. The general architecture of the surveying and analyzing system.

Automated response classification (Problem 1, Sect. 2.4) is based on the proposed A-QDX (*Automated Questionnaire Data eXtraction*) algorithm, in which two major phases can be distinguished: (1) analysing, extracting, transforming and loading of header data, and (2) extracting and loading of response data into relational structures. During the first stage content of each cell in the header row is examined using PCRE regular expressions [8] in search of occurrences of: classes data, course name, class format; academic title or degree, first and last name, position (optional) of a faculty member, survey question text. Typical structure of a header row cell is shown below.

> *Classes – common teachers for all students. [Databases – lecture(dr Jan Nowak, Assoc Prof)][Effectiveness of teaching methods.]*

Particular header data are extracted and inserted into appropriate database tables. The procedure also stores an index of each column, which contains answers concerning a given faculty member and his or her classes. In the second phase the A-QDX algorithm reads answer values from the CSV file and unequivocally assigns them to classes and teachers – based on the column indexes determined and saved during the first phase. The A-QDX algorithm is specified below using the notation, which was introduced in the Sect. 2.2 *Formal Model*.

Algorithm A-QDX. Automated Questionnaire Data eXtraction.

Input. A survey schema Q_m; a source CSV file with semi-structural responses.

Output. A set of unequivocal structural survey responses $R \equiv \{r : r \equiv \{r_c, r_f\}\}$.

Procedure.

0. Start.
1. Read the first row (header) from the CSV file; compute its column number k.
2. For each header column $j \in [1; k]$ repeat the steps 3–7.
3.   Read content of the column j.
4.   Extract teacher's $t \in T$ data: title, first name, surname, position (optional).
5.   Extract classes data: course name $s \in S$ and classes format $f \in F$.
6.   Extract a survey question – a closed one $q_c \in Q_c$, a free-form one $q_f \in Q_f$ or an auxiliary one (e.g. announcing optional classes).
7.   If teacher's, classes and question data have been extracted in the steps 3–6, save them in the database tables (without duplicates).
8. For each data row $i \in [2; p]$ read from the CSV file, where p is the number of all rows, repeat the step 9.
9.   For each column $j \in [1; k]$ of a row i, containing an answer to a question q, store the answer in the database tables with respect to an appropriate column index.
10. Return database tables – unequivocal structural questionnaire responses R.
11. End.

Automated classification of free-form comments (Problem 2) is based on the proposed A-QCC (*Automated Questionnaire Comment Classification*) algorithm, which assigns such notes to faculty members $t \in T$, who taught classes evaluated within a given survey. It is done in two phases through text pattern matching of: (1) teachers' last names, (2) course names together with class format names. The resulting set of assignments Y is stored in the *CommentsClassification* database table (Fig. 3).

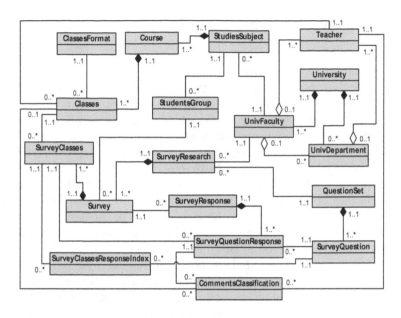

Fig. 3. Data structures of the questionnaire analyzer program (UML class diagram).

Algorithm A-QCC. Automated Questionnaire Comment Classification.

Input. A survey schema Q_m; a questionnaire response set $R \equiv \{r : r \equiv \{r_c, r_f\}$, $r_f \equiv (a_{qf1}, \ldots, a_{qfm})\}$

Output. A set Y of assignments (a_{qfi}, t), where $r_f \equiv (a_{qf1}, \ldots, a_{qfm})$ and $t \in T$.

Procedure.

0. Start.
1. Determine a set T of staff members, whose classes were evaluated in the response set R.
2. For each teacher $t \in T$ repeat the steps 3–5.
3.   Retrieve a set of comments A_t containing the last name of the teacher t.
4.   For each comment $a \in A_t$ repeat the step 5.
5. $Y := Y \cup \{(a, t)\}$
6. Determine a set C of classes, which were evaluated in the set R and were taught by single faculty members.
7. For each class $c \in C$, such that $c = (s, f, t)$, where $s \in S$ is a course title, $f \in F$ – class format, and $t \in T$ – teacher's data, repeat the steps 8–10.
8.   Retrieve a set of comments A_c containing combined names s and f.
9.   For each comment $a \in A_c$ repeat the step 10.
10. $Y := Y \cup \{(a, t)\}$
11. Return Y.
12. End.

4   Experimental Results

The proposed model of surveying and methods of automated response processing were experimentally verified with real-life data, which were acquired using the *LimeSurvey* system in the years 2014–2015 within teaching evaluation, which engaged students of computer science at the President Stanislaw Wojciechowski Higher Vocational State School in Kalisz. A total of 341 survey responses were obtained. At first they were processed in a traditional way (using spreadsheets) and then they were imported, transformed and analysed using the methods and software described in the Sect. 3.2. Both sets of results were compared with each other and the following measures were calculated (Table 1): (a) completeness and accuracy of closed question data extraction and classification using the A-QDX algorithm (with respect to the Report 1 for particular student participant groups – see the Sect. 2.3 *Processing of Survey Results*); (b) completeness and accuracy of free-form comment (open question notes) classification using the A-QCC algorithm as compared to classification done by a human expert.

As we can see in Table 1, perfect completeness and accuracy of the A-QDX algorithm were achieved, while completeness of the A-QCC algorithm was significantly lower as a result of variety and complexity of students' free-form comments. Dashed lines in some cells mean that mathematical expressions were inadequate to given situations (e.g. no comments within a survey). The A-QCC failed to properly classify a comment whenever a class name or teacher's last name were missing, or when they were given not in the nominative case.

Table 1. Results of the survey research and response data processing.

Students participant group	Survey date completed	Responses acquired	Number of students	Response rate	A-QDX complet.	A-QDX accuracy	A-QCC complet.	A-QCC accuracy
2 INF S	2014-11-24	21	30	70 %	100.00 %	100.00 %	100.00 %	100.00 %
4 INF S	2014-11-21	33	34	97 %	100.00 %	100.00 %	0.00 %	- - - - -
6 INF S	2014-11-24	25	45	56 %	100.00 %	100.00 %	100.00 %	- - - - -
2 INF N	2014-11-22	9	10	90 %	100.00 %	100.00 %	- - - - -	- - - - -
4 INF N	2014-11-22	13	15	87 %	100.00 %	100.00 %	0.00 %	- - - - -
6 INF N	2014-11-22	15	20	75 %	100.00 %	100.00 %	0.00 %	- - - - -
1 INF S	2015-04-14	21	32	66 %	100.00 %	100.00 %	100.00 %	- - - - -
3 INF S	2015-04-10	22	30	73 %	100.00 %	100.00 %	0.00 %	- - - - -
5 INF S	2015-04-10	19	35	54 %	100.00 %	100.00 %	100.00 %	100.00 %
7 INF S	2015-01-23	40	44	91 %	100.00 %	100.00 %	100.00 %	100.00 %
3 INF N	2015-04-11	4	10	40 %	100.00 %	100.00 %	- - - - -	- - - - -
5 INF N	2015-04-11	9	15	60 %	100.00 %	100.00 %	71.43 %	100.00 %
7 INF N	2015-01-24	15	16	94 %	100.00 %	100.00 %	0.00 %	- - - - -
2 INF S	2015-06-09	17	25	68 %	100.00 %	100.00 %	- - - - -	- - - - -
4 INF S	2015-06-12	26	30	87 %	100.00 %	100.00 %	100.00 %	100.00 %
6 INF S	2015-06-12	28	35	80 %	100.00 %	100.00 %	71.43 %	100.00 %
4 INF N	2015-06-13	9	11	82 %	100.00 %	100.00 %	0.00 %	- - - - -
6 INF N	2015-05-30	15	15	100 %	100.00 %	100.00 %	44.12 %	100.00 %
Total		**341**	**452**	**75 %**	**100.00 %**	**100.00 %**	**47.44 %**	**100.00 %**

5   Conclusions

In the paper an effective, semi-supervised model of university surveying, based on concise electronic matrix questionnaires was presented. It enables high response rates as compared to alternative, unsupervised methods, and thus it can be recommended to other universities. The contribution of this work includes the formal model of matrix-based surveys (the first such one due to the author's knowledge), novel methods and the related computer system for automated questionnaire data extraction, processing and classification. Two original algorithms were proposed: A-QDX for extracting and classifying responses to closed questions, and A-QCC for classification of students' free-form comments. Experiments on real-life university survey data proved outstanding performance of the A-QDX algorithm, while the other one showed good accuracy, but clearly limited completeness – caused by complex and multi-variant nature of input text, written in natural language without any structural restrictions. Some improvements of the A-QCC algorithm could be achieved by using a thesaurus with synonyms and different syntactic variants of words (especially teachers' names), but further progress would need employing more sophisticated natural language processing and understanding.

References

1. APR System: ProAkademia Software (2016). http://www.aprsystem.com.pl
2. Basaran, M.A., Kalayci, N., Atay, M.T.: A novel hybrid method for better evaluation: evaluating university instructors teaching performance by combining conventional content analysis with fuzzy rule based systems. Expert Syst. Appl. **38**, 12565–12568 (2011)
3. Cano-Hurtado, J.J., Carot-Sierra, J.M., Fernandez-Prada, M.A., Fargueta, F.: An Evaluation Model of the Teaching Activity of Academic Staff. Valencia University of Technology, Valencia (2009)
4. Ciesielski, P.: Remarks for students' questionnaire evaluation of teaching. Jagiellonian University in Krakow (2012) (in Polish). http://ww2.ii.uj.edu.pl/~wladek/ankiety/OZD.1011.podsumowanie.sprawozda%F1.pdf
5. Cork, D.L., Cohen, M.L., Groves, R., Kalsbeek, W. (eds.): Survey Automation: Report and Workshop Proceedings. The National Academies Press, Washington, DC (2003). doi:10.17226/10695
6. Dudek, D.: Studies in Computer Science. Students' Evaluation Survey of Teachers and Classes in the Summer Semester 2014–2015. The President Stanislaw Wojciechowski Higher Vocational State School, Kalisz, Poland (2015) (in Polish)
7. Golding, C., Adam, L.: Evaluate to improve: useful approaches to student evaluation. Assess. Eval. High. Educ. **41**, 1–14 (2014). doi:10.1080/02602938.2014.976810
8. Hazel, P.: PCRE–Perl Compatible Reg. Expressions (2015). http://www.pcre.org
9. Inter-University Computerization Center: USOS–University Study-Oriented System (2016) (in Polish). http://muci.edu.pl
10. Kiang, M.Y.: A comparative assessment of classification methods. Decis. Support Syst. **35**, 441–454 (2003)

11. Kirchner, A., Norman, A.D.: Evaluation of electronic assessment systems within the USA and their ability to meet the National Council for Accreditation of Teacher Education (NCATE) Standard 2. Educ. Assess. Evalu. Account. **26**(4), 393–407 (2014). doi:10.1007/s11092-014-9204-3
12. Kotsiantis, S.B.: Supervised machine learning: a review of classification techniques. Informatica **31**, 249–268 (2007)
13. Larman, C.: Applying UML and Patterns: An Introduction to Object-Oriented Analysis and Design and Iterative Development, 3rd edn. Pearson Education, Upper Saddle River (2004)
14. Leung, D.Y.P., Kember, D.: Comparability of data gathered from evaluation questionnaires on paper and through the internet. Res. High. Educ. **46**(5), 571–591 (2005)
15. Mistry, R., Misner, S.: Introducing Microsoft SQL Server 2014. Technical Overview. Microsoft Press, Redmond (2014)
16. Moreno-Murcia, J.A., Torregrosa, Y.S., Pedreno, N.B.: Questionnaire evaluating teaching competencies in the university environment. Evaluation of teaching competencies in the university. J. New Approaches Educ. Res. **4**(1), 54–61 (2015). doi:10.7821/naer.2015.1.106
17. Moskal, A.C.M., Stein, S.J., Golding, C.: Can you increase teacher engagement with evaluation simply by improving the evaluation system? Assess. Eval. High. Educ. **41**, 286–300 (2015). doi:10.1080/02602938.2015.1007838
18. Nulty, D.D.: The adequacy of response rates to online and paper surveys: what can be done? Assess. Eval. High. Educ. **33**(3), 301–314 (2008)
19. Phyu, T.N.: Survey of classification techniques in data mining. In: Ao, S.I., Castillo, O., Douglas, C., Feng, D.D., Jeong-A Lee, J.-A. (eds.) Proceedings of the International MultiConference of Engineers and Computer Scientists (IMECS 2009), Hong Kong, Newswood Limited, pp. 727–731 (2009)
20. Polish Ministry of Science and Higher Education: Higher Education Reform. Warsaw, Poland (2011)
21. Reddy, T.A.: Classification and Clustering Methods. In: Reddy, T.A. (ed.) Applied Data Analysis and Modeling for Energy Engineers and Scientists. Springer, New York, Dordrecht, Heidelberg, London (2011)
22. Republic of Poland: Act of 27 July 2005 - Law on Higher Education. Official Journal of Laws of 2005, No. 164, item 1365, Warsaw, Poland (2005)
23. Schmitz, C.: LimeSurvey: An Open Source Survey Tool. LimeSurvey Project Team, Hamburg, Germany (2016). http://www.limesurvey.org
24. Stein, S.J., Spiller, D., Terry, S., Harris, T., Deaker, L., Kennedy, J.: Unlocking the impact of tertiary teachers' perceptions of student evaluations of teaching. Research Report. Ako Aotearoa National Centre for Tertiary Teaching Excellence, Wellington, New Zealand (2012)
25. Stein, S.J., Spiller, D., Terry, S., Harris, T., Deaker, L., Kennedy, J.: Tertiary teachers and student evaluations: never the twain shall meet? Assess. Eval. High. Educ. **38**, 892–904 (2013). doi:10.1080/02602938.2013.767876
26. The PHP Group: PHP page. http://php.net
27. The Polish Accreditation Committee. http://www.pka.edu.pl/en/
28. University of Exeter: Evaluating teaching: guidelines and good practice. Teaching Quality Assurance Manual (2014). http://as.exeter.ac.uk/academic-policy-standards/tqa-manual/lts/evaluatingteaching/
29. Vapnik, V.N.: An overview of statistical learning theory. IEEE Trans. Neural Netw. **10**(5), 988–999 (1999)

Cloud Computing

Evaluating Raft in Docker on Kubernetes

Caio Oliveira$^{(\boxtimes)}$, Lau Cheuk Lung, Hylson Netto, and Luciana Rech

Universidade Federal de Santa Catarina, Florianopolis, Brazil
caio.po@grad.ufsc.br, {lau.lung,luciana.rech}@ufsc.br,
hylson.vescovi@blumenau.ifc.edu.br

Abstract. In computing systems, some applications require high availability. The creation of copies improves availability, but keeping the copies synchronized requires the replication of the application state. Raft is a consensus algorithm that emerged with an easy understanding logic and a consequently well accepted solution. At infrastructure level, containers offer an alternative for replacing traditional virtual machines in cloud providers. This paper (This project was supported by CNPq proc. 401364/2014-3) evaluates the execution of Raft in physical machines and in Kubernetes, a container management system developed by Google and other companies. Results show similar performance for Raft in both environments.

Keywords: Raft · Performance · Kubernetes · Docker · Containers

1 Introduction

Data centers can manage resources dynamically in an efficient manner with the use of Virtual Machines [12]. This characteristic matches the nature of *cloud computing*, defined by NIST [7] as "a model for enabling ubiquitous, convenient, on-demand network access to a shared pool of configurable computing resources (...) that can be rapidly provisioned and released with minimal management effort or service provider interaction". Virtualization at level system, known as Containers [1], provides faster resource allocation, in comparison with virtual machines [3]. Docker is an example of container implementation [11]. Some companies interested in create standards for adopting containers as technology for improvement in resource management founded the *Cloud Native Computing Foundation* (CNCF) [14].

Google has a large experience with containers [16]. Kubernetes is an open source management system for Docker containers [1] which was presented as an initial result from CNCF. Some engineers of Borg [16] (the current container management system at Google) worked in the construction of Kubernetes. Kubernetes replicate containers with the aim of improving availability. Failed containers are recreated by Kubernetes, but the state of the application inside the container is not restored. External volumes can persist the state. However, the volumes should be protected against failures and concurrent access to the volume have to be controlled by the application.

© Springer International Publishing AG 2017
J. Świątek and J.M. Tomczak (eds.), *Advances in Systems Science*, Advances in Intelligent Systems and Computing 539, DOI 10.1007/978-3-319-48944-5_12

Raft [10] is an algorithm derived from Paxos [6]. It can be used to implement replicated state machines (SMR) [13] in local area networks (LAN). Raft emerged as a understandable algorithm, when compared to Paxos. Consequently, many Raft implementations became available[1] in various programming languages. With Raft, all requests send to any replicated container will be executed on all replicas, in the same order. Raft can be applied in Kubernetes at application level, i.e., inside containers. There are some evaluations of Raft [5,9] but its characterization of terms of latency and throughput are still incipient.

This paper brought an evaluation of Raft to provide state machine replication. We compare its execution in containers implemented by Docker on the Kubernetes environment with the execution directly in bare metal (i.e. physical machines). Although throughput is quite different and higher in physical machines, we found that clients observe similar latencies in both environments. Kubernetes provides many management features that could compensate the overhead perceived in the throughput measurements.

The remainder of this paper is organized as follows. Section 2 presents concepts about containers in Kubernetes. In Sect. 3 we take an overview in Raft. Section 4 brings the adaptions make to run Raft in Kubernetes. In Sect. 5 we evaluated the executions of Raft and finally in Sect. 6 we conclude the paper.

2   Containers and Kubernetes

Virtualization at system level appeared in the FreeBSD operational system as an extended version of the *chroot* command called *jail* [1]. Next, Solaris improved this resource, which was called *zone*. Containers are instantiated from static images. The state of a container can be defined by all data stored inside the container since its instantiation from the image. When a container is destroyed, its state is lost. It is not common to maintain session data inside containers. Thus, containers can be considered stateless virtual machines. Images of containers are usually small because they use a layered file system, e.g. `aufs`. That way, only files that do not exist in its host are effectively stored in the container image. With a level system virtualization and a layered file system, containers can be created and destroyed faster than traditional virtual machines [3]. It brings a more dynamic resource management capability to data centers. An example of container implementation is Docker [11].

Google created Kubernetes [1] as an evolution of its current management system called Borg [16]. Kubernetes has many features originally designed in Borg. For example, considers a web server and its logs, which are consumed by a log analyzer. Host these applications in different containers is recommended because it benefits the individual software maintenance. There containers should remain on the same machine, to avoid the effort of sending the log over the network. A feature of Borg called *Alloc* maintains containers together in the same machine. Kubernetes has a component called *POD*, which also maintains containers together on the same machine.

---

[1] raft.github.io.

Kubernetes is composed of machines (virtual or physical) called nodes (Fig. 1). The component *POD* contain one or more containers. Each POD receives a network address. Containers inside a POD can share resources such as external data volumes. A *firewall* forwards requests from clients to nodes in the Kubernetes cluster. Each request will be delivered to only one node. The load balance policy is defined by the rules of the firewall, which is a component external to the Kubernetes cluster. The *proxy* component forwards requests to PODs, which can be replicated or not. When PODs are replicated, requests are distributed in a round-robin fashion. PODs are managed by the *kubelet* component. The *kubelet* also sends data about monitoring of containers to the main node.

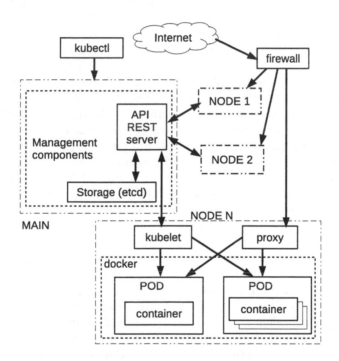

Fig. 1. Kubernetes architecture.

The *main* node of Kubernetes contains the management components. All information about the Kubernetes cluster is persisted in a storage component. The tool *etcd* [2] implements the storage in Kubernetes. Components interact via *REST* APIs. Components use the API Server to save and retrieve information. A human operation can interact with the cluster using the *kubectl* command interface. Some examples of commands are the creation of PODs, checking the health of the cluster and getting the description of PODs that are running.

3   Raft

Raft [10] is a consensus protocol designed to be of easy understanding, in comparison with Paxos [6] which is one of the most known consensus protocol. As consequence of the effort to create a simple protocol, many implementations of Raft are available [4]. Raft is already available in the *etcd* component[2] and can distribute the storage across the network, making it fault tolerant [15]. A set of replicas that have to maintain the same state is called a *configuration*. Raft ensure that replicas members of a configuration execute requests in the same order, what enables the maintenance of a replicated state on the replicas [13]. The system still works even if a minority of replica fail by crash. Formally, the system must have $n = 2f + 1$ replicas in the system, on which up to f replicas can fail. In Raft, crash failures are tolerated.

Replicas can act as leader, follower or candidate. An unique replica acts as a leader, determining the order of requests. Clients interact only with the leader. If a client interact with a follower replica, the client will be informed about the address of the leader replica. Follower replicas obey the proposal of the leader, running the requests in the established order. When a leader fails, elections are started and replicas can become candidate. The most updated replica in the system wins the election and becomes the new leader, sending heartbeat messages to other replicas to establish its authority and prevent new elections.

Each replica has to receive periodically answers from the leader, through a heartbeat mechanism. If leader does not answer after a timeout, replicas become candidate and start elections. Replicas use randomized timeouts (in a fixed interval, e.g., 150–300 ms) to prevent split votes. Initially, a candidate replica sends **RequestVote** messages to all replicas, which answer by **Vote** messages. A replica only does not accept a leader if this replica has an state more updated than the state of the proposer, and when a new leader was not elected yet. A new leader is elected when it is accepted by a majority of replicas.

An elected leader receives requests, assigns order to them and send a list of ordered requests to all replicas. The message which contains this list is called **AppendEntry**. On receiving this list, a follower replica updates its own list and answers the leader about this update. When the leader acknowledges that a majority of replicas (including itself) have new updated requests, the leader executes these new requests, sends answers to the clients (replies) and sends to replicas a list with the executed requests. The follower replicas also execute these requests and update their states. In Raft site [4] an animation allows the simulation of some actions (e.g. faults, request sending) in a quorum of five servers. In scenario the system can tolerate two faults, because the crash fault tolerance rule ($n = 2 * f + 1$) requires $n = 5$ servers to tolerate $f = 2$ faults.

---

[2] github.com/coreos/etcd/tree/master/raft.

4 Modifying Raft for Kubernetes

We choose one[3] of the many available implementations of Raft [4]. We choose an implementation in *Go*, the language on which Kubernetes was developed. Some modifications were done to allow the execution of Raft in the Kubernetes environment. The source code is available in GitHub[4]. We modified the original chosen implementation about the following characteristics:

- *Replica discovery.* Replicas communicate to each other via the IP address. When a replica enters in the system, it have to know the IP of the other replicas. To get this addresses, each replica was provided with a call to the *endpoints*[5] Kubernetes API, which returns a list with the IP of all replicas. This action is necessary because each container receives a dynamic IP during its instantiation. The API call has the format of an URL like `http://ip-of-main-node:port-of-api-server/api/v1/endpoints`. Usually, the port of the API server is the 8080 port. Replicas query the *endpoints* API periodically to keep updated with new replicas that entered in the system. These modifications are in the `caiopo/raft` repository, inside the `raft.go` program. This program also contains a hard-coded IP prefix like "18.16." (this IP prefix was defined inside the Flannel configuration (see Sect. 5.1). The *endpoints* API returns the IP of all containers. So, we compared the prefix with each returned address and consider only the IP of the our Raft replicated containers. It is possible to specify metadata in the creation of the container and get endpoints of an specific tag, but we did use this feature on our container.
- *Accepting of requests by any replica.* In Raft, only the leader can receive requests from clients. When a non-leader replica receives requests, they are ignored and the client is informed with the address of the leader. We modify Raft to allow each follower replica answer to clients about requests that they receive. It is mandatory because Kubernetes does an internal load balance (via `proxy` component, Sect. 2), delivering requests to any replicated container. This is done with a direct verification like "`if t.node.State == Leader`", being the block of code inside the *else* clause responsible by sending the requests to the leader and waiting for the answer. Modifications of this default behavior are in the `caiopo/pontoon` repository, inside the `http_transport.go` program. Another modifications in this program included the addition of commands like `hash`, which was used to get the hash over the log of executed commands in the container. We could compare the hash of all replicas in order to check if the state remains equals in all replicas after the execution of each test.

The modified source code of Raft is installed in a Docker container and is available in Docker Hub[6] under the tag `caiopo/raft`.

---

[3] github.com/mreiferson/pontoon.
[4] github.com/caiopo/pontoon and github.com/caiopo/raft.
[5] kubernetes.io/docs/api-reference/v1/definitions/#_v1_endpoints.
[6] hub.docker.com.

5   Evaluation

In this section we evaluate the performance of Raft in physical machines and in Kubernetes. Containers are virtual machines at system level. Therefore, the system will provide similar performance when running Raft in physical machines and in Docker containers on Kubernetes [3].

5.1   Experiment

An experiment was conducted to investigate the presented conjectures. We used a cluster with four computers Intel i7 3.5 GHz, QuadCore, cache L3 8 MB, 12 GB RAM, 1TB HD 7200 RPM. Machines are connected through an Ethernet 10/100 MBits network. Ubuntu Server 14.04.3 64 bits with kernel 3.19.0-42 was installed in the computers. The Kubernetes 1.1.7 was used. Docker containers communicate among them between physical machines via Flannel virtual network[7].

Raft was executed in Kubernetes and directly on the physical machines, using 3 replicas. The number of clients that simultaneously accessed Raft was varied following the sequence 4, 8, 16, 32 and 64. When using only Kubernetes, the number of replicas was three. To simulate the access of clients to Raft, the Apache HTTP server benchmarking tool[8] was used.

5.2   Results and Discussion

The evaluated latencies presented stable results, with low standard deviation (except for 4 and 8 clients), as presents Table 1. The performance of Raft is faster when running in physical machines. Although the overhead of Kubernetes starts high (111.1 %) for few clients, it decreases as more clients start to make requests, stabilizing around 21 % for 16 or more simultaneous access. Also represented in Fig. 2, latencies increase as more clients enter in the system.

Table 1. Measurements of Raft

Cli.	Req.	Kubernetes			Physical machine			Ku/Phy (%)
		Latency (ms)	St.Dev. (ms)	Time of test (s)	Latency (ms)	St.Dev. (ms)	Time of test (s)	
4	16000	19	62.8	75.3	9	1.0	34.5	111.1
8	12000	27	57.1	40.2	17	1.4	26.2	58.8
16	8000	42	2.3	12.1	35	2.1	17.3	20.0
32	8000	84	4.1	21.0	69	3.3	17.3	21.7
64	8000	169	9	21.2	139	7.7	17.5	21.6

---

[7] coreos.com/flannel.

[8] httpd.apache.org/docs/2.4/programs/ab.html.

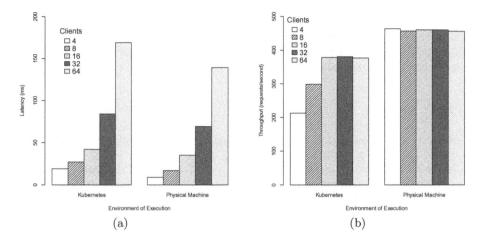

Fig. 2. (a) Latency and (b) Throughput of Raft in the evaluated environments.

Throughput of Raft increases as more clients access Raft in Kubernetes. For more than 16 clients, it remains around 380 requests per second. When Raft runs in physical machines, throughput is stable and presents the better value among all measurements. However, it does not scale as more clients send requests. This superior cut point can occurs because of a limitation in our chosen Raft implementation, which does not implement batch of messages.

Ongaro [9] have preliminary tests of performance in which throughput scales as more replicas enter in the system, but latency is evaluated considering only one client accessing the system. Another work [5] repeat the Raft authors' performance analysis, not considering latency perceived by clients nor throughput. Although executing Raft in Kubernetes presents throughput lower than when running in physical machines, we argue that the reduction of approximately 17.4 % is compensated by the benefits offered by the Kubernetes features.

6   Conclusion

This paper presents an evaluation of the Raft algorithm running on physical machines and in Docker containers managed by Kubernetes. We modified Raft to be capable of dynamically discover replicas and to allow clients send requests to any replica of Raft (i.e., not only the leader). Results show that Kubernetes provides competitive throughput when compared with the execution in bare metal. As more clients enter in the system, the latency perceived by the clients is similar in both environments.

Improvements in this work includes the usage of other consensus algorithms. For example, EPaxos [8] did not use leader, which could make better use of the load balance provided by Kubernetes. A more robust implementation of Raft can be used in order to achieve more efficient results. With containers, we can easily increase the number of replicas, when comparing with the context of traditional virtual machines. New experiments could consider scenarios with more faults. Applications that require high availability but are exposed to many risks (like sites exposed on Internet) are expected to suffer more faults than applications executed in local area networks or inside organizations.

References

1. Bernstein, D.: Containers and cloud: from lxc to docker to kuber-netes. IEEE Cloud Comput. **1**(3), 81–84 (2014)
2. CoreOS. etcd (2016). https://coreos.com/etcd. Accessed 12 May 2016
3. Felter, W., et al.: An updated performance comparison of virtual machines, Linux containers. In: International Symposium on Performance Analysis of Systems and Software, pp. 171–172. IEEE (2015)
4. GitHub. Raft (2016). http://raft.github.io. Accessed 12 May 2016
5. Howard, H., et al.: Raft refloated: do we have consensus? ACM SIGOPS Oper. Syst. Rev. **49**(1), 12–21 (2015)
6. Lamport, L.: The part-time parliament. ACM Trans. Comput. Syst. **16**(2), 133–169 (1998)
7. Mell, P., Grance, T.: The NIST definition of cloud computing. Technical report Spp. 800–145. Gaithersburg, MD, United States: National Institute of Standards & Technology (2011)
8. Moraru, I., Andersen, D.G., Kaminsky, M.: There is more consensus in egalitarian parliaments. In: Proceedings of the Twenty-Fourth Symposium on Operating Systems Principles, pp. 358–372. ACM (2013)
9. Ongaro, D.: Consensus: bridging theory and practice. Ph.D. thesis. Stanford University (2014)
10. Ongaro, D., Ousterhout, J.: In search of an understandable consensus algorithm. In: USENIX Annual Technical Conference, pp. 305–320 (2014)
11. Peinl, R., Holzschuher, F., Pfitzer, F.: Docker cluster management for the cloud - survey results, own solution. J. Grid Comput. **14**, 1–18 (2016)
12. Popek, G.J., Goldberg, R.P.: Formal requirements for virtualizable third generation architectures. Commun. ACM **17**(7), 412–421 (1974). ISSN:0001–0782
13. Schneider, F.B.: Implementing fault-tolerant services using the state machine approach: a tutorial. ACM Comput. Surv. **22**(4), 299–319 (1990)
14. Sill, A.: Emerging standards, organizational Patterns in cloud computing. IEEE Cloud Comput. **2**(4), 72–76 (2015). ISSN:2325–6095
15. Toffetti, G., et al.: An architecture for self-managing microservices. In: Proceedings of the 1st International Workshop on Automated Incident Management in Cloud, pp. 19–24. ACM (2015)
16. Verma, A., et al.: Large-scale cluster management at Google with Borg. In: European Conference on Computer Systems, p. 18. ACM (2015)

Performance Evaluation of MPTCP Transmission of Large Data Objects in Computing Cloud

Robert R. Chodorek[1][(✉)] and Agnieszka Chodorek[2]

[1] Department of Telecommunications, The AGH University of Science and Technology, Al. Mickiewicza 30, 30-059 Krakow, Poland
chodorek@agh.edu.pl
[2] Department of Information Technology, Kielce University of Technology, Al. Tysiaclecia Panstwa Polskiego 7, 25-314 Kielce, Poland
a.chodorek@tu.kielce.pl

Abstract. This paper discusses a problem with the multipath TCP transmission of large data objects in a computing cloud using the MPTCP protocol. The problem presented itself when a file transfer and the migration of a live virtual machine was observed. Experiments were carried out in a private cloud, built using the OpenStack software tool. Nodes of the test network were connected by Gigabit Ethernet links. Results show that while usage of the MPTCP gives an efficient file transfer and migration, the protocol is not able to fully exploit the potential of unloaded, high-speed paths due to restrictive congestion control.

Keywords: Computing cloud · Performance evaluation · MPTCP

1 Introduction

Cloud computing is a computing service (e.g. data processing, data storing), based on shared computer resources (e.g. servers, storage, applications), delivered to a wide range of users via an externally accessed network. Although the computing cloud (cloud) is a group of software and hardware devices, connected through an internal network, external users treat it as another, single endsystem. Cloud computing is becoming a key element of IT in many companies [1]. Cloud architecture is used for storage and processing of large data objects [1], complex computations and other services. Many companies run their services on virtual machines (VMs) that are hosted inside the cloud. Cloud computing increases the flexibility of service provisioning, scalability (if necessary, services may be cloned and run on the next VMs) and, primarily, reliability. Scalability and reliability is assured, amongst other things, by the possibility of the migration of a VM. For example, migration is required when one physical host running multiple VMs gets overloaded.

Many services, currently operating in the cloud, require the transmission of large data objects between components of a computing cloud. This is done both

© Springer International Publishing AG 2017
J. Świątek and J.M. Tomczak (eds.), *Advances in Systems Science*, Advances in Intelligent Systems and Computing 539, DOI 10.1007/978-3-319-48944-5_13

when a service works with large data [1] and during the migration of running VM instances [2]. For both of the above tasks, one of the critical factors is the performance and reliability of a network that connects the components of a cloud. The increase in performance is possible through the use of high-speed links (e.g. 40 and 100 Gigabit Ethernet in a data center [3]). Connection reliability is increased by creating multiple paths between components of a cloud.

The presence of multiple physical paths between the components of a computing cloud enables to apply multipath communication, where the transmission required by an application is performed by the simultaneous usage of multiple network connections. The transfer of large data objects between components of a computing cloud is performed typically using the Transmission Control Protocol (TCP). The TCP can be relatively easily replaced by the Multipath TCP (MPTCP) protocol, which was designed as a multipath version of TCP and allows the usage of the same socket interface functions. Thus the replacement of the TCP by the MPTCP does not require the rebuilding of applications and the realization of such a replacement in the cloud environment can be relatively fast. The concept of the applicability of the MPTCP for transmission in the cloud can be found in the [4]. However, this paper does not include tests of real MPTCP implementations. In [5] the authors present simulation analysis of the potential usage of MPTCP for data transmission in data centers and the dependence of MTCP performance on the topology and path load. In opinion of the Authors of [4], "multipathed communications are becoming essential to augment cloud communications". Cloud communications must also to stand up to the problems of the coexistence of elastic and inelastic traffic in heterogeneous networks [6].

The aim of this paper is to analyze the impact of the usage of the MPTCP protocol on the transmission of data inside a computing cloud. The problem is to transfer a large data object in as short a time period as possible. In our paper, large data objects are understood to be data objects large enough to be transmitted in a time period that - from a given cloud service point of view - is not negligible. There are three main types of transmission of large data objects between components of a cloud: data transmission (transfer of photographs, recorded videos, data files, etc.), migration of virtual machines, and real-time data streaming. The first two use reliable transport protocols (as TCP or MPTCP). The third uses real-time transport protocols.

In the paper, the impact of the usage of the MPTCP protocol on the transmission of large data objects inside a computing cloud is presented using the example of two experiments, representative of operations performed by a cloud. There is the transfer of a large file and the migration of a live VM instance. A private cloud, used for these experiments, was created using OpenStack software. In the paper we focus on two factors that affect the performance of the MPTCP - a congestion control method and the number of simultaneously transmitted flows.

The rest of this paper is organized as follows. Section 2 describes the MPTCP protocol. Section 3 shows the test environment. Sections 4 and 5 present the results obtained during the transmission of a large data objects (transfer of large files and migration of live VM instances). Section 6 concludes the paper.

2   MPTCP Protocol

The MPTCP protocol, specified in the document RFC 6824 [7], is an extension of the typical TCP protocol with transmissions via multiple paths. The protocol is able to send TCP packets using different network interfaces. Each path from a sender to a receiver is associated with one TCP flow. Using that concept, TCP flows can be understood as subordinates of the MPTCP. The MPTCP occupies an upper sublayer of the transport layer, while TCP instances (one for each path) are located at a lower sublayer.

The mechanisms of the MPTCP are an adaptation of typical TCP mechanisms for multi-path transmission. Minor modifications were made to the connection setup mechanism, error control (now based on two-lever sequence numbering) and flow control. Mayor modifications affected the congestion control.

The MPTCP congestion control is an extension of TCP's congestion control for multiple flows. The TCP's mechanism for congestion control is based on the congestion window. In the first phase of transmission, when the probability of congestion is low, the Slow Start algorithm increases the congestion window exponentially from 1 to a threshold value. If the congestion window exceeds the threshold value, it is increased linearly. The Slow start threshold is a conventional boundary between the low and high probability of congestion.

The basic method of the MPTCP'a congestion control - the Linked-Increases Algorithm (LIA), specified in the document [8] - defines an additional, total congestion window as a sum of the TCP congestion windows created for each flow. The LIA modifies the phase of the linear incrementation of the TCP congestion window. The speed of growth of a single TCP congestion window depends now on the reciprocal of the total congestion window and on the coefficient of the overall aggressiveness.

The modified LIA method - the Opportunistic Linked-Increases Algorithm (OLIA) - presented in [9] deals with a LIA tradeoff between optimal congestion balancing and responsiveness. The OLIA removes the total congestion window from the formula used for the calculation of the growth of the TCP congestion window. It also changes the definition and formula of the coefficient of aggressiveness, which now is calculated for each path.

While the LIA and the OLIA methods were focused on the growth of the congestion windows, the Balanced linked adaptation (Balia) [10] method describes the full evolution of the congestion window when there is a high probability of congestion. This method makes the congestion window a function of the maximum sending rate of a path and the sending rate of the given path.

In the above methods, the growth of the congestion window depends on the estimated round trip times (RTT). The Cubic method, presented in [11], decouples the window from the RTT, which can be helpful in multipath transmissions.

3   Test Environment

A test cloud was build using real network equipment (Fig. 1) and the OpenStack open source cloud operating system [12] enabling for running virtual servers.

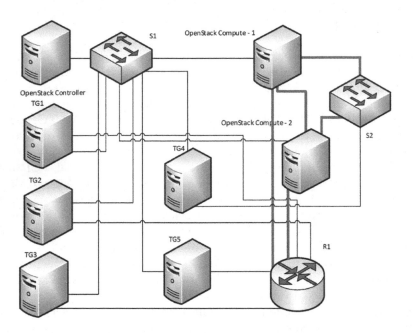

Fig. 1. Topology of test network (Color figure online)

The test system includes a computing cloud (OpenStack Controller and Open-Stack Computes) and a set of trafic generators (from TG1 to TG5). The imple-mentation [13] of the MTCP was used in the experiments. As a traffic generator, iPerf measurement tool [14] was used. The receiver part of the iPerf tool was implemented in the OpenStack Compute 2 node. Generated data were conveyed to this node from TG1...TG3 and TG5 via the R1 router and from TG4 via the S2 switch.

Nodes are connected using a gigabit Ethernet network that is built using twisted pair cables. All nodes are high-performance personal computers, equipped with one Ethernet card built into a motherboard and at least one dual-port external Ethernet cards. All build-in cards are connected with the common S1 switch. They are used for management purposes and transmit computing cloud signalling and experiment signalling. In the traffic generators, dual-port external cards are used for sending background traffic. These cards are connected to the R1 router (except the card mounted into the TG4, which is connected to the S2 switch). OpenStack Computer software runs on computers with two dual-port external network cards. The first cards are connected to the R1 router. Additional cards support virtualization. They are used both for making direct connection between OpenStack Compute 1 and OpenStack Compute 2 nodes and to connect these nodes via the S2 switch.

Signalling traffic and background traffic is transmitted using the TCP pro-tocol. The MPTCP transmission is carried out between OpenStack Compute 1 and OpenStack Compute 2 (four links, marked in Fig. 1 in red, are available to the MPTCP). Thus, the MPTCP protocol can use three different paths: a direct path (card-to-card connection), path via the S2 switch, and path via the router R1. The direct path is dedicated for MPTCP transmission and the other two paths are shared with TCP transmissions. Paths vary in length, presence of background traffic and used intermediate devices.

4   Transmission of Bulk Data

During the experiments, a large data file was transmitted from OpenStack Compute 1 node to OpenStack Compute 2 node using multi-path transmission. Paths available for multi-path transmission are marked in red (Fig. 1). The size of the file (10 GB) assured transmission time long enough (about 80 s at 1 Gbps) to eliminate the impact of the beginning and ending of transmission. Experiments were conducted using four alternative methods of congestion control, a variable number of paths (from 1 to 3), with and without background traffic.

Tables 1 and 2 present throughput of the MPTCP transmissions, calculated as the amount of transmitted data divided by transmitted time. Overall throughput was evaluated using total amount of transmitted data, while throughput of a single data flow covers only data conveyed by this flow. Values shown in Tables 1 and 2 are obtained by repeating each test 10 times and averaging the results. The confidence level is 95 %.

Overall throughput of MPTCP transmission in unloaded links (without any background traffic) is shown in Table 1. During tests, the method of the congestion control that was used was changed and the number of paths used for MPTCP transmission varies from 1 to 3. Paths were connected in a fixed sequence. The initial path was the one through switch S2. The second was the path via router R1. At the end, the direct path was switched on. The tests show that the changes to the sequence did not cause noticeable changes to throughput.

The results show that the usage of two paths (two simultaneously transmitted MPTCP flows) instead of one increases the MPTCP throughput by about 30 % in the case of 3 out of 4 congestion control methods (Cubic, LIA, OLIA) and by nearly 35 % in the case of the Balia method. Although the third path leads to the best direct connection (dedicated for internal traffic of the computing cloud), addition of the third path improves overall performance by only 10–15 %.

Note that only a single MPTCP flow utilizes more than 90 % of available bandwidth. Two and more MPTCP flows were able to occupy, respectively, 63 % of 2 Gbps and 48 % of 3 available bandwidth. It was experimentally verified that neither the MPTCP's receive window, nor PC performance caused that phenomena. This limitation is caused by the congestion control mechanism of the MPTCP protocol. Moreover, instantaneous throughput of the MPTCP transmission with the absence of background traffic (Fig. 2) shows repeatable, short-term collapses of transmission, caused by the congestion control mechanism implemented in the MPTCP.

As is shown in Table 1, the impact of the method of congestion control (Cubic, LIA, OLIA or Balia) on the performance of the transmission of bulk data in an

Table 1. Overall throughput of MPTCP transmission (no background traffic)

MPTCP congestion control	No. of MPTCP flows		
	1 flow	2 flows	3 flows
Cubic	0.97 ± 0.00 Gbps	1.26 ± 0.00 Gbps	1.39 ± 0.01 Gbps
LIA	0.95 ± 0.00 Gbps	1.24 ± 0.03 Gbps	1.38 ± 0.04 Gbps
OLIA	0.93 ± 0.00 Gbps	1.20 ± 0.01 Gbps	1.40 ± 0.01 Gbps
Balia	0.94 ± 0.00 Gbps	1.23 ± 0.02 Gbps	1.41 ± 0.03 Gbps

Table 2. Overall throughput of MPTCP transmission (two paths, presence of background traffic)

MPTCP congestion control	Shared paths		
	1st (initial)	2nd	both
Cubic	0.92 ± 0.00 Gbps	0.92 ± 0.00 Gbps	0.31 ± 0.00 Gbps
LIA	0.64 ± 0.01 Gbps	0.83 ± 0.01 Gbps	0.36 ± 0.01 Gbps
OLIA	0.69 ± 0.01 Gbps	0.85 ± 0.01 Gbps	0.36 ± 0.01 Gbps
Balia	0.93 ± 0.01 Gbps	0.93 ± 0.01 Gbps	0.33 ± 0.01 Gbps

unloaded network wasn't large. The difference between the largest and the smallest throughput is about 0.04 Gbps for one MPTCP flow, 0.06 Gbps for transmission consisted of two MPTCP flows and 0.03 Gbps in the case of three MPTCP flows. The best results are obtained for Cubic and Balia methods of congestion control. The Cubic method gives the best performance for one (0.97 Gbps) and two (1.26 Gbps) flows and average performance (1.39 Gbps) in the case of three flows. The Balia method gives maximal observable throughput for three flows (1.41 Gbps) and average for one and two flows (0.94 Gbps and 1.23 Gbps, respectively). Both methods also show the shortest collapses of overall instantaneous throughput (Fig. 2), although at the cost of a large irregularity of instantaneous throughput of the flows (the Cubic method) or the longest collapse of one of the flows (the Balia method). However, the collected data doesn't allow, in the Authors opinion, for any general and unambiguous conclusion concerning the applicability of any particular method of congestion control to the transmission of bulk data in an unloaded high-speed network.

Table 2 presents the overall throughput of MPTCP flows working with the LIA congestion control method, when the MPTCP competes for bandwidth with the TCP that conveys the data generated by traffic generators. During tests, MPTCP flows were transmitted using two paths. The first, initial path led from OpenStack Compute 1 to OpenStack Compute 2 via the S2 switch. The link from S2 to the OpenStack Compute 2 could be shared with one TCP flow, sent by the TG4. The second path led from OpenStack Compute 1 to OpenStack Compute

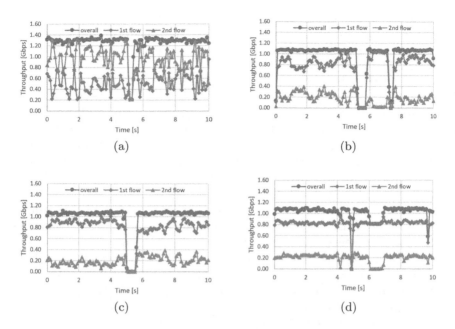

Fig. 2. Instantaneous throughput of MPTCP transmission (two paths, no background traffic): (a) Cubic, (b) Balia, (c) LIA, (d) OLIA

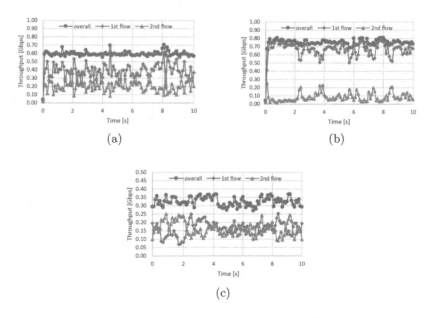

Fig. 3. Instantaneous throughput of MPTCP transmission (the LIA method of congestion control, two paths, presence of background traffic): (a) 1st path shared with the TCP traffic, (b) 2nd path share with the TCP traffic, (c) both paths shared with the TCP traffic

2 via the R1 router. The link from R1 to OpenStack Compute 2 could be shared with four TCP flows, sent by TG1, TG2, TG3 and TG4 traffic generators. TCP flows, generated by the iPerf software, were large enough to fill the shared links[1].

If the MPTCP competes for bandwidth in only one link, and the other is dedicated to the MPTCP traffic, the best performance is achieved using the Cubic (0.92 Gbps) and the Balia (0.93 Gbps) congestion controls, regardless of which path was shared with the TCP traffic. The throughput achieved with the use of those two methods was 30 to 45 % larger than the throughput achieved when the other two methods were used. If both paths were shared with the TCP traffic, the maximum throughput (0.36 Gbps) was observed when the LIA or the OLIA methods of congestion control were used, which give performance improvements of about 10–15%.

Figure 3 depicts the instantaneous throughput of the MPTCP transmission in two paths, if the LIA method of congestion control is used. The LIA method is the one which presents the longest collapse of overall throughput in the unloaded links (Fig. 2c). Competition with one TCP flow in the initial path (Fig. 3a) decreases throughput in that path by more than two times (0.4 Gbps) when compared with the results obtained in previous tests (0.89 Gbps). This effect was amplified by a decrease of throughput in the second, unloaded link, where total throughput achieved 0.24 Gbps (instead of 0.35 Gbps, observed in the test series described above). As a result, the overall throughput decreased by about half (from 1.24 to 0.64 Gbps) when compared with two unloaded links.

If the initial link was unloaded and the second was shared with four TCP flows (Fig. 3b), throughput of the initial flow is the greatest (more than 85 %) of the overall throughput of the MPTCP. The second flow occupies only 11 % of a 1 Gbps link. Overall throughput (0.83 Gbps) is larger than that achieved in the case of a shared initial link, and only a little smaller than 0.89 Gbps.

Note that in both tests the overall throughput of two MPTCP flows (when one of them was transmitted in an unloaded link), was smaller than the throughput of a single MPTCP flow in unloaded initial link (0.95 Gbps for the LIA congestion control). In the case of the MPTCP transmission in two shared links (Fig. 3c), the throughput measured in the initial link (0.17 Gbps) is close to throughput of the second link (0.19 Gbps). Overall throughput of the MPTCP flow achieved 0.36 Gbps. For comparison, in the same circumstances, a TCP flow achieved 0.47 Gbps in the first (initial) path, 0.185 Gbps in the second path and 0.91 Gbps in the path without background traffic.

It is worth noting that graphs of instantaneous throughput (Fig. 3a, b, c) don't show the short-term collapses of transmission that were observed with unloaded links (Fig. 2c).

---

[1] Link utilization, measured during homogeneous TCP transmissions (no MPTCP traffic), exceeded 90 % (93 % for single TCP flow, for four competing TCP flows).

5   Migration of Virtual Machines

In the OpenStack-based cloud, default VM migration is a block live migration. The performance of such migration depends on, amongst other things, network performance, disk performance and CPU load. Such factors, including actually occupied disk space in the virtual storage systems and currently occupied RAM, also have an influence on the performance of the migration. In our experiments, the virtual machine occupied 5.7 GB of virtual hard drive space. The virtual disk had declared a higher capacity (40 GB). During the migration, only allocated disk space was transferred. The same applies to a memory - only actually allocated memory was transferred to the destination node.

The OpenStack Compute node, used in the experiments, was built based on a dual-processor multi-core computer system. During tests, processors were lightly loaded, and they performed tasks associated with migration and functions which support the cloud. The load on all cores and all processors was monitored, and in all tests the sum of the loads of all processor cores was significantly below 100 %. Thus, the migration wasn't limited by the processors. Moreover, the average data transfer rate of the hard drive, measured using the hdparm software tool [15], achieved 381 MB/s (more than 3 Gbps), which implies the ability to potentially use three (i.e. all available for MPTCP connections) Gigabit Ethernet interfaces.

Migration time, measured during the MPTCP transmission, is shown in Tables 3 and 4. During the VM migration about 5.7 GB data of VM disk image was transferred and about 1 GB data of RAM memory was allocated in the VM. That gives about 6.7 GB of data to transfer during the migration. In our experiment, four methods of congestion control (Cubic, Balia, LIA and OLIA) dedicated to the MPTCP were tested, and number of MPTCP paths varied from 1 to 3. The sequence of paths was the same, as presented in the Sect. 5. Each test was repeated 10 times and measured times were averaged (confidence level 95 %). The impact of the OpenStack virtual environment on the obtained results can be seen as constant overhead of performed operations.

The average migration time of the MPTCP transmission in unloaded paths, presented in the Table 3, weakly depends on the method of congestion control. In the case of one or two used paths, the worst time (64 s and 52 s, respectively) was only one second longer than the typical time of migration. In the case of three paths, average migration time ranged from 45 s to 48 s. In all cases, the longest migration time was measured for the MPTCP protocol that used the LIA method of congestion control.

The migration time of the MPTCP transmission in the unloaded paths strictly depends on the number of paths available for the MPTCP transmission. The average migration time, measured during the experiment, varies from 45 s (3 paths) to 64 s (1 path). For sake of comparison, a VM migration in the direct link using the TCP transport protocol was tested. Measured migration time (also averaged over 10 times), obtained for the TCP, was equal to 65 s.

The presence of background traffic significantly increases average migration time. Migration using the 2-paths MPTCP transmission (Table 4) lasted about three times longer, when the protocol shared both links with the TCP. If only

Table 3. Average migration time (MPTCP transmission, no background traffic)

MPTCP congestion control	No. of MPTCP flows		
	1 flow	2 flows	3 flows
Cubic	63.0 ± 0.0 s	51.2 ± 0.0 s	44.8 ± 0.3 s
LIA	63.0 ± 0.0 s	51.8 ± 0.9 s	48.0 ± 1.1 s
OLIA	64.1 ± 0.0 s	51.0 ± 0.3 s	46.4 ± 0.3 s
Balia	63.4 ± 0.0 s	51.1 ± 0.6 s	45.3 ± 0.9 s

Table 4. Average migration time (MPTCP transmission in two paths, presence of background traffic)

MPTCP congestion control	Shared paths		
	1st (initial)	2nd	both
Cubic	63.6 ± 0.0 s	62.1 ± 0.0 s	162.1 ± 0.0 s
LIA	86.2 ± 0.8 s	68.9 ± 0.5 s	142.0 ± 1.3 s
OLIA	79.1 ± 0.6 s	68.1 ± 0.5 s	141.1 ± 1.1 s
Balia	63.0 ± 0.4 s	63.1 ± 0.4 s	154.1 ± 1.2 s

one link was shared with the TCP, the Cubic and Balia methods of congestion control enabled migration times comparable with the times measured for a single, unloaded path. In the case of LIA and OLIA methods, if the shared path was the initial path, measured migration times were about 25 to 35 % longer when compared with times obtained in a single, unloaded path. If the second path was the shared one, measured migration times were 4 to 6 s longer than those measured for a single, unloaded path.

For comparison, when the TCP transport protocol was used during migration, the process of migration via the S2 switch (initial path of the MPTCP) lasted 109 s (migration's flow competed for bandwidth with one TCP flow). Migration via R1 router (2nd path of the MPTCP) lasted 212 s (migration's flow competed for bandwidth with four TCP flows).

6 Conclusions

In the paper, the performance evaluation of the MPTCP transmission of large data objects inside a computing cloud was presented. The large data objects consisted of big files and live VM instances (consisting of a disk and memory image and the state of the VM machine). Experiments were carried out changing the congestion control method and varying the number of paths in multi-path transmissions. File transfer and migration tests showed that in the case of strong asymmetry in path load (paths with and without a background traffic) more efficient transmission is achieved when LIA and OLIA methods are used, while in the case of all links shared with TCP flows, Cubic and Balia methods gave better performance. However, all the above methods of congestion control seem to be too restrictive to allow the multipath transmission to achieve high performance in high-speed, unloaded paths. Solutions for overcoming the shortcomings of the investigated protocols versions will be the subject of further research.

Acknowledgment. The work was supported by the contract 11.11.230.018.

References

1. Hashem, I.A.T., Yaqoob, I., Anuar, N.B., Mokhtar, S., Gani, A., Khan, S.U.: The rise of big data on cloud computing: review and open research issues. Inf. Syst. **47**, 98–115 (2015)
2. Cerroni, W., Esposito, F.: Optimizing live migration of multiple virtual machines. IEEE Trans. Cloud Comput. **PP**(99), 1 (2016)
3. Jiang, W., Ren, F., Lin, C.: Phase plane analysis of quantized congestion notification for data center ethernet. IEEE/ACM Trans. Netw. (TON) **23**(1), 1–14 (2015)
4. Coudron, M., Secci, S., Maier, G., Pujolle, G., Pattavina, A.: Boosting cloud communications through a crosslayer multipath Protocol architecture. In: 2013 IEEE SDN for Future Networks and Services (SDN4FNS), Trento, pp. 1–8 (2013)
5. Tariq, S., Bassiouni, M.: Performance evaluation of MPTCP over optical burst switching in data centers. In: International Telecommunications Symposium (ITS), pp. 1–5. IEEE (2014)
6. Chodorek, A., Chodorek, R.R., Krempa, A.: An analysis of elastic and inelastic traffic in shared link. In: Proceedings of IEEE Conference on Human System Interaction, Cracow, Poland, pp. 873–878 (2008)
7. Raiciu, C., Handley, M., Bonaventure, O.: TCP extensions for multipath operation with multiple addresses, RFC 6824 (2013)
8. Raiciu, C., Handley, M., Wischik, D.: Coupled congestion control for multipath transport protocols, RFC 6356 (2011)
9. Khalili, R., Gast, N., Popovic, M., Le Boudec, J.: Opportunistic linked-increases congestion control algorithm for MPTCP, Internet Draft draft-khalili-mptcp-congestion-control-05 (2014)
10. Walid, A., Peng, Q., Hwang, J., Low, S.: Balanced linked adaptation congestion control algorithm for MPTCP, Internet Draft draft-walid-mptcp-congestion-control-04 (2016)
11. Ha, S., Rhee, I., Xu, L.: CUBIC: a new TCP-friendly high-speed TCP variant. ACM SIGOPS Oper. Syst. Rev. **42**, 64–74 (2008)
12. https://www.openstack.org
13. http://multipath-tcp.org/pmwiki.php/Main/HomePage
14. https://iperf.fr/
15. http://sourceforge.net/projects/hdparm/

A Decentralized System for Load Balancing of Containerized Microservices in the Cloud

Marian Rusek[✉], Grzegorz Dwornicki, and Arkadiusz Orłowski

Faculty of Applied Informatics and Mathematics, Warsaw University of Life Sciences,
Nowoursynowska 166, 02–787 Warsaw, Poland
marian_rusek@sggw.pl

Abstract. Microservice architecture is a cloud application design pattern which shifts the complexity away from the traditional monolithic application into the infrastructure. Each microservice is a small containerized application that has a single responsibility in terms of functional requirement, and that can be deployed, scaled and tested independently using automated orchestration systems. We propose a simple swarm-like decentralized load balancing system for microservices running inside OpenVZ containers. It can potentially offer performance improvements with respect to the existing centralized container orchestration systems.

Keywords: Container orchestration · Swarm algorithm · Microservices

1 Introduction

A typical monolithic enterprise systems we build today are difficult to scale, difficult to understand and difficult to maintain. Written in a monolithic way, these systems tend to have strong coupling between the components in the service and between services. A system with the services tangled and interdependent is harder to write, understand, test, evolve, upgrade and operate independently. Strong coupling can also lead to cascading failures: one failing service can take down the entire system, instead of allowing you to deal with the failure in isolation. Popular application servers (e.g., WebLogic, WebSphere, JBoss or Tomcat) encourage this monolithic model.

Microservices-based architecture is free of these problems [2,16,18,24,28]. It advocates creating a system from a collection of small, isolated services, each of which owns their data, and is independently isolated, scalable and resilient to failure. Services integrate with other services in order to form a cohesive system thats far more flexible than a typical monolithic system. One of the key principles in employing a microservices-based architecture is the decomposition of the system into discrete isolated subsystems communicating over well defined asynchronous protocols and decoupled in time (allowing concurrency) and space (allowing distribution and mobility – the ability to move services around).

© Springer International Publishing AG 2017
J. Świątek and J.M. Tomczak (eds.), *Advances in Systems Science*, Advances in Intelligent
Systems and Computing 539, DOI 10.1007/978-3-319-48944-5_14

Some developers and researches believe that the concept of microserices is a specific pattern of implementation of service-oriented architecture (SOA). However the microservice pattern has the following unique specifics: microservices use lightweight HTTP mechanisms for communication, they are independently deployable by fully automated machinery, and there is only a bare minimum of centralized management [24]. Enterprise service bus (ESB) is a typical software model used for designing and implementing communication between mutually interacting software applications in SOA. ESB provides all of the routing and data transformation required to get the parts of an application talking to each other. In the microservices-based architecture there is no central unit like ESB which does the routing. The accidental complexity is shifted from inside of an monolithic application into the infrastructure. It is possible because now we have many more ways to manage that complexity: programmable infrastructure, infrastructure automation, and the movement to the cloud [28].

Today we have a much more refined foundation for isolation of services, using virtualization, Linux Containers (LXC), Docker, and Unikernels [15, 19]. This has made it possible to treat isolation as a necessity for resilience, scalability, continuous delivery and operations efficiency. It has also paved the way for the rising interest in microservices-based architectures, allowing you to slice up the monolith and develop, deploy, run, scale and manage the services independently of each other. The value of microservices and containers lies in how they enable smaller, faster, more frequent change [5, 14]. While cloud computing changed how we manage "machines," it didnt change the basic things we managed. Containers, on the other hand, promise a world that transcends our attachment to traditional servers applications and application components. One might claim that represent the fruition of the object-oriented, component-based vision for application architecture.

So how do you build a smart system from a data center filled with dumb servers? This is where tools like Google Kubernetes [4] and open source Apache Mesos [13] data center operating system come in. Also of note is Dockers platform, using its Machine, Swarm and Compose tools [26]. The role of orchestration and scheduling within these container platforms is to match applications to resources. Google developed Kubernetes for managing large numbers of containers. Instead of assigning each container to a host machine, Kubernetes groups containers into pods. For instance, a multi-tier application, with a database in one container and the application logic in another container, can be grouped into a single pod. The administrator only needs to move a single pod from one compute resource to another, rather than worrying about dozens of individual containers. Apache Mesos is a cluster manager that can help the administrator schedule workloads on a cluster of servers. Mesos excels at handling very large workloads, such as an implementation of the Spark or Hadoop data processing platforms. Docker Swarm is a clustering and scheduling tool that automatically optimizes a distributed applications infrastructure based on the applications lifecycle stage, container usage and performance needs. All these container orchestration systems are monolithic applications running as daemons on dedicated nodes of the cloud. They orchestrate containers in a centralized fashion.

Usually decentralized orchestration systems offer performance improvements. For example decentralized orchestration of composite web services yields increased throughput, better scalability, and lower response time [6]. In this paper a decentralized system for load balancing of containerized microservices is proposed. In Sect. 2 the internals of a virtualization container are analyzed. In addition to cloud application it can run an additional process implementing the mobile agent intelligence. In Sect. 3 the swarm-like algorithm of container migration in the cloud is introduced. The number of containers on each host plays a role analogous to a pheromone in colonies of insects or simple transceivers mounted on autonomous robots. In Sect. 4 some preliminary experimental results for a simple cloud consisting of 18 hosts are presented. We finish with a summary and brief remarks in Sect. 5.

2 Container Internals

The startup time for a container is around a second. Public cloud virtual machines take from tens of seconds to several minutes, because they boot a full operating system every time. Thus recently the cloud industry is moving beyond self-contained, isolated, and monolithic virtual machine images in favor of container-type virtualization [17,25]. Containers introduce autonomy for applications by packaging apps with the libraries and other binaries on which they depend. This avoids conflicts between apps that otherwise rely on key components of the underlying host operating system. Containers do not contain a operating system kernel, which makes them faster and more agile than virtual machines. Container-type virtualization is an ability to run multiple isolated sets of processes, each set for each application, under a single kernel instance. Having such an isolation opens the possibility to save the complete state of (in other words, to checkpoint) a container and later to restart it. This feature allows one to checkpoint the state of a running container and restart it later on the same or a different host, in a way transparent for running applications and network connections [11,17].

In this paper container-type virtualization is used to build a swarm of tasks in a cloud. Each container in addition to the application and their libraries contains an separate process representing the mobile agent [1,12,27]. It deals with sensing the neighboring containers and initiating live migration of its container to another host. A typical modern server can run only about 10 virtual machines or about 100 containers. Therefore the density of container based mobile agents in a cloud can be 10 times higher. Typical migration time of a virtual machine is about 10 s—container can be migrated in time of order of 1 s [29]. Thus container migration is about 10 times faster. Containers can call system functions of the operating system kernel running on the server. Therefore in principle they can initiate container migration without help of a separate daemon process running on the host server.

The years from 2002 to 2010 represented a time of experimentation, when two projects in particular moved the needle on virtualization containers in Linux.

VServer project patched the Linux kernel in order to split things up into virtual servers, an early version of what today we would call containers. The second project was OpenVZ, which transformed the Linux kernel, so that you could run containers in production. Despite its success, OpenVZ never managed to get the containerization technology merged into the stock Linux kernel and always required a custom patch to make it possible. At later time, control groups and namespaces [22] were introduced, making LXC containers a functionality available within the stock Linux kernel. It thus became possible to use something that looked like a container without patching the kernel. At the time, Salomon Hykes was leading dotCloud, an infrastructure platform as a service (PaaS) company that was committed to applying standards in the deployment of distributed architecture for applications. They spent three years running a cloud platform production using LXC, so they had a lot of operational experience. They learned that this technology was not practical, so they wrote a tool that was more stable and allowed to deliver container-based application deployment for large-scale hybrid cloud environments. In this way a popular Docker technology based on a libcontainer format was born.

Docker containers cannot be live migrated between hosts—they can only be snapshotted and restored on the same or other host. The generally accepted method for managing Docker container data is to have stateless containers running in the production environment that store no data on their own and are purely transactional. Stateless containers store processed data on the outside, beyond the realm of their container space, preferably to a dedicated storage service that is backed by a reliable and available persistence backend. Another class of container instances are these that host storage services, like upon pattern is to use data containers. The runtime engines of these stateful services get linked at runtime with the data containers. In practical terms, this would mean having a database engine that would run on a container, but using a "data container" that is mounted as a volume to store the state. Therefore to run a cloud hosting environment, it is important to have a distributed storage solution, like Gluster and Ceph, to provide shared mount points. This is useful if the container instances move around the cloud based on availability.

Parallels®Virtuozzo is another widely deployed container-based virtualization software for Linux and Windows operating systems. As opposed to Docker Virtuozzo allows for live migration of containers. The results presented in this paper were obtained using an open source version of this software called OpenVZ [17]. OpenVZ is available for Linux operating system only and runs on a custom kernel. There have been several studies on various optimizations of container migration algorithms [17]. Two best known examples are lazy migration and iterative migration. Lazy migration is the migration of memory *after* actual migration of container, i.e., memory pages are transferred from the source server to the destination on demand. In the case of an iterative migration iterative migration of memory happens *before* actual migration of container. In our experiments with stripped down OpenVZ containers with a size of 50 MB in a test system consisting of two nodes connected by a 100 Megabit Ethernet network we measured the migration time seen by the host $T = 6.61\,$s and the migration time seen by the container $\tau = 2.25$ s. The later is three times smaller than the former due to optimizations described above.

We have altered the OpenVZ kernel by adding a system function allowing the container to ask the host to migrate it to another host. By calling this function a container is placed in a queue in the kernel - a dedicated daemon reads this queue and migrates all the containers waiting in it. Thus containers can leave the host only in a sequence. Our studies indicate that a parallel migration is possible, but the performance gain is negligible—migration speed increases only by 8 %. In addition as shown in our previous paper dealing with two hosts only sequential migration helps to stabilize the swarm algorithm [23].

Processes running inside an OpenVZ container have their own disk with partitions and file systems. In reality this is a virtual disk and its image is stored as a file on the physical disk of the host. This solution makes the migration of the container's data to another host is very easy—only a single file needs to be copied between servers. The network of the container is isolated in a way that allows the container to have they own IP address on the network. This is not the IP address of host but it can be reached from the other containers and hosts. Each container maintains its own state: network connections, file descriptors, memory usage etc. Containers share only the kernel with the host operating system. Thanks to state isolation from other containers a container can be migrated to another system and resumed.

When we launch our container for the first time it does not know on what host it was started. However it can use an ICMP echo/reply mechanism to detect the IP address of the host. Each ICMP packet has a TTL (Time-to-Live) value. When this packet is routed trough router this value is decreased. When it reaches 0 the packet is destroyed and an error ICMP packet is send back. This ICMP error packet will have last router IP address. Thus to detect the IP address of the host our container can send an ICMP echo packet with value 1 of TTL to some arbitrary external IP address. The host system acts as a router for container's network. When this special packet is sent by the container it will never reach the destination but the host system will send back an ICMP error packet with its IP address.

Host system keeps all the containers filesystems mounted on its local file system. Each container's file system is visible as a folder located in `/var/lib/vz/root/CID` where CID is an unique container identification number. The location can be exported trough an network file system like NFS. Our container will mount it locally. To do this it needs to know the export path `/var/lib/vz/root` and the IP address of its host system. By counting the number of entries in this folder it can detect other containers running on the same hosts and count their number N. By calling a custom system function written by us it can also check for the number Q of containers queued for migration in the kernel. A container can also log in via ssh to another host and ask for these parameters there. Each container knows how many hosts we have H and knows their IP addresses. It also knows how many other containers are there C in the cloud. These numbers can be updated dynamically at runtime by probing other containers and hosts using ICMP echo/reply protocol either by the container itself or by the host.

3   Swarm Algorithm

There many examples in biology how complex global behaviors can arise from simple interactions between large numbers of relatively unintelligent agents. Examples of self-organized processes of natural aggregation are nest construction, foraging, brood sorting, hunting, navigation, and emigration. All involve only local interactions between individuals and between individuals and their environment. For example the ants rely only on physical contact and pheromone communication, but simple individual ant behaviors result in group behaviors that are thought to be optimal for the entire colony. Emerging technologies are making it possible to cheaply manufacture small robots with sensors, actuators and computation. Swarm approaches to robotics, involving large numbers of simple robots rather than a small number of sophisticated robots, has many advantages with respect to robustness and efficiency. Such systems can typically absorb many types of failures and unplanned behavior at the individual agent level, without sacrificing task completion [3,7–9,20,21]. These properties make swarm intelligence an attractive solution also for other problem domains. In this paper we use this approach for task scheduling in a complex distributed system—the cloud.

Let us now propose a swarm-like decentralized algorithm for container migration inspired by pheromone robots [20,21]. The proposed approach threats the containers as mobile agents and is also capable of automatic self-repair; the system can quickly recover from most patterns of agent death and can receive an influx of new agents at any location without blocking problems. Each host is described by a pheromone p which can be either repulsive $0 < p < 1$ or attractive $p < 0$. The complete algorithm executed by a dedicated process running inside each container reads as follows:

1: Use the method described in Sect. 2 to get the IP address of the host.
2: Mount the `/var/lib/vz/root` folder exported by it.
3: **loop**
4:    **repeat**
5:      Obtain the pheromone value p of the host.
6:      Generate a random number $0 < r < 1$.
7:    **until** $r < p$
8:    Randomly choose a host with an attractive pheromone $p < 0$.
9:    Ask the host to migrate to it.
10:   **repeat**
11:     Get the IP address of the host.
12:   **until** It's different from the previous one
      {Migration to another host is complete}
13:   Unmount `/var/lib/vz/root` from the old host.
14:   Mount it from the new host.
15: **end loop**

Thus the pheromone p can be viewed as a migration probability of a container. The simplest choice for p is the fraction of the number of containers on a host:

$$p = \frac{N - Q - n}{N - Q} \tag{1}$$

above the equilibrium value where the containers are equally distributed between the hosts:

$$n = \frac{C}{H} \tag{2}$$

In this case the tasks are migrated between the nodes until the number of tasks on each node is the same and equal to n. Using an analogy with physics the nodes can be imagined to be gas containers, and the tasks running on them— gas molecules. The network links between the nodes are tubes connecting the containers between each other. The pressure of the gas equilibrates until it is the same in each container. Similar analogy is used in a self-repairing formation of mobile agents [8]. Note that the subtraction of Q in Eq. (1) is necessary in order to avoid oscillations of the containers [23].

In realistic cloud environments, tasks often differ regarding CPU and I/O load. Hence, optimization towards an equal distribution of tasks across available hosts does not seem to be optimal in each case. Instead of using Eq. (2) the desired number of containers of a given type n could be computed on each server separately using Dominant Resource Fairness (DRF) algorithm [10]. For example consider a server with 9 CPU cores and 18 GB of RAM. Container of type A needs 1 CPU core and 4 GB of RAM, and container of type B—3 CPU cores and 1 GB od RAM. DRF gives $n_A = 3$, and $n_B = 2$. The pheromone value p from Eq. (1) needs to be calculated separately for each container type. There are separate migration queues Q for containers of different types.

4   Experimental Results

The experiments were performed on $H = 18$ servers equipped with Intel®i5-3570 Quad-Core CPU, 8 GB of RAM each connected by a dedicated 100 Megabit Ethernet network. All servers were running Debian GNU/Linux operating system with OpenVZ software installed. The Linux kernel was modified by adding new system functions as described in Sect. 3. At the initial time $C = 18 \cdot 17 = 306$ identical containers are launched on the first 17 hosts, and the last one was empty:

$$N_i = 18, \quad i = 1, \ldots 17, \quad N_{18} = 0 \tag{3}$$

Each container has a size of about 100 MB and its migration to another host takes $T = 16$ s. Each container is running a Python script implementing the algorithm from Sect. 3. It starts by scanning the network using nmap to find the number of containers C, and the number of hosts $H = 18$ and their IP addresses. Than the mean number of containers $n = 17$ is calculated and the script enters a loop in which it periodically checks the pheromone value p, and decides with probability p whether to migrate to another host. In addition on the host server a monitor program was started which periodically (period 5 s) checked the number of containers N in the filesystem and the number of containers queued for migration Q in the kernel queue (access to this data from a user process was possible by a custom system function added to the kernel).

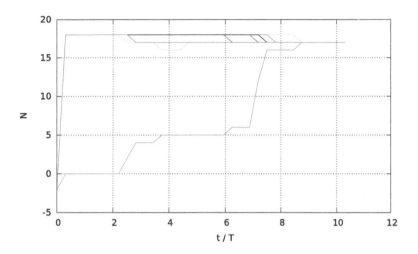

Fig. 1. Number of containers on each host versus time.

In Fig. 1 we have the numbers of containers N on each host plotted versus time t. It is seen from inspection of this plot that the containers can arrive to the destination host in parallel thus network bandwith was apparently not a problem during this experiment. Notice that around $t \simeq 4T$ the yellow line drops below the equilibrium value of $N = 17$—this happens because the migration process is inherently a probabilistic one. The migration probability is $p = 1/18$ but sometimes more than $pN = 1$ container can decide to jump to another host. Also the containers do not move independently but interact with each other. If more than one excess container asks the host for migration, then one of them must wait in a queue until the first one leaves the host. The system reaches equilibrium and migration stops around $t \simeq 9T$:

$$N_i = 17, \quad i = 1, \ldots 18 \qquad (4)$$

Thus at average two containers arrive to the destination host during time T.

Containers startup is not instantaneous. The experiment was arranged in such a way that the migration agent processes inside the containers were started in a loop—therefore some were started later than the other. In Fig. 2 we have the numbers of containers waiting for migration Q on each host plotted versus time t. Indeed we see a small delay in entering the queue. The first container leaves the migration queue around $t \simeq 1.5T$ (red line) but is deleted from the file system with some delay only after $t > 2T$ (c.f., Fig. 1).

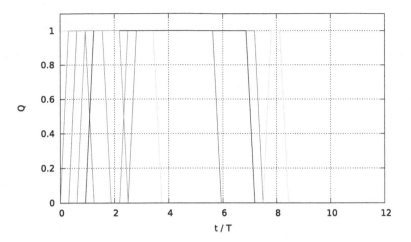

Fig. 2. Number of containers waiting for migration on each host versus time. (Color figure online)

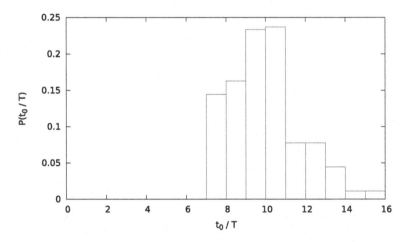

Fig. 3. Histogram of times needed to reach equilibrium.

To investigate the container self-organization process even further in Fig. 3 we have a histogram of times needed to reach equilibrium t_0 obtained from 300 runs of the algorithm. It is seen that the case discussed earlier is a fairly typical one.

5   Summary

In summary, the OpenVZ containerization software was used to implement a swarm of tasks executing in a cloud. Each task includes a mobile agent process which governs its migration to another nodes of the cloud. A variant of the Contained Gas Model known from self-repairing formations of autonomous robots is used. The tasks running on the nodes of the cloud self-organize to maintain a constant load among the servers. The system automatically adapts to creation and destruction of tasks as well as extension of the cloud by new servers. It can be easily adopted to react on server failures: a failing server can produce an artificial pheromone by creating entries in the /var/lib/vz/root directory of its filesystem. This will cause all the tasks running on it to migrate away from the pheromone. The performance of the swarm-like algorithm proposed to control the containers was experimentally tested on a simple "cloud" consisting of 18 nodes.

References

1. Aversa, R., Di Martino, B., Rak, M., Venticinque, S.: Cloud agency: a mobile agent based cloud system. In: 2010 International Conference on Complex, Intelligent and Software Intensive Systems (CISIS), pp. 132–137. IEEE (2010)
2. Balalaie, A., Heydarnoori, A., Jamshidi, P.: Migrating to cloud-native architectures using microservices: an experience report. In: Celesti, A., Leitner, P. (eds.) ESOCC Workshops 2015. CCIS, vol. 567, pp. 201–215. Springer, Heidelberg (2016). doi:10.1007/978-3-319-33313-7_15
3. Berman, S., Halász, A., Kumar, V., Pratt, S.: Bio-inspired group behaviors for the deployment of a swarm of robots to multiple destinations. In: 2007 IEEE International Conference on Robotics and Automation, pp. 2318–2323. IEEE (2007)
4. Brewer, E.A.: Kubernetes and the path to cloud native. In: Proceedings of the Sixth ACM Symposium on Cloud Computing, pp. 167–167. ACM (2015)
5. Calinciuc, A., Spoiala, C.C., Turcu, C.O., Filote, C.: Openstack and docker: building a high-performance iaas platform for interactive social media applications. In: 2016 International Conference on Development and Application Systems (DAS), pp. 287–290. IEEE (2016)
6. Chafle, G.B., Chandra, S., Mann, V., Nanda, M.G.: Decentralized orchestration of composite web services. In: Proceedings of the 13th International World Wide Web Conference on Alternate Track Papers & Posters, pp. 134–143. ACM (2004)
7. Cheah, C.C., Hou, S.P., Slotine, J.J.E.: Region-based shape control for a swarm of robots. Automatica 45(10), 2406–2411 (2009)
8. Cheng, J., Cheng, W., Nagpal, R.: Robust and self-repairing formation control for swarms of mobile agents. In: AAAI, vol. 5, pp. 59–64 (2005)
9. Dorigo, M., Trianni, V., Şahin, E., Groß, R., Labella, T.H., Baldassarre, G., Nolfi, S., Deneubourg, J.L., Mondada, F., Floreano, D., et al.: Evolving self-organizing behaviors for a swarm-bot. Auton. Robots 17(2–3), 223–245 (2004)
10. Ghodsi, A., Zaharia, M., Hindman, B., Konwinski, A., Shenker, S., Stoica, I.: Dominant resource fairness: fair allocation of multiple resource types. In: NSDI, vol. 11, p. 24 (2011)
11. Hacker, T.J., Romero, F., Nielsen, J.J.: Secure live migration of parallel applications using container-based virtual machines. Int. J. Space Based Situat. Comput. 12(1), 45–57 (2012)
12. Haichun, N., Yong, L.: A mobile agent-based task seamless migration model for mobile cloud computing. In: 2014 IEEE Workshop on Advanced Research and Technology in Industry Applications (WARTIA), pp. 241–246. IEEE (2014)

13. Hindman, B., Konwinski, A., Zaharia, M., Ghodsi, A., Joseph, A.D., Katz, R.H., Shenker, S., Stoica, I.: Mesos: a platform for fine-grained resource sharing in the data center. In: NSDI, vol. 11, p. 22 (2011)
14. Kratzke, N.: About microservices, containers and their underestimated impact on network performance. In: Proceedings of CLOUD COMPUTING 2015 (2015)
15. Madhavapeddy, A., Scott, D.J.: Unikernels: rise of the virtual library operating system. Queue 11(11), 30 (2013)
16. Malavalli, D., Sathappan, S.: Scalable microservice based architecture for enabling DMTF profiles. In: 2015 11th International Conference on Network and Service Management (CNSM), pp. 428–432. IEEE (2015)
17. Mirkin, A., Kuznetsov, A., Kolyshkin, K.: Containers checkpointing and live migration. Proc. Linux Symp. 2, 85–90 (2008)
18. Namiot, D., Sneps-Sneppe, M.: On micro-services architecture. Int. J. Open Inf. Technol. 2(9), 39 (2014)
19. Pahl, C.: Containerisation and the paas cloud. IEEE Cloud Comput. 2(3), 24–31 (2015)
20. Payton, D., Estkowski, R., Howard, M.: Progress in pheromone robotics. Intell. Auton. Syst. 7, 256–264 (2002)
21. Payton, D., Estkowski, R., Howard, M.: Compound behaviors in pheromone robotics. Robot. Auton. Syst. 44(3), 229–240 (2003)
22. Pike, R., Presotto, D., Thompson, K., Trickey, H., Winterbottom, P.: The use of name spaces in plan 9. In: Proceedings of the 5th Workshop on ACM SIGOPS European Workshop: Models and Paradigms for Distributed Systems Structuring, pp. 1–5. ACM (1992)
23. Rusek, M., Dwornicki, G., Orłowski, A.: Swarm of mobile virtualization containers. In: Świątek, J., Borzemski, L., Grzech, A., Wilimowska, Z. (eds.) Information Systems Architecture and Technology: Proceedings of 36th International Conference on Information Systems Architecture and Technology – ISAT 2015 – Part III. AISC, vol. 431, pp. 75–85. Springer, Heidelberg (2016). doi:10.1007/978-3-319-28564-1_7
24. Savchenko, D., Radchenko, G., Taipale, O.: Microservices validation: mjolnirr platform case study. In: 2015 38th International Convention on Information and Communication Technology, Electronics and Microelectronics (MIPRO), pp. 235–240. IEEE (2015)
25. Scheepers, M.J.: Virtualization and containerization of application infrastructure: a comparison. In: 21st Twente Student Conference on IT, pp. 1–7 (2014)
26. Stubbs, J., Moreira, W., Dooley, R.: Distributed systems of microservices using docker and serfnode. In: 2015 7th International Workshop on Science Gateways (IWSG), pp. 34–39. IEEE (2015)
27. Thant, H.A., San, K.M., Tun, K.M.L., Naing, T.T., Thein, N.: Mobile agents based load balancing method for parallel applications. In: APSITT 2005 Proceedings. 6th Asia-Pacific Symposium on Information and Telecommunication Technologies, pp. 77–82. IEEE (2005)
28. Thönes, J.: Microservices. IEEE Softw. 32(1), 116 (2015)
29. Zhao, M., Figueiredo, R.J.: Experimental study of virtual machine migration in support of reservation of cluster resources. In: Proceedings of the 2nd International Workshop on Virtualization Technology in Distributed Computing, p. 5. ACM (2007)

Layered Reconfigurable Architecture for Autonomous Cooperative UAV Computing Systems

Grzegorz Chmaj[(✉)] and Henry Selvaraj

University of Nevada, Las Vegas, Las Vegas, USA
{grzegorz.chmaj,henry.selvaraj}@unlv.edu

Abstract. Cooperative processing in UAV swarms requires efficient architectural and algorithmic solutions to maximize the operational speed and life time span. Solutions need to include communication, data gathering, data processing and general management. In this paper, we address all these needs and we also present a reconfigurable approach to the UAV swarm cooperative processing with maximum extent of decentralization. UAVs with multiple devices (such as cameras, radars), and multiple processing units (CPUs and reconfigurable FPGAs) onboard are considered. We present main elements of the architecture and two reconfiguration algorithms used in the process of management of FPGA chips. Results are evaluated using the dedicated simulation framework and demonstrate the efficiency of the proposed solutions.

Keywords: Reconfigurable · Autonomous · UAV · Cooperative · Computing

1 Introduction

One of the main needs that appear in today's world of interconnected devices is to be able to throw the application onto some processing structure that might be unknown at that time, and let the structure do the processing according to the application specification. This type of process exists both in stationary powered and battery powered systems. In both cases minimizing the operational electrical energy expenditure is important, however in the latter case it is often critical. Application Specific Integrated Circuits are most efficient when it comes to processing power and efficiency, however they are tied to the specific functions or applications that makes them hard to use in universal systems. Reconfigurable FPGA (Field Programmable Gate Array) chips offer the way to program their structure with user-defined design, therefore becoming application-specific chip with very good efficiency. In our research, we focus on the UAV-based cooperative computing systems [1, 2]. They operate on battery power or onboard combustible engines, what makes their energy consumption critical, especially during missions. Decentralization is another issue that must be considered [3]. These are systems working in the field that are distant from the base station, thus they should not rely on one main component, as if it would become inoperable and render the entire swarm down [4]. Next key factor is the flexibility of the UAV swarm. Typically each vehicle is equipped with some extra sensing devices such as cameras, radars etc., they

© Springer International Publishing AG 2017
J. Świątek and J.M. Tomczak (eds.), *Advances in Systems Science*, Advances in Intelligent Systems and Computing 539, DOI 10.1007/978-3-319-48944-5_15

also must have computing capabilities. Use of general purpose CPUs is not efficient for heavy computations required to process the massive amount of data coming from the sensing devices. Therefore, we are using FPGAs to program them with the designs that are currently required, to get efficient processing. They can be reprogrammed for other uses anytime, also *on the fly* reprogramming is supported and is one of key factors of the electrical efficiency of our proposed architecture.

Multiple reconfigurable systems exist. A comprehensive survey of them, including features, challenges, concepts, and applications is presented in [5]. Work presented in [6] shows the development methods and tools used in embedded reconfigurable systems. The reconfigurable computing design patterns are described in [7]. [8] provides a general overall description of reconfigurable systems, together with the characteristics of reconfigurable logic. Spatial computation, configurable data path, distributed control and distributed resources were stated as the fundamental differences between traditional processing architectures and reconfigurable logic approach. Authors of [9] present the methods of workload distribution over processing resources, which are considered to be multi-core CPUs, GPUs (Graphical Processing Units), PPU (Physics Processing Unit) and FPGAs, among others. Presented work focuses on handling applications' requirements in the high-level design process, including the identified concurrencies used to achieve load balancing and efficient task distribution. The behavior of distributed reconfigurable systems is often researched using simulation frameworks. Results in [10] include modeling complex reconfigurable nodes, processor configurations and tasks along with general purpose processors and offers multiple metrics for the evaluation. [11] presents the Open Control Platform for reconfigurable distributed control systems that provides the coordination of distributed interaction among hierarchically organized units, and supports the dynamic reconfiguration.

2   System Operation

2.1   Layered Architecture

In order to achieve maximum flexibility in operation, and also in possible redesign of the system – the presented architecture is proposed to be using the layered approach. The following layers are used: (1) *application* (2) *processing* (3) *units* (4) *communication*. The application is defined only at its level, and does have no knowledge about lower layers. The processing layer analyzes the application and manages the processing (control, coordination, results). The processing layer uses units as processing resources: sends the tasks and collects results related to these tasks. Units use the underlying communication layer to exchange information and do any communication. Each layer is connected to others with interfaces. Such layered structure allows replacing any layer without modifying the remaining layers (e.g. using communication layer as TCP/IP, Bluetooth, ZigBee etc., or using the same application on various systems). The proposed layered structure is shown in Fig. 1.

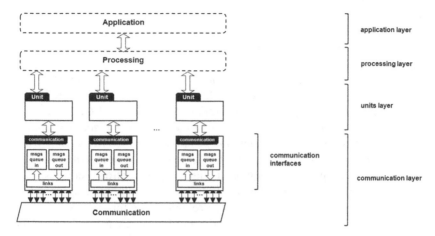

Fig. 1. Layered structure

2.2 General Structure

The cooperative computing system consists of V multiple units v that are autonomous with their decisions, i.e. the role of the centralized algorithm is minimized. Therefore, there is a need to design autonomous local algorithms that will control the local unit operation to the maximum extent. The decision making includes two cases: (1) decisions are taken just based on the local data, without getting the remote data (2) decisions are taken based on local and remote data (remote data must be fetched). The general operation of the system is depicted in Fig. 2. Two main phases are indicated: DP_PHASE_INIT and DP_PHASE_OPERATION.

During the DP_PHASE_INIT stage, the initial setup is done: the entry point is created, roles are assigned to units with the particular special role ROLE_CONTROL. This makes the cooperative system set up and ready for DP_PHASE_OPERATION. During 2nd phase, each unit can register an application that will be further processed on multiple units that participate in the group. The system allows each unit to register its own application, therefore there may be multiple applications running in the system concurrently, and each unit may process many different tasks belonging to different applications. System operates until the defined *end condition* is satisfied. During DP_PHASE_OPERATION roles of the units can be changed and/or reassigned.

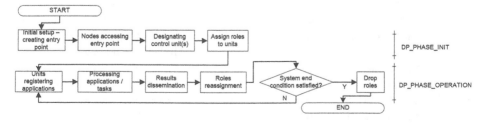

Fig. 2. The general system operation

Application is defined as a set of tasks, functions and data $A = \{T, F, D\}$. Functions operate on data, and are coordinated by operations. Data is not continuous and divided into data parts (if possible). This way such application can be executed in a distributed manner, where multiple functions can be executed on multiple units, and operate on various data parts. An example of such application A is the target location application (*TLA*) for UAV mission team. Set of functions: f_1 = *capture camera image*, f_2 = *capture IR camera image*, f_3 = *read radar data*, f_4 = *read GPS data*, f_5 = *request information from the other unit*, f_6 = *analyze terrain image*, f_7 = *analyze IR terrain image*. Set of tasks: t_1 = *determine position*, t_2 = *analyze terrain fragment*, t_3 = *validate terrain fragment with peers*, t_4 = *determine the next search area for the team*. UAVs are the units and are equipped with various electronic devices (cameras, GPS, radars etc.). As mentioned earlier, the ROLE_CONTROL must be present at least at one unit that serves as the database and control during the mission. A unit that starts the *target location* application (not necessarily the ROLE_CONTROL), registers the TLA at ROLE_CONTROL and therefore gets the ROLE_TASK_OWNER assigned. Remaining units that have ROLE_BASIC by default, also know the ROLE_CONTROL unit and get the information about TLA application from it. Further, they communicate directly with the task owner of the TLA and get the functions to execute according to their local resources (e.g. function f_2 can be executed only at the unit that has the IR camera installed). The proposed DPRS architecture allows dynamic (re) assignment of the functions during the operation. Given unit v_1 can execute task t_2 using functions f_1, f_4 and f_6 at the time T, finish t_2 execution at the time $T + T_1$ and then start the execution of task t_4 at the time $T + T_2$. This minimal set of operational elements provides a great flexibility required for the heterogeneous environment of cooperating units.

Each unit is considered to be a device with the processing capability (e.g. embedded system, PC computer, IoT device etc.). The processing capabilities are nowadays appearing in the form of multiple processors controlled by one operating system (server hardware, but also other architectures), processors with multiple cores, or single-chip single-core processor that is logically divided into processing threads by the operating system. Therefore, the unit v considered in this work is modeled as the device equipped with multiple processing devices – processors p such as CPUs, DSPs etc. These processors are general purpose – i.e. they are not designed for any specific application. The proposed approach allows using the reconfigurable FPGA chips, which can be programmed with some specific functions and thus becoming the application-specialized chips then. Hardware FPGA chips available on the market are programmed with bitstreams that are compiled hardware designs and model the hardware structure. This way, programming an FPGA reflects modeling its internal hardware structure. Market solutions also allow reprogramming on the fly, multiple functions can be programmed at the same time (if the internal chip space is sufficient) and chips handle thousands of reprogramming cycles. All these features are used in the proposed architecture.

Tasks are modeled as XML files that describe the relations between functions and data, also how the results are used. XML is also used to model any data structure in the system and all the relations. Functions f can appear in two forms: (1) programs compiled to the binary executable code (multiple platforms allowed) (2) definition in the form of script program (Ruby, Python, and similar). If a unit executes the processing of task t, it must fetch the functions required by this task – in the form that matches unit's architecture. For the FPGA processing devices, executing a function f in such device p requires programming the device p first, using the bitstream representing the function f. Application-specific programmed FPGAs provide more efficient processing, however require time and other resources to get the bitstreams and program them into the chip structure. Each unit is autonomous, but is participating in the team and cooperates with others units. This mechanism, along with the application structure as described above provides a cooperative processing structure with maximum flexibility, able to process multiple applications in the distributed manner with minimum centrality (thus more reliability for failures). The proposed system is especially useful for systems where: (1) several units have different capabilities and need to part and join the team structure constantly (UAV missions); (2) structures with multiple concurrent applications that need to be often switched / turned off / resumed, and benefit from efficient matching the structure to current application / needs (Internet of Things structures).

2.3 Reconfiguration

Each unit runs multiple processes, also each role assigned to the unit is served by one or more process. The very top-level scheme of the node operation is shown in Fig. 3. Two phases already described before (NODE_PHASE_INIT and NODE_PHASE_R-OLES) present the operations that unit performs when the entire system remains in these stages. The NODE_PHASE_SCHEDULER is used to execute all the scheduled operations that are not part of schemes /algorithms. Such operations include, among others: leaving the system, new role assignment, role drop, etc. Operations may be scheduled anytime in the advance. The focus of the presented work is the reconfiguration schemes their energy cost.

Figure 4 presents the operational scheme for ROLE_RECONFIGURABLE, that is assigned to every unit that is equipped with FPGA chip(s). The scheme assigned to ROLE_RECONFIGURABLE works as follows for processor p. The reconfiguration algorithm AL_RECONFIGURE_FPGA is executed regularly for p, and the period of the execution is determined by RECONFIGURATION_PERIOD value. RECONFI-GURATION_PERIOD value can be either determined once, before the system

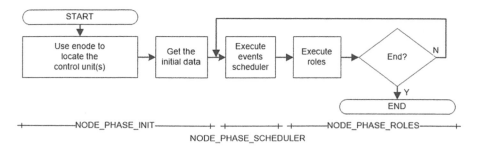

Fig. 3. Top-level operation of the unit

operation starts, or changed dynamically during system operation. In the work presented in this paper, the former way is used. The variable *reconf_counter* is set to RECONFIGURATION_PERIOD and then decremented each time the ROLE_RECONFIGURE_FPGA scheme is executed. Once the value reaches zero, the algorithm performs the *reconfiguration attempt*.

If *reconf_counter* = 0, but the related FPGA processor still executes a function, then the procedure waits till the function execution ends. Once the reconfiguration attempt is started, the AL_RECONFIGURE_FPGA algorithm is executed to determine if the reconfiguration should be done for the processor. This determination is done based on local knowledge and additional metrics that can be requested from other nodes (especially from ROLE_CONTROL and ROLE_TASK_OWNER. Based on all the inputs, the decision is taken whether the reconfiguration process should be started immediately or the next attempt should be taken next time the *reconf_counter* reaches zero.

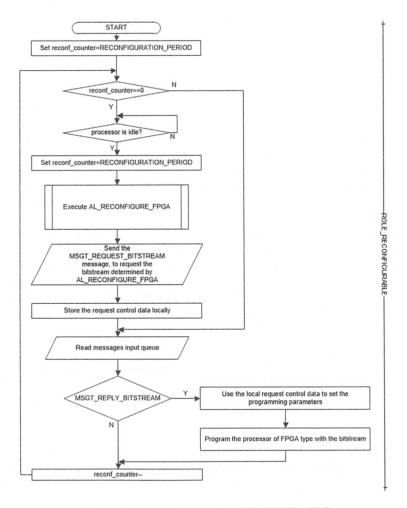

Fig. 4. Operation of ROLE_RECONFIGURABLE

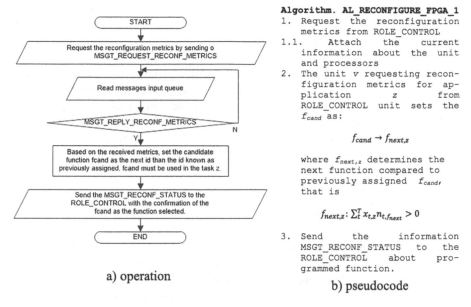

a) operation

Algorithm. AL_RECONFIGURE_FPGA_1
1. Request the reconfiguration metrics from ROLE_CONTROL
1.1. Attach the current information about the unit and processors
2. The unit v requesting reconfiguration metrics for application z from ROLE_CONTROL unit sets the f_{cand} as:

$$f_{cand} \rightarrow f_{next,z}$$

where $f_{next,z}$ determines the next function compared to previously assigned f_{cand}, that is

$$f_{next,z} : \Sigma_t^T x_{t,z} n_{t,f_{next}} > 0$$

3. Send the information MSGT_RECONF_STATUS to the ROLE_CONTROL about programmed function.

b) pseudocode

Fig. 5. AL_RECONFIGURE_FPGA_1 algorithm

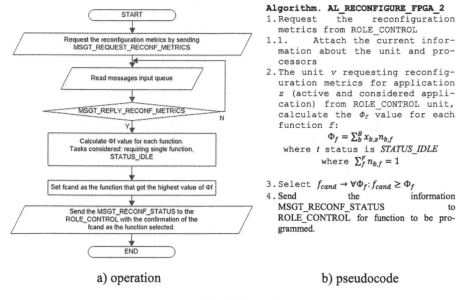

a) operation

Algorithm. AL_RECONFIGURE_FPGA_2
1. Request the reconfiguration metrics from ROLE_CONTROL
1.1. Attach the current information about the unit and processors
2. The unit v requesting reconfiguration metrics for application z (active and considered application) from ROLE_CONTROL unit, calculate the Φ_f value for each function f:

$$\Phi_f = \Sigma_b^B x_{b,z} n_{b,f}$$
where t status is *STATUS_IDLE*
where $\Sigma_f^F n_{b,f} = 1$

3. Select $f_{cand} \rightarrow \forall \Phi_f : f_{cand} \geq \Phi_f$
4. Send the information MSGT_RECONF_STATUS to ROLE_CONTROL for function to be programmed.

b) pseudocode

Fig. 6. AL_RECONFIGURE_FPGA_2 algorithm

Two reconfiguration algorithms are used and compared. The following notation is used to describe their operation: $x_{t,z} = 1$ if task t belongs to application z, 0 otherwise; $n_{t,f} = 1$ if task t requires function f, 0 otherwise; AL_RECONFIGURE_FPGA_1 works according to the diagram Fig. 5(a) and pseudocode Fig. 5(b). The main idea of this algorithm is to provide the even distribution of the functions programmed in the FPGAs over the system. AL_RECONFIGURE_FPGA_2 works according to the diagram Fig. 6 (a) and pseudocode Fig. 6(b). This algorithm takes the popularity of the requirement into consideration (how many tasks require a function *f*), considering tasks that require single function. Descriptions are using multiple message types: MSGT_REQUES-T_RECONF_METRICS – message to request all the data, known to the ROLE_-TASK_OWNER and/or ROLE_CONTROL that can be useful for determining reconfiguration; MSGT_REPLY_RECONF_METRICS – the reply for MSGT_RE-QUEST_RECONF_METRICS; MSGT_RECONF_STATUS – an update message to the ROLE_CONTROL unit.

3   Experimentation Results

The most important efficiency metric for the cooperative processing system is the electrical energy efficiency, as the UAV systems often are built using devices that rely on the battery power. Thus lowering the electrical energy consumption increases the operational timespan, also for the specific UAV systems reduces the frequency of a vehicle to be forced to come back to the ground base for recharging. The following elements of the electrical energy expenditure were included: E_GETTING_TASKS – process of getting the task definition to the unit, including control data and incoming transmission costs; E_GETTING_FF – getting the definitions of the functions from ROLE_TASK_OWNER; E_GETTING_EXT_DATA – acquiring any values that are needed for task processing, such as remote data from other units; E_PRO-CESSING_TASK – computational process (executing task); E_BASIC_COST – all costs including configuration, idle operations, and others not included in the remaining elements; E_OUTBOUND_TRAFFIC – cost of outgoing data transmission; E_FPGA_PROGRAMMING_BITSTREAM – cost of programming the bitstream into FPGA; E_FPGA_DOWNLOAD_BITSTREAM – cost of downloading the bitstream files. E_FPGA_DECIDE_RECONF – cost of ROLE_RECONFIGURATION except AL_RECONFIGURE_FPGA and programming bitstream;

The total cost is then:

$$E = \sum_v^v \sum_s^s \sum_p^p (v_{E\_GETTING\_TASKS}^s + v_{E\_GETTING\_FF}^s + v_{E\_GETTING\_EXT\_DATA}^s$$
$$+ p_{E\_PROCESSING\_TASK}^s + v_{E\_OUTBOUND\_TRAFFIC}^s$$
$$+ v_{E\_FPGA\_DOWNLOAD\_BITSTREAM}^s + p_{E\_FPGA\_DECIDE\_RECONF}^s$$
$$+ p_{E\_FPGA\_PROGRAMMING\_BITSTREAM}^s) + T p_{E\_BASIC\_COST}$$

$$C_{AL2}^{AL1} = \frac{AL2 - AL1}{AL2} \cdot 100\%$$

obj_{elem}^s indicates the energy that object *obj* spends on operations executing element *elem* during time slot *s*. C_{AL2}^{AL1} compares two E values.

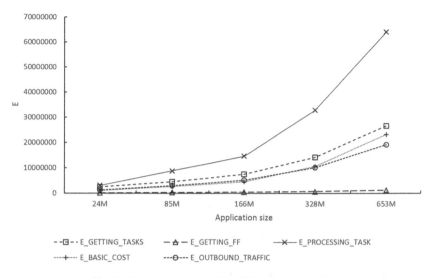

Fig. 7. Energy elements for different data volumes

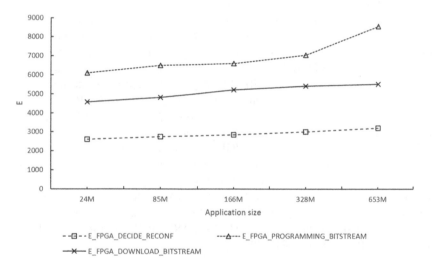

Fig. 8. Energy elements related to reconfiguration

Experiments were done for the *target search* application *A* described in Sect. 2, modified to incorporate five different total tasks sizes (24 M, 85 M, 166 M, 328 M, 653 M respectively). Different task sizes reflect different search area and different data load coming from data sensing equipment. Figures 7 and 8 show energy elements for five cases. Results show, that for the same application the size of the processing data has the most impact to E_PROCESSING_TASK, moderate impact to E_GET-TING_TASKS, E_BASIC_COST, E_OUTBOUND_TRAFFIC and minimal to E_GETTING_FF. Impact on functions is small as the same application logic contains

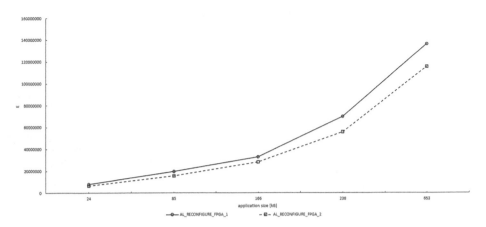

Fig. 9. Energy consumption for two AL_RECONFIGURE_FPGA algorithms

the same set of function, just differs with how many these functions will be sent to units. Regarding the reconfiguration-related energy elements, the E_FPGA_DECI-DE_RECONF depends on the timespan of the system operation, so its cost is linear. The largest impact of the size can be observed to the process of programming the bitstream into FPGA (also, being the most costly element of reconfiguration). The even distribution of the functions selected to be programmed onto FPGA, used in AL_RECONFIGURE_FPGA_1 exposed high operational cost, and the differences compared to the AL_RECONFIGURE_FPGA_2 differ depending on the size of processed tasks (Fig. 9, up to 22 % calculated as C). AL_RECONFIGURE_FPGA_2 uses the information about tasks' requirements and returns f_{cand} being most universal for the current task resources and suitable for most tasks.

The advantage of AL_RECONFIGURE_FPGA_2 comes from the periodic reconfiguration not limited by remote resources matching. Thus, AL_RECONFIGURE_FPGA_2 is much more flexible and adapts better to the current processing needs of the system. AL_RECONFIGURE_FPGA_1 required the longest time of processing, and the difference compared to the remaining two algorithms increased with the increase of the task(s) size. AL_RECONFIGURE_FPGA_2 is 31 %–38 % faster than AL_RECONFIGURE_FPGA_1.

4 Conclusions

Development of cooperative processing systems that operate on battery power creates demand for efficient algorithms for management. The centralized management is not that much desired for its single point of failure design. At the same time the flexibility of cooperative systems is demanded. In this work, we propose an architecture that addresses all of the above challenges, among others. The use of FPGA chips and the proposed efficient management algorithms create an effective and flexible solution. Research showed, that the proposed algorithms can save both electrical energy and time, while using the same resources for the same application. We also describe the proposal of the general architecture with multiple layers, multiple roles and various types of resources, along with the universal format of defining the cooperative application suitable for reconfigurable environment. Therefore, it creates a basis for further research, especially in the areas of advanced scheduling algorithms for reconfigurable systems, efficient applications management and improving the life of battery power for decentralized cooperative systems with autonomous nodes.

References

1. Chmaj, G., Selvaraj, H.: Distributed processing applications for UAV/drones: A Survey. In: Selvaraj, H., Zydek, D., Chmaj, G. (eds.). AISC, vol. 366, pp. 449–454Springer, Heidelberg (2015). doi:10.1007/978-3-319-08422-0_66
2. Chmaj, G., Selvaraj, H.: UAV cooperative data processing using distributed computing platform. In: Selvaraj, H., Zydek, D., Chmaj, G. (eds.). AISC, vol. 366, pp. 455–461. Springer, Heidelberg (2015). doi:10.1007/978-3-319-08422-0_67
3. Chmaj, G., Latifi, S.: Decentralization of a multi data source distributed processing system using a distributed hash table. Int. J. Commun. Netw. Syst. Sci. 6(10), 451–458 (2013)
4. Department of Defense, Unmanned Systems Integrated Roadmap FY2013-2038 (2013)
5. Jóźwiak, L., Nedjah, N.: Modern architectures for embedded reconfigurable systems a survey. J. Circuits Syst. Comput. 18(2), 209–254 (2009)
6. Jóźwiak, L., Nedjah, N., Figueroa, M.: Modern development methods and tools for embedded reconfigurable systems: a survey. Integr. VLSI J. 43(1), 1–33 (2010)
7. DeHon, A., et al.: Design patterns for reconfigurable computing. In: 12th Annual IEEE Symposium on Field-Programmable Custom Computing Machines, FCCM 2004, pp. 13–23 (2004). doi:10.1109/FCCM.2004.29
8. Bondalapati, K., Prasanna, V.: Reconfigurable computing systems. Proc. IEEE 90(7), 1201–1217 (2002). doi:10.1109/JPROC.2002.801446
9. Freitas, E., Binotto, A., Pereira, C., Stork, A., Larsson, T.: Dynamic reconfiguration of tasks applied to an UAV system using aspect orientation. In: International Symposium on Parallel and Distributed Processing with Applications, ISPA 2008, Sydney, NSW, pp. 292–300 (2008). doi:10.1109/ISPA.2008.69
10. Nadeem, M., Ostadzadeh, S., Nadeem, M., Wong, S., Bertels, K.: A simulation framework for reconfigurable processors in large-scale distributed systems. In: 2011 40th International Conference on Parallel Processing Workshops (ICPPW), Taipei City, pp. 352–360 (2011)
11. Wills, L., Sander, S., Kannan, S., Kahn, A., Prasad, J., Schrage, D.: An open control platform for reconfigurable, distributed, hierarchical control systems. In: 2000 Proceedings of the 19th Digital Avionics Systems Conference, DASC, Philadelphia, PA, pp. 4D2/1–4D2/8 (2000)

A Practical Verification of Protocol and Data Format Negotiation Methods in ComSS Platform

Łukasz Falas, Patryk Schauer[✉], Radosław Adamkiewicz, and Paweł Świątek

Department of Computer Science and Management,
Wroclaw University of Science and Technology,
Wybrzeże Wyspiańskiego 27, 50-370 Wrocław, Poland
patryk.schauer@pwr.wroc.pl

Abstract. The main aim of the research presented in this paper was to perform a practical verification of protocol and data format negotiations between streaming services with the use of Future Internet research infrastructure providing network isolation and tools for proper measurement of the key parameters. The paper presents a brief introduction to the ComSS Platform and automated protocol and data format negotiation methods, followed by in depth analysis of the used negotiation scheme and management message exchange protocol efficiency, as well as network traffic analysis related to these communication protocols.

Keywords: Data stream processing · Composite data stream processing services · Communication protocol negotiation · Data format negotiation

1 Introduction

The concept of composite streaming services extends the well-known Service Oriented Architecture paradigm by introducing tools and techniques for data stream processing, which are compliant with the general vision of service orientation. One of most common use cases for such services is the online processing of multimedia streams. For instance, composite streaming service can be used for video and audio data streams (gathered from cameras connected to a surveillance system) processing, aimed at detection of unauthorized access to a certain area or at detection of other incidents requiring reaction. In this example, composite service could consist of a service responsible for gathering data from the cameras, from services responsible for video and audio analysis focusing on anomaly detection, from a decision making service which would raise an alarm or contact system operator on the basis of analysis services output and from data storing service which would save gathered data in the cloud storage.

One of the key requirements for composite streaming services execution is the assurance of proper and compatible communication protocols and data formats used for data stream transfer between different services. In order to ensure the required compatibility, we are introducing tools and mechanisms for automated negotiation between

© Springer International Publishing AG 2017
J. Świątek and J.M. Tomczak (eds.), *Advances in Systems Science*, Advances in Intelligent Systems and Computing 539, DOI 10.1007/978-3-319-48944-5_16

different atomic services, during which the communication between them is established. The negotiations can be used, for instance to choose a common communication protocol and stream bitrate considering the available communication resources.

The main goal of the experiments presented in this article was to evaluate different negotiation methods, necessary during composite streaming services execution, which we have implemented in our solution for Composition of Streaming Services (ComSS) platform. Depending on the amount of available resources different methods of establishing connections between atomic streaming services were applied. The various methods differed in execution time and stream quality parameters like bitrate. Performed tests of different negotiation methods in various network and computing environments (differing in their infrastructure and in the amount of available resources), allowed us to verify resource requirements for each of the tested methods in different environments. Additionally, due to utilization of network traffic generator available in one of the testbeds, we were able to verify the efficiency of our solution and to determine its vulnerability to extreme network conditions. Also, the utilization of distributed environment offered by PLLAB2020 infrastructure allowed us to perform tests in environment which can be configured to mimic real environment conditions for using composite streaming services.

Experimental scenarios which were conducted focused on measuring the time and quality of the negotiation process, executed during establishing connections between atomic services of composite streaming service. The planned scenarios differed in:

- composite service structure (depending on number of serial and parallel sub-structures present in the general composite service structure the quality parameters can differ significantly),
- resources availability:
  - network connections parameters (and characteristic of background traffic),
  - computing resources.

In each scenario different negotiation methods were tested. Additionally influence of background traffic on quality and time of negotiations was also evaluated.

2 Related Works

Such services, especially from multimedia and Internet of Things domain, run constantly basing on processing an ingoing stream of data. Work on the distribution of streaming services can be found in [1], where authors describe methods for processing sensor data or in [2, 3] introducing specialized middleware for data stream processing. The natural distribution of the source of the stream and its destination was extended via introduction of more computing services in between [4, 5], introducing composite stream processing services in eHealth, rehabilitation and recreation fields. Many application from eHealth domain like sportsman and patient monitoring may work as streaming services [6]. Also some examples can be found in computational science and meteorological applications, where there are multiple data sources and multiple recipients interested in the processed data stream [7]. This work is a continuation of our previous works and presents results concerning the efficiency of ComSS Platform communication [8].

3   ComSS Platform

3.1   Platform Architecture

SOA paradigm assumes the existence of many atomic services, realising defined functionalities. Such services are executed in different locations within the infrastructure, which allows them to communicate and exchange data. It is necessary to use some middleware software to provide composite functionality of processing data streams (presented in Fig. 1) [9]. Such a tool should provide the following functions:

- service structure generation,
- accurate atomic services selection,
- services preparation for execution.

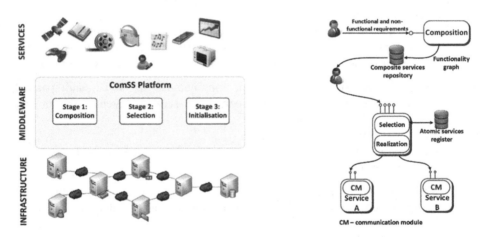

Fig. 1. ComSS Platform environment and process stages and ComSS Platform typical use

All features, mentioned in the previous section, describe a prototype environment called ComSS Platform (COMposition of Streaming Services). Figure 1 shows also a general use of the proposed tool.

Construction of a composite streaming service using discussed platform, consists of three steps, composition, selection and initialisation, which were described in [10]. Each stage corresponds to one function of previously described middleware application. Based on SOA architecture, all of them were implemented as a services [11].

In this article we focused on the last stage of composite streaming designing - service initialisation (also referred to as service set-up). Each atomic service has to be configured and initialised. Because of the distributed environment, this process bases on communication. Procedure prepares all services to realise their functions with other services as one composite service. The input of this stage is a services graph. Except the connection structure and input/output configuration of each service, the graph contains identifiers (addresses) of each services and formats available on every interface.

Each service class is described by input and output data types, what is needed in the composition stage. Each service, which belongs to the class can operate on different formats of a stream. For instance, data type stream from accelerometer sensors can be represented by different formats, such as Manufacture A Acc Stream or Manufacture B Acc Stream. Both streams contain the same information, but they are represented in a different way. This differences may impact non-functional parameters, especially connected with quality of service. Additionally some services could have functionality of converting one formats into another, but other could not do that. This can result in the failure of negotiation, when two services operate on the same type of a data, but they do not handle common format. During service configuration, data formats should be determined and all possible problems with connections should be solved.

3.2   Automated Communication Protocol and Data Format Negotiation Methods

Service connections parameters, especially formats, can be chosen by two general methods. First is based on auto-negotiation between services. Second assumes, that formats are selected centrally with the use of planning methods. In both cases composite service orchestrator is needed to control the whole initialisation process, which is divided into three steps: resources reservation, service configuration and composite services execution.

We propose two methods of auto-negotiation and two methods of planning [12]. First auto-negotiation method is ad-hoc negotiation. These methods focused only on pairs of services, which have to be connected. In some cases, whole connection graph analysis is necessary because of relations between formats availability. For this reason we propose method of sequential negotiations, in which services communicate with each other step by step.

When using planning based method, parameters selection is realised during the resources reservation. In such case the decision about using formats is taken centrally. Two planning algorithms which base on graph search were implemented.

In our experiment we were focusing on analysis of network communication issues of these methods. Due to the fact, that all of these methods use exactly the same communication protocol, which was presented in our previous works, all of the experiments were conducted for the ad-hoc negotiation method.

4   Experiment Set-up

The experiment was conducted on the PLLAB2020 research infrastructure [13]. In our experiment we have utilized 21 virtual machines (20 machines for services used during tests, 1 machine for our ComSS management software). In order to perform the test and measure the network traffic in proper isolation we have set up 38 dedicated virtual networks:

- 20 networks were used for our solution's control messages transmissions managing the execution of composite data stream processing services
- 18 networks were used for inter-service communication and communication protocol and data format negotiations

For our experiment we have prepared 4 different composite service workflows which cover majority of communication patterns which we are expecting to encounter in production environments. The aim of the experiment was to test the communication modules of services built on our framework for data stream processing services, which enable automated protocol and data format negotiation and provide interfaces for remote management of execution of composite data stream processing services with our management software (ComSS). For each of the services in the workflow we were measuring the parameters of negotiation message exchange with other services participating in the composite data stream processing workflow and the parameters of management message exchange with the management software responsible for orchestrating the composite service execution.

The schematic overview of the composite service workflows are presented below (Figs. 2, 3, 4 and 5).

As it is depicted on the diagrams, in each of the workflows we selected links on which we were analysing the network traffic with the support of a dedicated analysing device (DAG) and a traffic generator (SPIRENT) to test how background traffic affects network connectivity and message exchange in our solution.

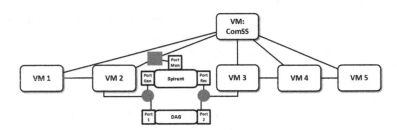

Fig. 2. General schema of the first composite data stream processing workflow

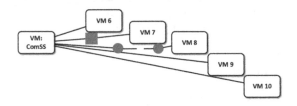

Fig. 3. General schema of the second composite data stream processing workflow

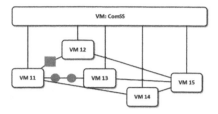

Fig. 4. General schema of the third composite data stream processing workflow

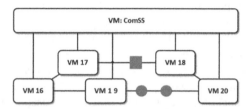

Fig. 5. General schema of the fourth composite data stream processing workflow

In order to thoroughly verify our communication modules and automated protocol and data format negotiation methods we have prepared a set of evaluation tests. The tests differed in following environmental parameters:

- selected composite data stream processing workflow (**g1, g2, g3, g4**)
- number of individual user requests sent during test (**i10**: 10 requests, **i50**: 50 requests, **i100**: 100 requests)
- number of feasible data formats, which were used during negotiations (**f03**: 3 formats, **f14**: 14 formats, **f21**: 21 formats)
- characteristic of background network traffic (**r1**: 0 % of background traffic, **r2**: 95 %, **r3**: 99 %, **r4**: 100 %)

The tests were conducted for all combinations of the described parameters which resulted in a total of 144 different test scenarios. Each of the tests was repeated 5 times (which gives a total number of 720 test runs), and the results from the repeated test runs were averaged.

During the performed tests we have measured the following parameters:

- Management message exchange time at 3 different stages of composite data stream processing service set up and execution process for each connection between services and our management software:
  - resource reservation and negotiation interface setup message exchange (reservation stage),
  - inter-service communication configuration message exchange (configuration stage),
  - composite service execution control message exchange (control stage).
- Protocol and data format negotiation message exchange time between each of the interconnected services.

- Message exchange traffic characteristics (data stream throughput variability) on management links (between management software and services).
- Message exchange traffic characteristics (data stream throughput variability) on negotiation links (between services participating in composite data stream processing workflow).

5 Experiment Results

In this section of the report we have describe the results obtained during the experimentation on PLLAB2020 infrastructure. Due to the large number of tests, which resulted in a relatively big set of test results, we are only presenting a selection of experimental results. Also, during the test we have noticed that the test results from different workflows does not differ considerably, hence in this report we only focused on presenting the results from workflow 1 (which results were similar to results of workflow 3 and 4, probably due to the fact that in all cases management and communication are independent, hence the structure of the workflow does not influence performance of point to point communication between services and between services and management component) and the results of workflow 2. It is also worth noting that in case of workflow 2 we are only analysing the communication with management software, due to the fact that this workflow is completely parallel and, due to this fact, it does not require any inter-service negotiations. In order to prove our solution usefulness in production environments, we assumed that that the communication time between services during the negotiations should, on average, be lower than 600 ms and we expected that the average aggregated time of communication during three phases of composite service initialization (reservation, configuration and control phases) should not exceed 1 s.

Table 1. Aggregated communication times required for composite data stream processing services set-up for a stream of 10, 50 and 100 consecutive user requests.

			Workflow 1.				Workflow 2.			Workflow 1.				Workflow 2.	
			max init	max neg	mean init	mean neg	max init	mean init		max init	max neg	mean init	mean neg	max init	mean init
i50	f03	r1	16,00	15,37	14,20	13,44	40,15	33,18	i100	31,00	29,67	28,70	27,72	72,09	68,14
		r2	32,00	31,08	29,10	28,24	76,58	75,27		57,00	56,56	46,50	45,83	152,45	148,36
		r3	32,50	30,53	28,20	27,37	77,27	74,91		58,00	57,67	55,40	54,79	152,90	121,37
		r4	29,50	28,54	27,90	27,37	77,71	75,78		60,00	58,68	56,60	55,80	151,31	148,93
	f14	r1	16,50	15,23	12,90	12,41	36,07	33,49		30,00	29,90	26,90	26,52	79,89	70,34
		r2	30,00	28,83	28,60	28,11	79,12	76,29		56,50	55,65	54,70	53,91	152,59	122,94
		r3	32,00	28,97	28,40	27,14	78,59	77,45		55,50	53,76	54,20	53,49	153,41	122,81
		r4	36,00	35,84	29,70	29,50	79,02	75,93		57,50	55,40	53,90	53,56	151,99	123,83
	f21	r1	16,50	15,07	14,40	13,37	35,82	33,05		65,50	60,75	48,50	43,71	70,16	65,59
		r2	6,00	4,98	3,80	3,30	79,68	75,90		9,50	8,78	7,80	7,00	150,85	121,63
		r3	5,00	5,18	3,50	3,32	80,67	76,93		10,00	7,75	7,10	6,34	153,05	149,54
		r4	29,50	29,40	27,80	27,31	77,27	75,87		59,50	57,45	55,20	54,72	149,92	148,18

We have started the analysis of experiment results with an examination of aggregated communication times required for proper set-up of the composite data stream processing services for a stream of 10, 50 and 100 consecutive user requests. All of the times were measured along with the analysis of network traffic characteristics performed by the DAG traffic analyser.

Table 1 shows the mean and maximum time of both, composite data streaming service set-up times and inter-service protocol and data format negotiation communication. The results presented in the table show that, as it was expected, the observed aggregated communication time required for composite data stream processing service set-up increases along with the increase of background traffic. However, it is worth noting that the increase in observed time is less significant than we expected (which is even more evident in case of average processing time for single request presented further). The measured communication time with r2, r3 and r4 background traffic still meets our QoS expectations. Also, there are no significant differences between times measured with r2, r3 and r4 background traffic, which proves that our communication modules and communication methods are reliable and retain required performance even at very high network loads. We can also observe that the increase in the number of format does not impact the performance of network communication significantly, even in case of high network load (Fig. 6).

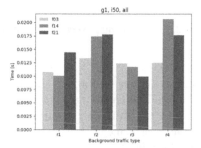

Fig. 6. Average aggregated communication time value of reservation, configuration and control stages times for workflow 1.

After the analysis of aggregated communication times required for proper set-up of the composite data stream processing services for a stream of 10, 50 and 100 consecutive user requests, we have started to analyse average request processing communication times for single request in different stages (reservation, configuration and control) of the set-up process as well as aggregated values of all three of these stages. Figure 7 presents selected times measured during the tests performed on workflow 1, which were measured by our software.

The results show that the measured average communication times for each of the stages are relatively low, as well as the aggregated time of all three stages and the communication time required for composite data stream processing service set-up meets our current requirements. It is also important that values in different cases are similar, hence the variability of network traffic and variety of network formats does not

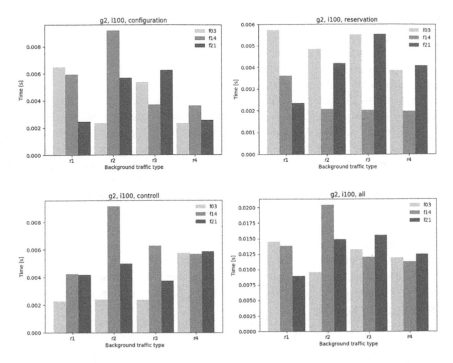

Fig. 7. Average reservation, configuration, control stage communication time and aggregate time of 3 stages for workflow 2.

have a direct performance impact on our solution. Some degree of small scale variability can be observed on the provided charts, however it may be a result of the fact that the testing infrastructure was not fully deterministic. Due to this fact we have encountered some outliers in our measurements, however we have decided that we will not exclude them from the final results.

The results obtained during the analysis of workflow 2 (as well as 3 and 4) were similar to workflow 1. The communication times obtained during the tests of 2 workflow were also relatively low and meet our expectations. Unfortunately, in this case we also encountered some small scale variability, due to the presence of outliers in our measurements, similarly to the workflow 1 (Figs. 8 and 9).

The utilization of SPIRENT traffic generator and DAG analyser allowed us to perform the analysis of traffic characteristics measured in different test cases. The measured characteristics were presented on charts above. Charts related to 1st workflow present the characteristics on both, management and negotiation links. For 2nd workflow measurements were performed only on management link.

In case of reference r1 background traffic (0 %) the maximum observed throughput reached approximately 15 kB/s (120 kb/s). This result is in line with our supposition that we should be able to achieve a relatively low network usage, due to utilization of our dedicated communication modules and communication protocols. As a result of our

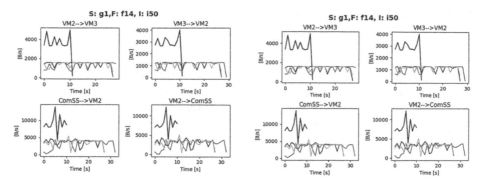

Fig. 8. Network characteristics for negotiation and management message exchange between negotiation modules of services for 14 data formats, 50 and 100 user requests (**black** – r1 background traffic, **dark_grey** – r2, **grey** – r3, **light_gray** – r4).

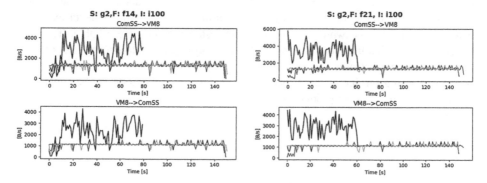

Fig. 9. Network characteristics for management message exchange between negotiation modules of services for 14 and 21 data formats and 100 user requests (**black** – r1 background traffic, **dark_grey** – r2, **grey** – r3, **light_gray** – r4).

optimization, even in case of heavy background traffic (95 %, 99 %, 100 %) observed throughput was similar. In majority of cases it did not exceed 5 kB/s (40 kb/s).

In case of the second workflow the results for management link were relatively similar. The measured throughput on configuration link without background network traffic, on average, did not exceed 6 kB/s (48 kb/s), while with background traffic message were exchanged with an average throughput of 2 kB (16 kb/s). All of these results met our expectations related to the QoS requirements for our solution. An interesting fact related to this experiment, is that in high network load cases the background traffic caused traffic shaping on the link. This result was related to the uniform characteristic of background traffic cause, which smoothed the traffic related to management message exchange.

The key conclusion of the network traffic characteristics analysis is that even in high network load scenarios, when the network links are almost completely reserved, our negotiation modules and protocols retain their required performance and are reliable, enabling efficient service set-up with composite data stream processing.

6 Conclusions and Lessons Learned

The possibility to perform the described experiment enabled us to fully evaluate and verify our protocol and data format negotiations in an isolated and controlled environment which simulates real world use cases. From strictly technical perspective, the conducted tests allowed usto improve our solution and fix some of critical flaws and bugs in the implementation of our methods in communication modules, which were decreasing our methods' performance and which, in some specific cases, resulted in a very high number of packet retransmissions.

The results of the experiments which we have performed after the removal of critical flaws in our software have shown, that the methods we have developed are performing in line with our expectations. The tests have also shown, that the structure of the workflow does not have a considerable impact on the performance of the communication modules and protocol and data format negotiation methods. Following that, the tested protocols and negotiation methods retain their reliability and performance even in case of larger number of communication formats and in case of high network load. Also, the average, as well as maximum, measured times never exceeded to required QoS levels, which proved that the tested methods are reliable. The tests have also shown that the introduction of adaptable protocol and data format negotiation methods does not introduce a considerable delay to the process of composite data stream processing service set-up, hence all of the developed methods can be considered as viable for utilization in production environments.

References

1. Gu, X., Yu, P., Nahrstedt, K.: Optimal component composition for scalable stream processing. In: 2005 Proceedings of the 25th IEEE International Conference on Distributed Computing Systems, ICDCS 2005, pp. 773–782, June 2005
2. Chen, L., Reddy, K., Agrawal, G.: Gates: a grid-based middleware for processing distributed data streams. In: 2004 Proceedings of the 13th IEEE International Symposium on High performance Distributed Computing, pp. 192 – 201, June 2004
3. Schmidt, S., Legler, T., Schaller, D., Lehner, W.: Real-time scheduling for data stream management systems. In: 2005 Proceedings of the 17th Euromicro Conference on Real-Time Systems, (ECRTS 2005), pp. 167–176, July 2005
4. Gu, X., Nahrstedt, K.: On composing stream applications in peer-to-peer environments. IEEE Trans. Parallel Distrib. Syst. 17(8), 824–837 (2006)
5. Rueda, C., Gertz, M., Ludascher, B., Hamann, B.: An extensible infrastructure for processing distributed geospatial data streams. In: 2006 18th International Conference on Scientific and Statistical Database Management, pp. 285 –290 (2006)
6. Świątek, P., Klukowski, P., Brzostowski, K., Drapała, J.: Application of wearable smart system to support physical activity. In: Advances in Knowledge-based and Intelligent Information and Engineering Systems, pp. 1418–1427. IOS Press (2012)
7. Liu, Y., Vijayakumar, N., Plale, B.: Stream processing in data-driven computational science. In: 7th IEEE/ACM International Conference on Grid Computing, pp. 160–167, September 2006

8. Stelmach, P., Świątek, P., Schauer, P.: Communication protocol negotiation in a composite data stream processing service. In: Swiątek, J., Grzech, A., Swiątek, P., Tomczak Jakub, M. (eds.) AISC, vol. 240, pp. 681–690. Springer, Heidelberg (2014). doi:10.1007/978-3-319-01857-7_65

9. Alamri, A.: Cloud-based e-health multimedia framework for heterogeneous network. In: 2012 IEEE International Conference on Multimedia and Expo Workshops (ICMEW), pp. 447–452 (2012)

10. Świątek, P., Schauer, P., Kokot, A., Demkiewicz, M.: Platform for building eHealth streaming services. In: 2013 IEEE 15th International Conference on e-Health Networking, Applications & Services (Healthcom), Lisbon, pp. 26–30 (2013)

11. Świątek, P., Stelmach, P., Prusiewicz, A., Juszczyszyn, K.: Service composition in knowledge-based SOA systems. New Gener. Comput. **30**, 165–188 (2012)

12. Stelmach, P., Świątek, P., Falas, Ł., Schauer, P., Kokot, A., Demkiewicz, M.: Planning-based method for communication protocol negotiation in a composition of data stream processing services. In: Kwiecień, A., Gaj, P., Stera, P. (eds.) CN 2013. CCIS, vol. 370, pp. 531–540. Springer, Heidelberg (2013). doi:10.1007/978-3-642-38865-1_53

13. Binczewski, A., et al.: Infrastruktura PL-LAB2020, Przegląd Telekomunikacyjny, Wiadomości Telekomunikacyjne R, 88(12), s. 1399–1404 (2015)

Transportation and Multi-Robot Systems

Reactive Dynamic Assignment for a Bi-dimensional Traffic Flow Model

Kwami Seyram Sossoe[1(✉)] and Jean-Patrick Lebacque[2]

[1] IRT SystemX, Palaiseau, France
`kwami.sossoe@irt-systemx.fr`
[2] IFSTTAR-COSYS-GRETTIA Laboratory, Champs-sur-Marne, France

Abstract. This paper develops a graph-theoretic framework for large scale bi-dimensional transport networks and provides new insight into the dynamic traffic assignment. Reactive dynamic assignment are deployed to handle the traffic contingencies, traffic uncertainties and traffic congestion. New shortest paths problem in large networks is defined and routes cost calculation is provided. Since mathematical modelling of traffic flow is a keystone in the theory of traffic flow management, and then in the traffic assignment, it is convenient to elaborate a good model of assignment for large scale networks relying on an appropriate model of flow related to very large networks. That is the zone-based optimization of traffic flow model on networks developed by [8], completed and improved by [9].

Keywords: Reactive assignment · Cross-entropy algorithm · Traffic control · Continuum anisotropic network · Zone-based model of flow

1 Introduction

The assignment is one of the recurring issues in respect of networks operators. Particular attention is taken in the case of transport since such networks allow people to move every day providing to them means of mobility. The government and cities are concerned. There are many static allocation models with respect to the assignment problem in the literature. There are also dynamic allocation models. We are concerning in these second type of assignment. Mainly algorithm for assignment derived from the algorithm of Dijkstra. Versatile algorithms such genetic algorithms, greedy algorithms, revolutionary algorithms, have been developed too in the early century addressing assignment issues in traffic controlling over networks.

These models incorporating such algorithms are of good quality depending on the purposes for which they are deployed, and on particular networks. Although, it is difficult that they represent accurately the dynamic aspects of network flows when very large transport networks are involved.

In this paper we propose a realistic model of dynamic assignment of vehicles flow for the prediction and the estimation of traffic on wide and dense networks. Since the vagaries of traffic in densely populated areas are very numerous, different and varied, we thought it would be more appropriate to use a template

© Springer International Publishing AG 2017
J. Świątek and J.M. Tomczak (eds.), *Advances in Systems Science*, Advances in Intelligent Systems and Computing 539, DOI 10.1007/978-3-319-48944-5_17

stream that already aggregates the roads and the network into zones, say two-dimensional grids or cells with a finite number of directions of the propagation of flow [8,9]. Using instantaneous travel times of users over networks, a reactive dynamic assignment developed by [4] and applied on networks allow to describe behavioral movement of users over the networks. It allows to compute accurately two-dimensional cells flow of such type of networks. Paths of users are ventilate along such directions of propagation.

2   The Bi-dimensional Flow Model

The bi-dimensional flow model is particularly timely responding to the traffic flow computing problematic over large and dense networks. We mean by dense network, a network with infinite secondary roads and very close to each other. It is the case of the road network of the city of Paris. Its road network forms a spiderweb, ranking in the type of anisotropic networks. For US cities, cities are new cities and their networks are rather orthotropic since roads are not gradually constructed as cities grow.

2.1   Concept of the Zone-Based Traffic Flow Modeling

At a very large scale, the area of a dense network is well approximated by a continuous medium where vehicles flow corresponds to fluid flow flowing on a surface, with a finite directions of preferred propagation. That is vehicles flow is viewed from a great elevation into the airspace, approximately 100 m to 500 m. Our case study concerns transport networks comprising major arteries, secondary urban/suburban roadways. The method is to decompose the surface of such networks in zones in such a way that the principal roads constitute frontiers of certain zones. The zone is meshed in two-dimensional cells. Within cells, are reduced the directions of flow propagation in four preferred outflow directions and four preferred inflow directions. Those directions of propagation of the flow will ventilate the generated traffic demands, from cell to cell. Cells represent edges and relations between cells represent vertices, when we are referring such obtained simplify network in graph theory.

Based on local behavior of flow at a macroscopic scale, a global two-dimensional behavior of flow is easily constructed [8]. It implies flow conservation for the two-dimensional zones both in Eulerian and Lagrangian coordinates. Every cell comprises of course many lanes in the preferred directions of propagation. For intersections of the simplified network that are formed in each zones or cells, intersection traffic flow model rules are applied [5]. It describes the interactions and the dynamic of incoming and outgoing flows at and out of intersections.

The conservation equation is constructed so that it takes into account turning rates at intersections within cells and interactions through interfaces of cells. Interfaces of cells are curved lines in the Euclidian space $\mathbb{R}^2$. We build a corresponding Lagrangian system of the two-dimensional traffic flow for large-scale networks. This allows the estimation of the flow of Lagrangian data. That is in particular the estimation of floating cars flow in any zone of the network over a relatively long time intervals. For sake of clarity, we applied this concept of network flow computing without taking into account difference between major roads and secondary roads. These difference shall be studied in a secondary paper. Every cell is setting by a maximum flow capacity (a free flow capacity), a critical density and a maximal density constraints labeled with directions of propagation. Variables are the 4-cell inflows and 4-cell outflows, traffic demand and traffic supply of cells with respect to the directions.

2.2   Semidiscretized Shape of the Two-Dimensional Flow Model

Traffic theory on dynamics of vehicles on highways and urban network and the analogy with fluids flowing within two-dimensional domain suggest the formulation of the following physical model (1) for traffic in a cell:

$$N_{c,i}(t + \delta t) = N_{c,i}(t) + \left(Q_{fc}(t) - R_{cg}(t)\right)\delta t + \left(r_{c,i}(t^+) - q_{c,i}(t^+)\right)\delta t \qquad (1)$$

with (i) the direction of propagation of flow inside the cell (c). (f) and (g) are respectively the indexes of zones located at left and right of the target-cell (c).

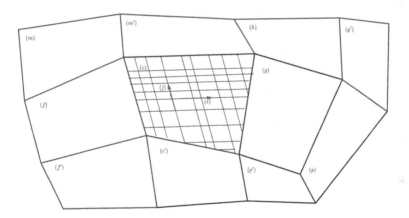

Fig. 1. Surface of the large-network (as the network-domain) is dis-aggregated in $2d$-zones. Zones are meshed in quadrangular cells. Each cell has certain parameters: a finite number of preferred propagation directions of flows, numbers of lanes in each direction, lengths of lanes in each direction.

Cell Internal Flows Control. Using intersection traffic flow model rules following [5], we find out that these functions (or variables) are solution of the below (2) linear-quadratic optimization problem:

$$\max_{(q,r)} \left(\sum_{i=1}^{4} \Phi(q_i) + \sum_{j=1}^{4} \Psi(r_j) \right)$$

$$\text{s.t.} \quad \begin{vmatrix} 0 \le q_i \le \Delta_{ci}^{t+1/2}, & \forall i \in \{1,2,3,4\}, \\ 0 \le r_j \le \Sigma_{cj}^{t+1/2}, & \forall j \in \{1,2,3,4\}, \\ -r_j + \sum_{i=1}^{4} q_i \Gamma_{c,ij}^{t} = 0, & \forall j \in \{1,2,3,4\}. \end{vmatrix} \qquad (2)$$

with $q = (q_1, q_2, q_3, q_4)$ and $r = (r_1, r_2, r_3, r_4)$ internal vectors flows which expresses traffic states inside cells. Functions $\Psi(\vartheta_\ell) \doteq \Phi(\vartheta_\ell)$ are defined by (3). They are assumed equal, concave, increasing and they describe interactions of vehicles inside cell. The optimization problem (2) results in an intersection model similar to the intersection models of [3, 10].

$$\Psi(\vartheta_\ell) \overset{\text{def}}{=} -\frac{1}{2}\vartheta_\ell^2 + \vartheta_\ell \cdot \vartheta_{\ell,max}. \qquad (3)$$

Notations and definitions are the following.

- $\forall i \in \{1,2,3,4\}$, q_i is the incoming vehicles flow in the direction i;
- $\forall j \in \{1,2,3,4\}$, r_j is the outgoing vehicles flow in the direction j;
- $\ell = i, j \in \{1,2,3,4\}$, ϑ_ℓ referring to q_i or r_j, $\vartheta_{\ell,max}$ denotes $q_{i,max}$ or $r_{j,max}$ which is the maximum flow constraint in the direction i or j;
- $\Gamma_{c,ij}^{t}$, assignment coefficients of flows within cell c, from direction i to direction j, at instant time t;
- μ_{ci}, number of lanes in cell (c) in the direction i;
- ν_{ci}, number of exiting lanes in the cell (c) respect to direction i;
- $\delta_i = \Delta_{ci}(\rho_{ci}^t)$, lane supply in direction i;
- $\sigma_i = \Sigma_{ci}(\rho_{ci}^t)$, lane demand in direction i;
- $\Delta_{ci}^{t+1/2} = \mu_{ci}\delta_i$, vehicles demand in c to direction i, at time t^+ $(t+1/2)$;
- $\Sigma_{cj}^{t+1/2} = \mu_{ci}\sigma_i$, cell c supply on line j, at time t^+ $(t+1/2)$;
- $r_{cj}^{t+1/2}, q_{ci}^{t+1/2}$, $\forall i,j$ denote the solution of the above convex optimization problem (2).

The obtained optimization model is easily solved with the python-cvxopt solver. The turning rates $\Gamma_{c,ij}^{t}$ can be considered as assignment coefficients; they are updated at each time-step t, $t \in [0,T]$ in the designed reactive dynamic assignment engine. Furthermore, having the number of vehicles in every 2d-cell, matrix of turning rates at time t, minimum travel paths, and directional outflows of zones of each path are easily identifiable and calculable.

Cell Inflows and Cell Outflows. Flows through cells denote by Q_{fc} and R_{cg} for $f \in \mathfrak{V}(c)$ are governed by a such semidiscretized model (1). $\mathfrak{V}(c)$ denotes the neighboring of the cell (c) comprising only adjacent cells that share an edge the cell (c). Q_{fc} and R_{cg} are respectively inflow and outflow of the cell (c) (see Fig. 1). There are define as following.

$$Q_{fc}(t) = \min\left(\delta_{f,i'}(t), \sigma_{c,i}(t)\right) \text{ and } R_{cg}(t) = \min\left(\delta_{c,k}(t), \sigma_{g,k'}(t)\right). \qquad (4)$$

where i lies in the sense of lanes on $f \rightarrow c$, from the cell (f); i' lies in the sense of lanes on $f \rightarrow c$, from the cell (f); k lies in the sense of lanes of (c) which flows flow directly into (g); k' lies in the sense of lanes on $c \rightarrow g$, within (g).

2.3   Computational Aspect: Methodology

Using a finite volume mesh of a transportation network area (which can be obtained easily by any mesh software for finite volume methods), we deduce a graph of the simplified network obtained at the two-dimensional scale. To compute bi-dimensional cells flows, the general structure of the algorithm is shown schematically in Fig. 2.

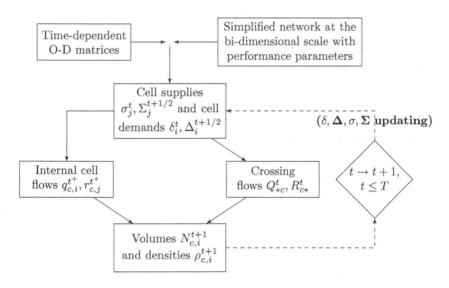

Fig. 2. Bi-dimensional network flows computing engine.

Flows Q_{*c}^t and R_{c*}^t denote the directional inflow $Q_{fc}(t)$ and directional outflow $R_{cg}(t)$ respectively. They are the flows that cross interfaces of cell (c). We name them crossing flows which are at the opposite of the cell internal flows $q_{c,i}^{t+}, r_{c,j}^{t+}$. These are computed with the linear-quadratic optimization problem (3). At each time step in the computation, the number of vehicles $N_{c,i}(t+1)$ in cell $c \in \mathfrak{C}$ is calculated by just applying the semidiscretized bi-dimensional formula (1), with $\mathfrak{C}$ the set of all nodes of the simplified network.

3   Reactive Dynamic Assignment

In this section we develop a reactive dynamic traffic assignment model applying to the semidiscretized traffic flow model (1). The model derives from works of

[2,11]. Thought we use rather instantaneous travel times instead of predictive travel time or experienced travel time. The instantaneous travel time allow to capture rapid changes in flow when traffic breakdown occurs. [1] give computational procedures for instantaneous travel times within macroscopic approximation of interrupted traffic flow. By analogy, we give definition of instantaneous travel time in the two-dimensional traffic flow modeling theory, and provide a diagram of the reactive assignment over large network.

Notation:

- π_c^d: the weight of path of minimum cost that reached the destination point (the node (d)) from the node (c).
- $\pi_c^{d,k}$: the weight of path of minimum cost that reached the destination point (the node (d)) from the node (c), consisting in k-arcs.
- $\Gamma_{c,ij}(t)$: turning rate movements of vehicles within cell (c), from direction (i) to direction (j), at time instant t.
- $\Gamma_{c,ij}^d(t)$: at time instant t turning rates, of incoming flow in the direction (i) in the cell (c), and that going to (c)-direction (j), in order to reach the destination cell (d).
- $\varpi_{cc'}$: the cost of the arc (c,c').

3.1   Instantaneous Travel Time

Let $(c) \in \mathfrak{C}$ a cell. For (i) a direction, we denote by $V_{cg,i}^t$ the cell exit speed of (c) in the direction $(i) = c \to g$. We are defining instantaneous travel time (ITT) for cell links: that is the links in $2d$-cell that lie in the preferred directions of flow propagation. It is a good approximation since flows will assign through these preferred directions of propagation. This is even the main expected in two-dimensional modeling: reduce the infinite links and nodes of dense network in a simplified network while still ensure a way of providing almost perfect information about network traffic states. Let us mention that instantaneous travel time in two-dimension space shall be describe as an integral along the path a user or vehicle will follow with respect to its velocity. A formal definition of ITT is the following (see [7]).

$$ITT(a,b,t) = \int_a^b d\chi/V(\chi,t) \tag{5}$$

$T(x,t) \overset{def}{=} ITT(x,b;t)$ is instantaneous travel time from x to b estimated at time t, labeled backward in [7]. This formula is valid in non-interrupted traffic flow, particularly when velocity is always bound by a strictly positive lower speed.

The authors of [7] have give clear computational definition of the ITT in interrupted traffic. The cell exit speed defined as $V_{cg,i}^t = R_{cg}L_{c,i}/N_{c,i}^t$ by authors is a CFL condition: It permits emulation of FIFO behavior within each $2d$-cell. Proper discretization constraint such $R_{cg}\delta t \leq N_{c,i}^t$ is set. Introducing the cell travel time $ITT_{c,i}^t = T_{c,i}^t - T_{f,i}^t$, these below (6) formulas hold:

$$\left[\begin{array}{l} ITT_{c,i}^{t+1} - L_{c,i}/V_{c,i}^t = \left(1 - \frac{\alpha_{c,i}\nu_{c,i}^t}{1-\nu_c^t}\right)\left(ITT_{c,i}^t - L_{c,i}/V_{c,i}^t\right) - \left(T_{f,i}^{t+1} - T_{f,i}^t\right), \\[2mm] \qquad\qquad\qquad\qquad\qquad \text{if } \nu_{c,i}^t \leq \frac{1}{1+\alpha_{c,i}} \\[4mm] ITT_{c,i}^{t+1} - L_{c,i}/V_{c,i}^t = -\frac{1-\alpha_{c,i}}{\alpha_{c,i}^t\nu_{c,i}^t}\left(T_{f,i}^{t+1} - T_{f,i}^t\right) \quad \text{if } \nu_{c,i}^t \geq \frac{1}{1+\alpha_{c,i}} \end{array} \right. \qquad (6)$$

where coefficients $\alpha_{c,i}$ and $\nu_{c,i}^t$ are defined such as:

$$\alpha_{c,i} \overset{def}{=} V_{c,i,max}\delta t/L_{c,i} \text{ and } \nu_{c,i}^t \overset{def}{=} V_{c,i}^t/V_{c,i,max} = R_{c,g}^t \delta t/\left(\alpha_{c,i}N_{c,i}^t\right). \qquad (7)$$

$V_{c,i,max}$ is the maximal exit speed of vehicles in the cell (c) and in the direction (i).

3.2 Travel Cost

The cost of an arc ϖ_{cf} can be estimated in the framework of the proposed model by the instantaneous travel time, which itself can be estimated at each time-step by the following:

$$\varpi_{cg} \approx N_{c,i}/R_{cg} \qquad (8)$$

if the cell (g) lies in direction (i) with respect to cell (c).

3.3 Logit Algorithm

Let us introduce a Logit model for the choice of neighboring nodes or cells, and address shortest paths computation. From a cell, vehicles have 4 possible choices for their next motion since there are 4 outflow directions. Due to target cell (the destination of vehicles), in a cell vehicles have generally just 2 possible directions that they may take when they are going out of the cell (a simplifying assumption). Therefore, the weight of path of minimum cost π_c^d can be decomposed as below (Fig. 3):

$$\pi_c^d \quad \rightarrow \quad \left\{ \begin{array}{l} \varpi_{cf} + \pi_f^d = C_f^d \\[1mm] \varpi_{cg} + \pi_g^d = C_g^d \end{array} \right. \qquad (9)$$

We can determine the probability of choice of users for choosing either the one cell between the neighboring cells of the cell they are located, at a fixed time t. The formulation of this probability of user cell-choice is given by (10):

$$\left[\begin{array}{l} P\left(\text{choice} = (f)/\text{Dest.} = d\right) = \dfrac{\exp(-\theta C_f^d)}{\exp(-\theta C_f^d) + \exp(-\theta C_g^d)} = \mathcal{F}_{cf}^d \\[4mm] P\left(\text{choice} = (g)/\text{Dest.} = d\right) = \dfrac{\exp(-\theta C_g^d)}{\exp(-\theta C_f^d) + \exp(-\theta C_g^d)} = \mathcal{F}_{cg}^d \end{array} \right. \qquad (10)$$

Parameters $\mathcal{F}_{cf}^d$ and $\mathcal{F}_{cg}^d$ allow the calculation of the coefficients of turning rates. We can easily compute π_c^d by the below algorithm (min,+ type, which can be improved as a Dijkstra algorithm):

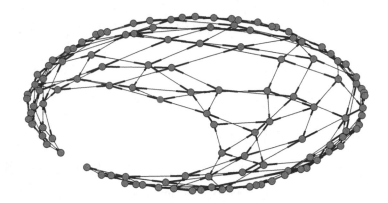

Fig. 3. Every node represents a zone. Every arc is a connection between two adjacent zones. The zones are two-dimensional computing cells, with at more 4-directions of propagation of the vehicles flow: (4-inflows and 4-outflows for each node/zone).

– $\pi_c^{d,1} = 0$.
– If $c \neq d$, $\pi_c^{d,1} = \varpi_{cd}$ if exists arc (c,d), or $= \infty$ if not.
– $\pi_c^{d,k+1} = \min \left(\pi_c^{d,k} , \min_{c' \in Succ(c)} \left(\varpi_{cc'} + \pi_{c'}^{d,k} \right) \right)$, $\forall t$.

Therefore, the traffic assignment model identifies the minimum cost travel paths, and the directional outflow within cells of each path. That is discussed in the Sect. 3. The general structure of the reactive algorithm is shown in Fig. 4.

The constructed assignment model enables flow assignment of big cities, if their transportation networks are approximated by two-dimensional media (orthotropic or anisotropic), completed by their main arteries network. For instance, we look at the road networks of Paris, Atlanta, Tokyo, Chicago, Manhattan, and Minneapolis. The presented assignment model allows computing paths of given OD pairs, and then successive cell-flows of the cells of the paths.

4   Recommendations

The main reasons of this bi-dimensional and zone-based approach is that it is difficult in practice to secure traffic data for all links of a dense network, even in a context of streaming data (portable, GPS), and that using traditional models (microscopic or macroscopic traffic models) requires cumbersome computational calculations. Further, very macroscopic management decisions do not necessarily require a very high level of detail of the traffic on the network.

Our zone-based approach results in a high level of aggregation of links flow. Hence the zone-based (or the two-dimensional scale framework) of traffic modeling requires less information than the traditional network approaches and makes possible to model traffic flow of transportation systems of large surface networks with few network sensors of traffic counts. It optimizes traffic zone-flows well and then provides a good traffic flow management.

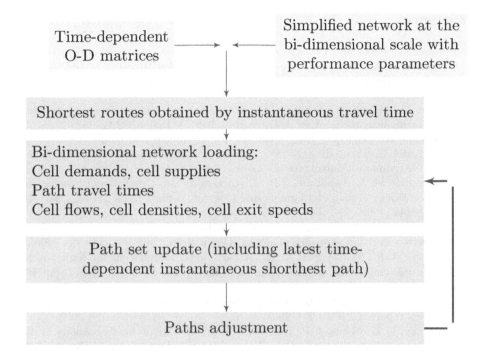

Fig. 4. Structure of solution algorithm to the RDA (Reactive Dynamic Assignment) problem.

Further issues are flow modeling and optimization per transportation mode within large-scale network. The bi-dimensional traffic flow model can be interfaced with a GSOM model [6] of the main motorways and arteries. In this perspective the bi-dimensional model will mainly describe dense networks of secondary roads. This modeling framework is also compatible with vehicular multimodality (distinguishing between private cars, taxis, electric vehicles, demand responsive systems etc.).

The reactive assignment introduced allows to calculate the exact flow on roads networks since dynamic user equilibrium is no longer valid in a very inhomogeneous transport network. A reactive assignment requires instantaneous travel time and capture perfectly vagaries of the traffic.

Acknowledgement. This research work has been carried out in the framework of the *Institute for Technological Research SystemX*, and therefore granted with public funds within the scope of the French Program *"Investissements d'Avenir"*.

References

1. Buisson, C., Lebacque, J.P., Lesort, J.B.: Travel times computation for dynamic assignment modeling. In: Transportation Networks: Recent Methodological Advances: Selected Proceedings of the 4th EURO Transportation Meeting, pp. 303–317 (1999)
2. Florian, M., Mahut, M., Tremblay, N.: Application of a simulation-based dynamic traffic assignment model. Eur. J. Oper. Res. **189**(3), 1381–1392 (2008)
3. Haj-Salem, H., Lebacque, J.-P., Mammar, S.: An intersection model based on the gsom model. In: Proceedings of the 17th World Congress the International Federation of Automatic Control, vol. 17(1), pp. 7148–7153 (2008)
4. Khoshyaran, M.M., Lebacque, J.P.: A reactive dynamic assignment scheme. In: Mathematics in Transport Planning and Control. Default Book Series, Chap. 13, pp. 131–143 (1998)
5. Lebacque, J.-P., Khoshyaran, M.M.: First-order macroscopic traffic flow models: intersection modeling, network modeling, Chap. 19, pp. 365–386. Elsevier (2005)
6. Lebacque, J.-P., Khoshyaran, M.M.: A variational formulation for higher order macroscopic traffic flow models of the gsom family. Procedia Soc. Behav. Sci. **83**, 370–394 (2013)
7. Lebacque, J.P.: Instantaneous travel times for macroscopic traffic flow models. Technical report, CERMICS (1996)
8. Saumtally, T., Lebacque, J.-P., Haj-Salem, H.: A dynamical two-dimensional traffic model in an anisotropic network. In: Networks and Heterogeneous Media, pp. 663–684, September 2013
9. Sossoe, K.S., Lebacque, J.-P., Mokrani, A., Haj-Salem, H.: Traffic flow within a two-dimensional continuum anisotropic network. Transp. Res. Procedia **10**, 217–225 (2015)
10. Tampère, C.M.J., Corthout, R., Cattrysse, D., Immers, L.H.: A generic class of first order node models for dynamic macroscopic simulation of traffic flows. Transp. Res. Part B Methodol. **45**(1), 289–309 (2011)
11. Ziliaskopoulos, A.K., Waller, S.T., Li, Y., Byram, M.: Large-scale dynamic traffic assignment: implementation issues and computational analysis. J. Transp. Eng. **130**(5), 585–593 (2004)

Using Repeated-Measurement Stated Preference Data to Investigate Users' Attitudes Towards Automated Buses Within Major Facilities

A. Alessandrini[1], P. Delle Site[2(✉)], D. Stam[3], V. Gatta[4],
E. Marcucci[4], and Q. Zhang[3]

[1] University of Florence, Florence, Italy
[2] University Niccolò Cusano – Telematica Roma, Rome, Italy
paolo.dellesite@unicusano.it
[3] University of Rome La Sapienza, Rome, Italy
[4] University Roma Tre, Rome, Italy

Abstract. The paper reports on the results of an investigation about users' attitudes towards automated and conventional minibuses for routes within major facilities. A common stated preference questionnaire has been used in four European cities. The econometric analysis is based on the estimation of three binomial logit models: one model considers all independent observations, a copula logit and an error component logit take into account the correlation among error terms of the observations by the same individual. The observed attributes are waiting time, riding time and fare. Of particular interest, is the estimation of the alternative specific constant (ASC) of the automated minibus, because this represents the mean of all the unobserved attributes of the automated alternative that affect the choice. With a common specification of the systematic utilities of the automated and conventional alternatives, the results show a positive value of the ASC, which is indicative, the observed attributes being the same, of a relatively higher preference for automation. The differences in policy implications among the three models estimated are negligible.

Keywords: Automated bus · Stated preference · Binomial logit · Copula · Error component

1 Introduction

Automated buses are the subject of the paper. The key advantage of automated buses is identified in the potential for offering a higher frequency of service in the off-peaks, provided the operating costs are lower than a conventional bus. Also, there is a potential for higher flexibility in adapting the supply to demand because of the lack of drivers' scheduling constraints. The technology is available and a few demonstrations have taken place. The Rivium in Rotterdam is the only system currently operated on a permanent basis (an extensive review of automation in road public transport is in NETMOBIL Consortium 2005). The key barriers to implementation are legal (on this aspect a review is in Csepinszky et al. 2015).

© Springer International Publishing AG 2017
J. Świątek and J.M. Tomczak (eds.), *Advances in Systems Science*, Advances in Intelligent Systems and Computing 539, DOI 10.1007/978-3-319-48944-5_18

The paper is based on results from the project CityMobil2 which has been funded within the European Commission's Seventh Framework Programme (a review of the project is in Alessandrini et al. 2014b; details on technological aspects of the automated vehicles that are tested within the project are in Guala et al. 2015).

The paper deals with the investigation about users' attitudes towards automated buses and conventional buses. A stated preference (SP) questionnaire has been administered to individuals with no experience of the automated bus. The econometric analysis has been based on the estimation of logit models which have considered the choice for two alternatives: automated and conventional minibus. The observed attributes are: waiting time, riding time and fare. Of particular interest, is the estimation of the alternative specific constant (ASC) of the automated bus, because this represents the mean of all the unobserved attributes of the automated system that affect the choice. With a common specification of the systematic utilities of automated and conventional minibus, the observed attributes being the same, a positive value of the ASC is indicative of a relatively higher preference for automation. The key results of this survey, which has been extended to the twelve cities where the planning of an automated bus has taken place within CityMobil2, are presented in Alessandrini et al. (2014a).

The present paper focuses on the cases where the route of the bus service is within major facilities, such as technology parks and university campuses. In such application cases a relatively higher preference for automation has been found. The original contribution of the paper consists in an extended econometric analysis which takes into account the correlations among the error terms that arise from repeated choices by the same respondent.

The approach that other authors have used to address the repeated measurement problem is based on error components. An autocorrelation structure for the random terms is assumed, with each random term decomposed into an individual-specific term, constant across choice tasks, independent across individuals and normally distributed, and an observation-specific term, independent across choice tasks and individuals, and Gumbel distributed (Abdel-Aty et al. 1995; Cantillo et al. 2007; Jensen et al. 2013).

Another option, yet unexplored with SP data, is to use the copula logit proposed by Bhat and Sener (2009) to model spatial correlation in binary logit. With copula logit, the marginal distribution (i.e. over the single observation) of each error term remains Gumbel.

In addition to the standard logit with all independent observations used in the previous analysis, the paper reports on the results of the estimation of an error component logit and a copula logit. The paper is organized as follows. Section 2 is about econometric methodology. Section 3 reports on the four case studies and the surveys. Section 4 on the results of the econometric analysis. Section 5 concludes.

2 Methodology

Two alternatives are considered. We denote by 1 the conventional bus alternative, by 2 the automated bus alternative. Let j be the index of alternative. Let $n = 1, ..N$ be the index of individual. Let $t = 1, ..T$ be the index denoting choice task. In the present case we have $T = 4$.

The utility $u_j^n(t)$ of an alternative is the sum of the systematic component $V_j^n(t)$ and the random term $\varepsilon_j^n(t)$:

$$u_j^n(t) = V_j^n(t) + \varepsilon_j^n(t) \quad j = 1, 2; \ t = 1, ..T; \ n = 1, ...N \tag{1}$$

The following is the basic specification of systematic utilities:

$$V_1^n(t) = \beta_1 \cdot WT_1^n(t) + \beta_2 \cdot RT_1^n(t) + \beta_3 \cdot FA_1^n(t) \tag{2}$$

$$V_2^n(t) = \beta_1 \cdot WT_2^n(t) + \beta_2 \cdot RT_2^n(t) + \beta_3 \cdot FA_2^n(t) + ASC \tag{3}$$

where WT is waiting time, RT is riding time, FA is fare, $\beta_1, \beta_2, \beta_3$ are the coefficients, ASC is the alternative specific constant of the automated alternative. For fare, effects coding ($-1/1$) was used instead of dummy coding ($0/1$) to avoid confounding with the ASC. The code -1 represents the case where an extra-fare is paid. The code $+1$ represents the case where the same fare as other public transport is paid.

2.1   Logit with All Independent Observations (LOGIT1)

For a given individual and choice task, the random terms are distributed independently and identically across alternatives according to a standard Gumbel distribution. Thus the difference between the random terms of the two alternatives is distributed according to a logistic distribution, which is denoted by Λ (having omitted the index of individual):

$$F(\varepsilon_1(t) - \varepsilon_2(t) < \eta(t)) = \Lambda(\eta(t)) = \frac{1}{1 + e^{-\eta(t)}} \tag{4}$$

This implies the marginal choice is binomial logit.

In LOGIT1 the random terms of each alternative are independent across choice tasks and individuals. The cumulative distribution of the difference of the random terms of a given individual is (having omitted the individual index):

$$F(\varepsilon_1(t) - \varepsilon_2(t) < \eta(t), \ t = 1, ...T) = \prod_{t=1}^{T} \Lambda(\eta(t)) \tag{5}$$

The likelihood function of LOGIT1 is:

$$L = \prod_{n=1}^{N} L^n = \prod_{n=1}^{N} \prod_{t=1}^{T} \frac{e^{\left[V_1^n(t) - V_2^n(t)\right] \cdot d^n(t)}}{1 + e^{\left[V_1^n(t) - V_2^n(t)\right]}} \tag{6}$$

where $d^n(t) = 1$ if alternative 1 is chosen, $d^n(t) = 0$ if alternative 2 is chosen. For estimation of LOGIT1 we have used the software NLOGIT.

2.2 Copula Logit (LOGIT2)

The copula logit (LOGIT2) is the model used by Bhat and Sener (2009). Equation (4) holds for each individual and choice task. LOGIT2 is obtained by considering that the random terms of each alternative are independent across individuals, while for each individual the differences between random terms are correlated across choice tasks according to the following multivariate version of the Farlie-Gumbel-Morgenstern (FGM) copula (Nelsen 2006; Karunaratne and Elston 1998) (having again omitted the individual index):

$$
\begin{aligned}
F(\varepsilon_1(t) - \varepsilon_2(t) &< \eta(t), t = 1, \ldots T) \\
&= \prod_{t=1}^{T} \Lambda(\eta(t)) \cdot \left[1 + \theta \cdot \sum_{t=1}^{T-1} \sum_{k=t+1}^{T} (1 - \Lambda(\eta(t))) \cdot (1 - \Lambda(\eta(k))) \right]
\end{aligned}
\tag{7}
$$

The parameter θ controls the pairwise correlation across choice tasks and is restricted to the interval between 0 (independence) and 1 (maximal correlation). This interval yields positive Pearson correlation coefficients. Notice that in the bivariate logistic distribution based on the FGM copula the maximal Pearson correlation coefficient corresponding to $\theta = 1$ is limited to 0.304. In bivariate FGM copulas the correlation between the two uniform marginals, i.e. the Spearman's correlation, cannot exceed 1/3. The requirement for a non-negative density function implies the following restriction on the values of θ:

$$
1 + \theta \cdot \sum_{t=1}^{T-1} \sum_{k=t+1}^{T} \Delta_t \cdot \Delta_k \geq 0 \quad \Delta_1, \ldots, \Delta_T \in \{-1, +1\}
\tag{8}
$$

The likelihood function of LOGIT2 takes the following closed-form expression:

$$
L = \prod_{n=1}^{N} L^n = \prod_{n=1}^{N} \left[\left(\prod_{t=1}^{T} \frac{e^{\left[V_1^n(t) - V_2^n(t) \right] \cdot d^n(t)}}{1 + e^{\left[V_1^n(t) - V_2^n(t) \right]}} \right) \right.
$$
$$
\left. \cdot \left[1 + \sum_{t=1}^{T-1} \sum_{k=t+1}^{T} (-1)^{d^n(t) + d^n(k)} \cdot \theta \cdot \left(1 - \frac{e^{\left[V_1^n(t) - V_2^n(t) \right] \cdot d^n(t)}}{1 + e^{\left[V_1^n(t) - V_2^n(t) \right]}} \right) \cdot \left(1 - \frac{e^{\left[V_1^n(k) - V_2^n(k) \right] \cdot d^n(k)}}{1 + e^{\left[V_1^n(k) - V_2^n(k) \right]}} \right) \right] \right]
\tag{9}
$$

The likelihood of LOGIT2 is neither concave nor log-concave. Therefore, a global optimization method is needed. We have implemented the Price's algorithm (Price 1978) using the Wolfram Mathematica programming environment.

The Price's algorithm is a controlled random search procedure. It requires knowledge of a compact set which contains in its interior the global maximum. Bounded intervals for $\beta_1, \beta_2, \beta_3$ and ASC have been defined using the results of the estimation of LOGIT1. The parameter θ is restricted to the bounded interval $[0, 1]$. The objective function is the natural logarithm of the likelihood. This choice is motivated by overflow problems, not by ease of differentiation since the algorithm does not use derivatives. The Price's algorithm is a heuristic which uses a set of initial points chosen at random. We have used different starting points and found the results robust.

For inference, the standard theorems on the asymptotic properties of maximum likelihood estimators, ensuring normality and providing the covariance matrix, require, in addition to some regularity conditions, unrestricted parameters (Greene 2012). In LOGIT2, a re-parameterisation is therefore needed for θ. If θ at maximum likelihood is in the interior of the interval $[0, 1]$, it is possible to consider the new unrestricted parameter z of the logit transformation:

$$\theta = \frac{e^z}{1 + e^z} \tag{10}$$

Thus the estimators of $\beta_1, \beta_2, \beta_3, ASC$ and z will be asymptotically normally distributed. We have obtained the estimate of the elements of the covariance matrix by using the outer product of gradients estimator (also known as BHHH estimator, based on Berndt et al. 1974) which uses the first-order derivatives of the loglikelihood. This computation also has been implemented in Mathematica.

2.3   Error Component Logit (LOGIT3)

In error component logit (LOGIT3) the utilities take the expression:

$$u_j^n(t) = V_j^n(t) + \alpha_j^n + \varepsilon_j^{\prime n}(t) \qquad j = 1, 2; \ t = 1, \ldots T; \ n = 1, \ldots N \tag{11}$$

with α_j^n independent across alternatives and normally distributed $\alpha_j^n \sim N[0, \sigma_\alpha^2]$, while $\varepsilon_j^{\prime n}(t)$ is independent across alternatives and tasks and Gumbel distributed with variance $\sigma_{\varepsilon\prime}^2 = 1$.

We have for the Pearson correlation of the error terms of a pair of tasks:

$$\rho = \frac{\sigma_\alpha^2}{\sigma_\alpha^2 + 1} \tag{12}$$

from which:

$$\sigma_\alpha = \sqrt{\frac{\rho}{1 - \rho}} \tag{13}$$

We have for the difference of the utilities:

$$u_1^n(t) - u_2^n(t) = V_1^n(t) - V_2^n(t) + \alpha_1^n - \alpha_2^n + \varepsilon_1^{\prime n}(t) - \varepsilon_2^{\prime n}(t) \tag{14}$$

and shortly:

$$\Delta u^n(t) = \Delta V^n(t) + \Delta \alpha^n + \Delta \varepsilon^{\prime n}(t) \tag{15}$$

with $\Delta \alpha^n \sim N[0, 2 \cdot \sigma_\alpha^2]$ and $\Delta \in^{\prime n}(t)$ distributed according to a logistic.

Now set $\Delta \alpha^n = \sqrt{2} \cdot \sigma_\alpha \cdot v$ with v standard normal.

We have the likelihood function:

$$L = \prod_{n=1}^{N} L^n = \prod_{n=1}^{N} \int_{-\infty}^{\infty} \frac{e^{-v^2/2}}{\sqrt{2 \cdot \pi}} \cdot \prod_{t=1}^{T} \frac{e^{\left[(\Delta V^n(t) + \sqrt{2} \cdot \sigma_\alpha \cdot v) \cdot d^n(t)\right]}}{1 + e^{\Delta V^n(t) + \sqrt{2} \cdot \sigma_\alpha \cdot v}} dv \qquad (16)$$

Make the change of variable: $x = v/\sqrt{2}$. We have:

$$L = \frac{1}{\sqrt{\pi}} \cdot \prod_{n=1}^{N} \int_{-\infty}^{\infty} e^{-x^2} \cdot \prod_{t=1}^{T} \frac{e^{\left[(\Delta V^n(t) + 2 \cdot \sigma_\alpha \cdot x) \cdot d^n(t)\right]}}{1 + e^{\Delta V^n(t) + 2 \cdot \sigma_\alpha \cdot x}} dx \qquad (17)$$

This integral can be solved numerically by Gauss-Hermite quadrature (simulation is not needed), yielding:

$$L = \frac{1}{\sqrt{\pi}} \cdot \prod_{n=1}^{N} \sum_{l=1}^{L} w_l \cdot Q^n(y_l) \qquad (18)$$

with w_l weight, y_l node and:

$$Q^n(y_l) = \prod_{t=1}^{T} \frac{e^{\left[(\Delta V^n(t) + 2 \cdot \sigma_\alpha \cdot y_l) \cdot d^n(t)\right]}}{1 + e^{\Delta V^n(t) + 2 \cdot \sigma_\alpha \cdot y_l}} \qquad (19)$$

For estimation of LOGIT3 we have used NLOGIT.

3   Case Studies and Surveys

The four routes chosen for the surveys are:

- CERN, Geneva, where it serves the main building area and the Restaurant No2 (2 km);
- Lausanne, where it runs in the EPFL (École Polytechnique Fédérale de Lausanne) campus and links the northern and the southern area (1.4 km);
- San Sebastian (Spain), where it runs inside the Miramón Technology park, linking the entrance of the park and the head offices of the different companies (2.1 km);
- Sophia Antipolis, where it links the areas of Trois-Mulins and Saint-Philippe (3.8 km).

Random samples from the population of potential users of the bus systems for the specific routes were drawn. A common SP questionnaire across the four cases was administered online. Data collection activities took place in Spring 2013.

The questionnaire includes the following parts. First, the route of the public transport service under planning is described. A brief description of two vehicle options, a conventional minibus and an automated minibus, is provided. It is specified that the two vehicles are equal in terms of propulsion, and of total and seating capacity. Both will run in mixed traffic. The difference is the presence or absence of the driver. In the second part of the questionnaire, respondents are asked to choose between a conventional minibus and an automated minibus in different supply scenarios for a trip of given length. The supply scenarios for the services provided by the two vehicles are defined according to different levels of waiting time, riding time and fare. The third part relates to the personal characteristics of the respondents: gender, age, income before taxes, education, occupation, car availability in the household, ownership of a public transport monthly ticket.

The attributes and corresponding levels of the SP design are shown in Table 1. The number of combinations in the full factorial design (eight combinations) has been reduced to four combinations using a within-alternative orthogonal design technique.

Table 1. SP design – Attributes and levels

Alternative	Attribute	Number of levels	Levels
Conventional/automatic	Waiting time	2	3/8 min
	Riding time	2	5/10 min
	Fare	2	As other public transport means in the city/extra-fare of 2 EUR per return journey

4 Results

The results of the estimation of the three logit models are in Table 2. The models have been estimated separately for each case study.

The coefficients of waiting time, riding time and fare have the right sign. The *ASC* is always positive denoting a relatively higher preference for automation.

The estimates of the copula logit are closer to the logit with all independent observations than the estimates of the error component logit. With copula logit large deviations appear only for non-significant coefficients.

The intuitive expectation that the logit with all independent observations overestimates the statistical significance of the coefficients on the logit models that take error correlation into account is not confirmed in all cases.

The results on the value of the error term correlation shown by the copula logit and the error component logit are consistent with each other. The correlation in the error component logit, computed with the delta method, is highly statistically significant.

For comparison with previous literature, Delle Site et al. (2011) is a relevant reference, because they have used logit models calibrated on SP data to assess the relative preference between a conventional and an automated bus service for a short connection with a parking area in the new trade fair district in Rome (Italy). They too have found a relatively higher preference for automation.

Table 3 shows the preference shares that are obtained under two policies: in policy 1 (P1) the same fare applies to both the automated and the conventional bus, in policy 2 (P2) there is an extra-fare (2 EUR per return journey) for the automated bus, on top of the current fare of public transport in the city. The travel time of the conventional bus is assumed lower than the travel time of the automated bus (80 % lower: 8 against

Table 2. Estimation of logit models

City	LOGIT1 coefficient (t-statistic)	LOGIT2 coefficient (t-statistic)	% deviation on LOGIT1	LOGIT3 coefficient (t-statistic)	% deviation on LOGIT1
CERN					
β_1	−.38293 (−13.24)	−.3947	−3 %	−0.4564 (−13.048)	−7.4 %
β_2	−.34816 (−17.28)	−.3578	−2.7 %	−0.4153 (−16.516)	−6.7 %
β_3	.99446 (11.62)	1.028	+3.3 %	1.1872 (11.73)	+19.4 %
ASC	1.22331 (12.14)	1.267	+3.5 %	1.4616 (10.546)	+19.5 %
θ	−	1.0	−	−	−
ρ	−	−	−	0.5786 (28.266)	−
Number of individuals: 482					
Lausanne					
β_1	−.3778 (−15.22)	.3792 (−14.09)	−0.3 %	−0.3658 (−14.408)	+0.4 %
β_2	−.3976 (−23.18)	−.3972 (−25.97)	+0.1 %	−0.3884 (−22.515)	+2.3 %
β_3	.55397 (8.19)	.5639 (7.91)	+1.7 %	0.5390 (8.136)	−2.7 %
ASC	.86415 (10.07)	.899 (9.61)	+4 %	0.8390 (9.053)	−2.9 %
θ	−	.889 (15.50)	−	−	−
ρ	−	−	−	0.4043 (22.831)	−
Number of individuals: 742					
San Sebastian					
β_1	−.2003 (−4.39)	.211	−5.3 %	−0.2349 (−3.839)	17.2 %
β_2	−.2078 (−6.72)	.2158	−3.8 %	−0.2461 (−6.197)	−18.4 %
β_3	.8358 (5.69)	.876	+4.8 %	0.9833 (5.085)	17.6 %
ASC	.28531 (1.85)	.364	+27.5 %	0.3409 (1.545)	19.5 %
θ	−	1.0	−	−	−
ρ	−	−	−	0.5816 (18.699)	−
Number of individuals: 200					

(*continued*)

Table 2. (*continued*)

City	Logit1 coefficient (t-statistic)	Logit2 coefficient (t-statistic)	% deviation on Logit1	Logit3 coefficient (t-statistic)	% deviation on Logit1
Sophia Antipolis					
β_1	−.3131 (−8.42)	−.3278	−4.6 %	−0.4033 (−8.116)	−28.9 %
β_2	−.2807 (−10.75)	−.2908	−3.5 %	-0.3615 (-10.72)	−28.8 %
β_3	.73213 (6.08)	.7759	+5.9 %	.9414 (6.244)	28.6 %
ASC	.16204 (1.24)	.2722	+67.9 %	.2230 (1.226)	37.6 %
θ	–	1.0	–	–	–
ρ	–	–	–	.6391 (25.326)	–
Number of individuals: 290					

10 min). This is because of safety reasons. Commercial speed of the automated bus will be the result of the design of the collision avoidance system. The application of an extra-fare reduces significantly the preference share for automation. The differences in the preference shares among the three logit models are in almost all cases negligible.

Table 3. Policy comparisons: preference share (%) for automated bus

City	P1			P2		
	Logit1	Logit2	Logit3	Logit1	Logit2	Logit3
CERN	62.8	63.4	62.2	18.8	18.2	19.9
Lausanne	51.7	52.6	51.4	26.1	26.4	22.2
San Sebastian	46.7	48.3	47.0	14.2	13.9	15.4
Sophia Antipolis	40.6	42.3	40.8	13.7	13.5	13.7

5 Conclusions

The results are representative of the attitudes of users who have been exposed to minimal information about automated buses, and who have had no experience of these innovative systems. Information and experience are, together with emotions, the key determinants of preference formation according to the standard behavioural choice model (McFadden 2014). The cases investigated relate to new services on routes where public transport is currently not existing. A common trait of the results across the cases where the route is within a major facility is the relatively higher preference for the automated minibus. The result confirms that of a previous study. It is not so when the automated minibus is used on routes serving different purposes, e.g. within city centre:

preference for either the automated or the conventional alternative is found (see Alessandrini et al. 2014a).

The paper contributes to the assessment of the impacts on coefficient estimation and on policy implications of different ways of taking into account, in binomial logit, the correlation of error terms across observations by the same individual. In terms of estimation, the copula logit provides results that are closer to logit with all independent observations than the error component logit. In terms of policy implications, both the copula logit and the error component logit provide, in the policies tested, preference shares that are comparable with the logit with all independent observations.

Acknowledgements. The survey activities have been carried out by the partners of the City-Mobil2 project involved in the feasibility studies in the four cities.

References

Abdel-Aty, M.A., Kitamura, R., Jovanis, P.P.: Investigating effect of travel time variability on route choice using repeated-measurement stated preference data. Transp. Res. Rec. **1493**, 39–45 (1995)

Alessandrini, A., Alfonsi, R., Delle Site, P., Stam, D.: Users' preferences towards automated road public transport: results from European surveys. Transp. Res. Procedia **3**, 139–144 (2014a)

Alessandrini, A., Cattivera, A., Holguin, C., Stam, D.: CityMobil2: challenges and opportunities of fully automated mobility. In: Meyer, G., Beiker, S. (eds.) Road Vehicle Automation. Lecture Notes in Mobility, pp. 169–184. Springer, Heidelberg (2014b)

Berndt, E.K., Hall, B.H., Hall, R.E., Hausman, J.A.: Estimation and inference in nonlinear structural models. Ann. Econ. Soc. Meas. **3**(4), 653–665 (1974)

Bhat, C.R., Sener, I.N.: A copula-based closed-form binary logit choice model for accommodating spatial correlation across observational units. J. Geogr. Syst. **11**(3), 243–272 (2009)

Cantillo, V., de Dios Ortúzar, J., Williams, H.C.W.L.: Modeling discrete choices in the presence of inertia and serial correlation. Transp. Sci. **41**(2), 195–205 (2007)

Csepinszky, A., Giustiniani, G., Holguin, C., Parent, M., Flament, M., Alessandrini, A.: Safe integration of fully automated road transport systems in urban environments: the basis for the missing legal framework. In: Proceedings TRB Annual meeting, Washington D.C (2015)

Delle Site, P., Filippi, F., Giustiniani, G.: Users' preferences towards innovative and conventional public transport. Procedia Soc. Behav. Sci. **20**, 906–915 (2011)

Greene, W.H.: Econometric Analysis, 7th edn. Pearson, Boston (2012)

Guala, L., Alessandrini, A., Sechi, F., Delle Site, P., Holguin, C., Salucci, M.V.: Testing autonomous driving vehicles in a mixed environment with pedestrians and bicycles. In: Proceedings 22nd ITS World Congress, Bordeaux, France, October (2015)

Jensen, A.F., Cherchi, E., Mabit, S.L.: On the stability of preferences and attitudes before and after experiencing an electric vehicle. Transp. Res. Part D Transp. Environ. **25**, 24–32 (2013)

Karunaratne, P.M., Elston, R.C.: A multivariate logistic model (MLM) for analyzing binary family data. Am. J. Med. Genet. **76**(5), 428–437 (1998)

McFadden, D.: The new science of pleasure: consumer choice behavior and the measurement of well-being. In: Hess, S., Daly, A. (eds.) Handbook of Choice Modelling. Edward Elgar (2014)

Nelsen, R.B.: An Introduction to Copulas, 2nd edn. Springer, New York (2006)

NETMOBIL Consortium: EU Potential for Innovative Personal Urban Mobility. Deliverable D7 of the NETMOBIL (New transport system concepts for enhanced and sustainable personal urban mobility) project. Fifth Framework Programme, European Commission (2005)

Price, W.L.: A controlled random search procedure for global optimization. In: Dixon, L.C.W., Szegö, G.P. (eds.) Towards Global Optimization 2. North Holland, Amsterdam (1978)

Comparing Signal Setting Design Methods Through Emission and Fuel Consumption Performance Indicators

S. de Luca[1], R. Di Pace[1(⊠)], S. Memoli[1], and L. Pariota[2]

[1] Department of Civil Engineering, University of Salerno,
Via Giovanni Paolo II 132, 84084 Fisciano, SA, Italy
{sdeluca,rdipace,smemoli}@unisa.it
[2] Department of Civil Building and Environmental Engineering,
University of Naples Federico II, Via Claudio 21, 80125 Naples, Italy
luigi.pariota@unina.it

Abstract. In order to address the Signal Setting Design at urban level two main approaches may be pursued: the coordination and the synchronisation approaches depending on the steps considered for the optimisation of decision variables (two steps vs. one step). Furthermore, in terms of objective functions mono-criterion or multi-criteria may be adopted. In this paper the coordination approach is implemented considering the multi-criteria optimisation at single junctions and mono-criterion optimisation at network level whereas the synchronisation is implemented considering the mono-criterion optimisation.

The main purpose of the paper is the evaluation of the performances of two strategies not only considering indicators such as the total delay, the queue length etc. but also considering other indicators such as the emissions and the fuel consumption. The methodological framework is composed by three stages: (i) the decision variables (green timings and offsets) computation through optimisation methods; (ii) the implementation of optimal signal settings in a microscopic traffic flow simulator ("Simulation of Urban MObility"-SUMO); (iii) the estimation of emissions and fuel consumption indicators.

Keywords: Network signal setting design · Macroscopic traffic flow model · Microscopic traffic flow model · Sustainable transportation

1 Introduction and Motivation

This paper aims to compare the results of two different Network Signal Setting Design methods not only by considering the usually adopted indicators such as the capacity factor (computed at network level), the total delay etc. but also considering other indicators such the total emissions and the fuel consumption.

The mitigation of environmental impact due to the traffic congestion is still a difficult challenge to be pursued; the considered strategies usually refer to the Intelligent Transportation Systems [1, 2] and in particular to the enhanced eco-driving technologies [6] or on the application of transport policies (demand or supply).

© Springer International Publishing AG 2017
J. Świątek and J.M. Tomczak (eds.), *Advances in Systems Science*, Advances in Intelligent Systems and Computing 539, DOI 10.1007/978-3-319-48944-5_19

As regards the supply strategies some researchers investigated the significance of the correlation between signalised junctions and emissions/fuel consumption [9] whilst others [11–13, 18] developed multi-criteria optimisation frameworks using these indicators as alternative criteria.

In fact even though optimisation strategies are usually based on total delay minimisation, or on the combination of total delay minimisation and capacity factor maximisation, none of these objective functions are strictly related to the emission/consumption evaluation; in particular these indicators are usually represented by the number of stops. Summing up the optimisation problem considering the trade-off between the network performances and the air pollution indicators estimation might be represented through multi-criteria method. Based on previous considerations, the paper aims to preliminarily investigate the effectiveness of two control strategies in terms of emission and fuel consumption indicators. The main contribution with respect to the current literature is on the enhanced traffic control strategies.

The research is organised in three sections as in following described in more details.

In the first section the results achieved through two optimisation strategies, coordination and synchronisation are shown and discussed; in particular, in case of coordination method the multi-criteria optimisation at a single junction (green timings are the decision variables) is adopted (capacity factor maximisation an delay minimisation, are the considered criteria) and mono-criterion optimisation is applied at network level (offsets are the decision variables), whilst in case of synchronisation method the mono-criterion optimisation (green timings and offsets are optimised together) is adopted (the total delay is the considered objective function); in terms of algorithms some meta-heuristic algorithms are adopted in both cases and in particular the Genetic Algorithms [4] (to get optimal green times at each single junction) and the Hill Climbing (to get optimal offsets) are combined for the coordination approach [7], whereas the Simulated Annealing is adopted for mono-criterion optimisation in the synchronisation approach. In both strategies the adopted traffic flow model is based on macroscopic approach.

In the second section the considered network and the given values of input variables computed through the optimisation strategies, are implemented in a microscopic simulator ("Simulation of Urban MObility"-SUMO).

Finally, in the third stage the total emissions and the fuel consumption are computed through HBEFA model [4] embedded in SUMO.

The reminder of this paper is organised as follows: Sect. 2 provides a description of the optimisation strategies focusing on decision variables, constraints and objective functions description and on the presentation of the macroscopic traffic flow model used for total delay computation; in Sect. 3 the emissions and fuel consumption estimations through SUMO is briefly discussed; the results of the numerical application are shown in Sect. 4; finally conclusions and further perspectives are presented in Sect. 5.

2 Optimisation Strategies

In this section the basic notations, the constraints and the objective functions are described. Furthermore, the considered traffic flow model is described.

2.1 Variables and Constraints

Assuming that the green scheduling is described by the stage matrix (i.e. the stage matrix composition and sequence), let

c be the cycle length, assumed known or as a decision variable (common to all junctions);

for each junction (not explicitly indicated)

t_j be the duration of stage j as a decision variable;

t_{ar}, be the so-called all red period at the end of each stage to allow the safe clearance of the junction, assumed known (and constant for simplicity's sake);

Δ be the approach-stage incidence matrix (or stage matrix for short), with entries $\delta_{kj} = 1$ if approach k receives green during stage j and 0 otherwise, assumed known;

l_k be the lost time for approach k, assumed known;

$g_k = \Sigma_j \delta_{kj} t_j - t_{ar} - l_k$ be the effective green for approach k;

$r_k = c - g_k$ be the effective red for approach k;

y_k be the arrival flow for approach k, assumed known;

s_k be the saturation flow for approach k, assumed known;

$(s_k \cdot g_k)/(c \cdot y_k)$ be the capacity factor for approach k;

and for each junction in the network

ϕ_i be the offset as the time shift between the start of the plan for the junction i and the start of the reference plan, say the plan of the junction number 1, $\phi_1 = 0$.

Some constraints were introduced in order to guarantee:
stage durations being non-negative

$$t_j \geq 0 \qquad \forall j$$

effective green being non-negative

$$g_k \geq 0 \qquad \forall k$$

this constraint is usually guaranteed by the non-negative stage duration, but for a too short cycle length with regard to the values of all-red period length and lost times, say

$$\sum\nolimits_j \mathrm{MAX}_k \left(\delta_{kj} l_k + t_{ar} \right) \geq c$$

consistency among the stage durations and the cycle length

$$\sum_j t_j = C$$

the minimum value of the effective green timing

$$g_k \geq g_{min} \qquad\qquad \forall k$$

A further constraint was included in order to guarantee that the capacity factor must be greater than 1 (or any other value)

$$((s_k \cdot g_k)/(c \cdot y_k) - 1) \geq 0 \qquad \forall k$$

Such a constraint may be added only after having checked that the maximum junction capacity factor for each approach k in the junction i is greater than 1, otherwise a solution may not exist whatever the objective function is.

Finally let assume

$$c \geq \phi_i \geq 0.$$

2.2 Objective Functions

At a single junction, the objective functions in the optimisation problems were:

– the junction capacity factor computed as

$$CF = MAX_k (s_k \cdot g_k)/(c \cdot y_k)$$

– the total delay computed
– for non-interacting approaches (isolated or external junctions) by the two terms Webster's formula [17] as

$$TD = \sum_k y_k \cdot (0.45 \cdot c \cdot (1 - g_k/c)^2/(1 - y_k/s_k)$$
$$+ \ y_k \cdot 0.45/(s_k \cdot g_k/c) \cdot ((g_k/c) \cdot (s_k/y_k) - 1))$$

– for the interacting approaches by evaluating vehicles queuing interval by interval and considering input as the flow obtained by cyclic flow profiles. A more detailed expression consistent with the traffic flow modelling will be described in Subsect. 2.3.

Further objective functions could be considered such as the queue length, the number of stops etc.

2.3   Traffic Flow Model

One of the considered objective functions is the total delay, as described in more details above; to compute total delay at single junction different analytical formulations may be applied (e.g. [16]) whereas at network level to represent total delay traffic flow modelling is required. With reference to the literature traffic flows may be described through

- microscopic models, modelling both the space-time behaviour of the systems' elements (i.e. vehicles and drivers) as well as their interactions;
- mesoscopic models, modelling traffic by groups of vehicles possibly small, the activities and interactions of which are described at a low detail level;
- macroscopic flow models, modelling traffic at a high level of aggregation as a flow without distinguishing its parts (i.e. the traffic stream is represented in an aggregate manner using characteristics as flow-rate, density, and speed).

In this paper traffic flow is modelled through macroscopic model and in particular the Cell Transmission Model (CTM; [8]) is implemented. Moreover, since CTM assumes the same speed for all the vehicles on a road, it cannot fully predict realistic traffic flow behaviour as the platoons keep the same density when moving from the upstream stop-line section to the downstream section, and all vehicles travel at the same free flow speed. The CTM includes the horizontal queuing at the cost of not considering the platoon dispersion then to overcame this limitation the CTM & PDM (see [3]) allowing horizontal queues and platoon dispersion modelling was adopted.

3   Total Emissions and Fuel Consumption Estimation

Different models have been developed in the literature for emissions and fuel consumptions estimation; among them some are based on traffic conditions such as stop-and-go or free- flow driving other on the estimation of emissions/consumptions produced via engine (e.g. HBEFA, Handbook of Emission Factors; [5]; Road Model, [15]) or are vehicle operating models thus various driving cycle variables are required as input (e.g. PHEM; Passenger car and Heavy Emission Model, [10]; CMEM; Comprehensive Modal Emissions Model; [14]).

In this paper emissions and fuel consumption have been estimated through TraCI4Matlab which is an implementation of the TraCI (Traffic Control Interface) protocol; through this protocol user is able to interact with SUMO (Simulation of Urban Mobility) in a client (Matlab)-server (SUMO).

4   Numerical Application

In this section it is shown an application to a network with four interacting
signalised junctions forming a loop (shown in Fig. 1 as a square for simplicity's sake). Two design strategies were considered:

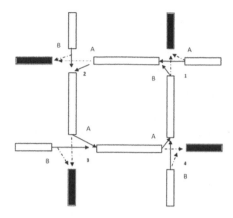

Fig. 1. Junction layout; stage matrix; characteristics of the junction

A coordination: once fixed a master junction in the network, say the junction 1, this
   consists in optimising the node offsets, say the time distance among the start of the
   signal plan of the other junctions in the network and the start of the signal plan of
   junction 1. The green times at each junction were previously computed through a
   multi-criteria. Genetic Algorithm which optimises simultaneously the junction total
   delay (to minimise) and capacity factor (to maximise);
B synchronisation: offsets and green times were simultaneously optimised considering the minimisation of network total delay.

All information related to the network, in terms of flow and saturation flow for each
approach, are summarised in Table 1. With regards to the links length L1 = 500 m,
L2 = 200 m; L3 = 600 m and L4 = 200 m.

Strategy A: In this case two optimisation steps are identified: first the green times of
each junction were optimised considering the trade-off effect between two criteria
(Total Delay, TD and Capacity Factor, CF); as expected, any solution was not dominant with respect to both criteria at the same time (see Table 2); then starting from the
timings obtained by the multi-criteria single junction signal setting design the network

Table 1. Flow and saturation flow for each approach

Junction	Stage	y_k [veic/h]	s_k [veic/h]
1	A	400	1750
	B	245	
2	A	390	
	B	227	
3	A	422	
	B	286	
4	A	375	
	B	191	

Table 2. Total delay, capacity factor and effective greens of each approach

Junction	Stream	TD [PCU-hr/hr]	CF	Effective green
1	A	1.27	1.66	31
	B	2.35	2.15	49
2	A	1.75	2.32	40
	B	1.91	2.47	40
3	A	1.49	1.88	33
	B	2.30	2.68	47
4	A	1.53	2.38	40
	B	1.79	2.44	40

Table 3. Offsets, TD and DOS

Coordination results

Offset 1–2[a] [sec]	Offset 2–3[a] [sec]	Offset 3–4[a] [sec]	Offset 4–1[a] [sec]	TD [PCU-hr/hr]	DOS [%]
54	31	58	37	8.48	60

[a]Offset i-j refers to the time distance between the start of signal plan of junction j with respect to the start of signal plan of junction i.

total delay, the degree of saturation (DOS) and the link offsets between the signal plans of the interacting junctions were carried out (see Table 3).

Strategy B: In this case the green times at each junction (see Table 4) and the offsets (see Table 5) are simultaneously optimised considering the minimisation of network total delay (as described in Sect. 2).

Table 4. Synchronisation results: effective greens

Synchronisation results

Junction	Stream	Effective green
1	A	38
	B	42
2	A	45
	B	35
3	A	43
	B	37
4	A	46
	B	34

The network parameters (i.e. the input flows and the link lengths) were fixed as equal to those adopted in previous implementations. Results shown in Table 5 make it clear that there is a greater efficiency in terms of level of service function by using a synchronisation strategy with respect to a coordination strategy; in fact

Table 5. Synchronisation results: offsets

Offset 1–2[b] [sec]	Offset 2–3[b] [sec]	Offset 3–4[b] [sec]	Offset 4–1[b] [sec]	TD [PCU-hr/hr]	DOS [%]
54	13	72	41	6.71	49

[b]Offset i-j refers to the time distance between the start of signal plan of junction j with respect to the start of signal plan of junction i.

$TD_{Coordination} = 8.48$ PCU-hr/hr whereas $TD_{Synchronisation} = 6.71$ PCU-hr/hr and $DOS_{Coordination} = 60$ % whereas $DOS_{Synchronisation} = 49$ %.

The signal settings obtained through Strategy A and Strategy B have been then implemented in the microscopic simulator SUMO, in order to get further indicators which were related to emissions and fuel consumption.

Results shown in Table 6 makes it clear, as expected, that the Synchronisation approach allows to improve the network performances also in terms of air pollution indicators. As a matter of fast such results highlight relevant insights into optimal traffic signal strategy for fuel consumption and emissions minimisation.

Table 6. Emissions and fuel consumption

		Coordination	Synchronisation
[ton/year]	CO2	52.822	50.695
	CO	0.575	0.455
	HC	0.024	0.014
	Nox	0.200	0.098
	PMx	0.008	0.004
	FuelConsumed	22.660	20.211

5 Conclusions and Research Perspectives

The main focus of the paper is on the comparison of the results performed through two different optimisation strategies for Network Signal Setting Design.

With reference two the decision variables (green timings and offsets) the considered optimisation strategies were the coordination and the synchronisation; the first one was carried out in two steps where during the first step the optimal values of the green timings at single junction were computed whereas in the second step the optimal values of the offsets were carried out; the second strategy was computed in only one step thus all decision variables were optimised at the same time.

In case of coordination multi-criteria optimisation was adopted at single junction whilst mono-criterion optimisation was adopted at network level; in case of synchronisation only mono-criterion optimisation was considered. Furthermore, in terms of algorithms, due to the nature of the optimisation problem, meta-heuristics algorithms were adopted; in particular in coordination approach Genetic Algorithms and Hill Climbing where respectively considered at single junction and at network level, whereas in synchronisation approach the Simulated Annealing was adopted.

Moreover, the objective functions considered in the optimisation procedure were: (i) in case of coordination, the capacity factor and the total delay at a single junction and the total delay at network level; (ii) in case of synchronisation the total delay.

The paper deals with the preliminary investigation of the effect of two optimisation strategies on alternative indicators in order to evaluate the significance of the introduction of other objective functions such as the emissions (to be minimised) and the fuel consumption (to be minimised).

To this aim, in order to compare the effectiveness of two strategies were considered common indicators such as the total delay and the degree of saturation, and were also introduced the emissions and the fuel consumption indicators.

The results point out the relevant effect of synchronisation with respect to the coordination thus highlighting the possibility to introduce in multi-criteria optimisation of further criteria based on emissions and fuel consumption. Furthermore it is expected that increasing the degree of complexity of the network the effect of two strategies in terms not only of performance indicators but also in terms of air pollution indicators, could be more significant and then relevant.

In future works researchers

- will investigate the relevance on air pollution indicators of optimisation strategies by increasing the traffic flows;
- will evaluate the effectiveness of the strategies for bigger grid network;
- will develop a multi-criteria optimisation based on performance criteria and air pollution indicators.

Acknowledgements. The research has been supported by the University of Salerno, Italy, EU under local grant n. ORSA151059 financial year 2015 and by 'APPS4SAFETY – PON03PE_00159_3'.

References

1. Bifulco, G.N., Cantarella, G.E., Simonelli, F.: Design of signal setting and advanced traveler information systems. J. Intell. Transp. Syst. Technol. Plan. Oper. **18**(1), 30–40 (2014)
2. Bifulco, G.N., Cantarella, G.E., Simonelli, F., Velonà, P.: Advanced traveller information systems under recurrent traffic conditions: network equilibrium and stability. In: Transportation Research Part B: Methodological (2016). doi:10.1016/j.trb.2015.12.008
3. Cantarella, G.E., de Luca, S., Di Pace, R., Memoli, S.: Network signal setting design: meta-heuristic optimisation methods. Transp. Res. Part C Emerg. Technol. **55**, 24–45 (2015)
4. Cantarella, G.E., de Luca, S., Di Pace, R., Memoli, S.: Signal setting design at a single junction through the application of genetic algorithms. In: de Sousa, J.F., Rossi, R. (eds.) Computer-Based Modelling and Optimization in Transportation. AISC, vol. 262, pp. 321–331. Springer, Heidelberg (2014)
5. Colberg, C.A., Tona, B., Stahel, W.A., Meier, M., Staehelin, J.: Comparison of a road traffic emission model (HBEFA) with emissions derived from measurements in the Gubrist road tunnel (Switzerland). Atmos. Environ. **39**, 4703–4714 (2005)

6. de Luca, S., Di Pace, R., Marano, V.: Modelling the adoption intention and installation choice of an automotive after-market mild-solar-hybridization kit. Transp. Res. Part C: Emerg. Technol. **56**, 426–445 (2015)
7. Di Gangi, M., Cantarella, G.E., Di Pace, R., Memoli, S.: Network traffic control based on a mesoscopic dynamic flow model. Transp. Res. Part C Emerg. Technol. **66**, 3–26 (2016)
8. Daganzo, C.F.: The cell-transmission model. Part 2: Network traffic, University of California, Berkeley, California (1994)
9. Guo, R., Zhang, Y.: Exploration of correlation between environmental factors and mobility at signalized intersections. Transp. Res. Part D Transp. Environ. **32**, 24–34 (2014)
10. Hausberger, S.: Simulation of real world vehicle exhaust emissions, VKM-THD Mitteilungen Technical University Graz, vol. 82, Graz (2003)
11. Kwak, J., Park, B., Lee, J.: Evaluating the impacts of urban corridor traffic signal optimization on vehicle emissions and fuel consumption. Transp. Plan. Technol. **35**(2), 145–160 (2012)
12. Lee, J., Park, B.B., Malakorn, K., So, J.J.: Sustainability assessments of cooperative vehicle intersection control at an urban corridor. Transp. Res. Part C Emerg. Technol. **32**, 193–206 (2013)
13. Osorio, C., Nanduri, K.: Urban transportation emissions mitigation: coupling high-resolution vehicular emissions and traffic models for traffic signal optimization. Transp. Res. Part B: Methodol. **81**, 520–538 (2015)
14. Scora, G., Barth, M.: Comprehensive modal emissions model (CMEM), version 3.01. User guide. Centre for Environmental Research and Technology. University of California, Riverside (2006)
15. Simulation of Urban MObility. www.sumo-sim.org
16. Sjödin, Å., Ekström, M., Hammarström, U., Yahya, M.-R., Ericsson, E., Larsson, H., Almén, J., Sandström, C., Johansson, H.: Implementation and evaluation of the ARTEMIS road model for sweden's international reporting obligations on air emissions. In: Proceedings of 2nd Conference Environment & Transport, Including 15th Conference on Transport and Air Pollution, Reims, France, 12–14 June 2006, vol. 1, no. 107. Inrets ed., Arcueil, France, pp. 375–382 (2006)
17. Webster, F.V.: Traffic signal settings (No. 39) (1958)
18. Zhou, Z., Cai, M.: Intersection signal control multi-objective optimization based on genetic algorithm. J. Traffic Transp. Eng. **1**(2), 153–158 (2014). (English Edition)

GSOM Traffic Flow Models for Networks with Information

Megan M. Khoshyaran[1,2(✉)] and Jean-Patrick Lebacque[1,2]

[1] ETC Economics Traffic Clinic, 35 avenue des Champs Elysées, 75008 Paris, France
megan.khoshyaran@wanadoo.fr
[2] UPE IFSTTAR Cosys/Grettia, 14-20 Boulevard Newton, Cite Descartes,
Champs sur Marne, 77447 Marne-la-Vallée, France

Abstract. This paper proposes a macroscopic model for managing networks in a context of inter-vehicular and system to vehicle information flow. The model is based on the GSOM methodology, supplemented with an instantaneous travel time estimator and a multilane extension. The instantaneous travel time estimator can be estimated with the GSOM model and can take into account multilane. Thus the model is compatible with information based reactive dynamic assignment based on a logit behavioral scheme. It is also applicable for transportation systems endowed with vehicular multimodality.

Keywords: Macroscopic traffic models · GSOM models · Multilane traffic · Information · Communication · Travel time · Dynamic traffic assignment · Lane assignment · Mean field

1 Introduction

Automation and communication for cars are being intensively developed by manufacturers and are expected to become an essential feature of traffic flow. They will impact the way traffic flows in networks and the way it will be managed. For instance VANETs for traffic management have been the object of intensive study: refer to [LW2, CF1, YL1]. Another important feature of recent evolution in traffic is the emergence of *vehicular modes*, induced by the availability of information and internet services and new technologies: demand responsive services, electrical cars, car-sharing systems ("autolib"), car pooling..., which compete with the more classical taxis and buses. All these modes share the same infrastructure and compete for the available ressources, and all are affected by the information flow. Thus the setting of the problem is multi-modal.

The main issue addressed by this paper is to provide a modelling framework for traffic flow in this context. Classical macroscopic traffic modeling is based on a finite information speed propagation (wave speed/car following). In contrast V2V (vehicle to vehicle) and V2S (vehicle to system) communication is instantaneous long-range, and non directional, and it impacts user decisions at the local

© Springer International Publishing AG 2017
J. Świątek and J.M. Tomczak (eds.), *Advances in Systems Science*, Advances in Intelligent
Systems and Computing 539, DOI 10.1007/978-3-319-48944-5_20

level (interaction with neighbouring vehicles) and at the global level (in the case of routing). It is liable to induce adverse effects (Braess-like paradoxes) at all scales.

The starting point of our model is constituted by the GSOM model ([LM1, KL2, LK4]), which combines a LWR-like kinematical representation of traffic, with dynamics of individual driver attributes. The driver attributes include modal and OD descriptors and thus enable the description of a vehicular multimodal system with assignment. They also include the impact of information on users and enable the modeling of this impact, particularly at the mesoscopic level. The paper focuses on travel time and traffic state information and proposes a model for the estimation of travel times in the context of information. The study of routing issues can be addressed following the ideas of [KL1] concerning reactive dynamic assignment, and [ML1] for behavioral aspects, assignment and general multimodal systems.

Micro-simulation of traffic in association of VANET is currently being given much attention in the literature. Questions addressed include network simulation, the design of information dissemination and management of VANET (see [MB1]), the assessment of the impact on traffic flow, urban congestion management and CO_2 emission (see for instance [TG1, YL1]), and the improvement of routing in the network [DW1]. General reviews are provided by [DR1, DB1]. Nevertheless we consider that macroscopic models combining macroscopic flow with information are better suited to address macroscopic issues concerning routing, dynamic network equilibrium, global traffic flow and aggregate dynamics. Indeed, macroscopic models integrate naturally macroscopic contraints, whereas microscopic models even when very precise at a local level, must achieve the satisfaction of macroscopic constraints by the aggregation of the local behaviour of the simulated particles.

The paper proposes an extension of the model towards multi-lane modeling, in order to describe more precisely the travel time and assignment aspects. Some multilane models describe only the overall impact of lane change on traffic flow [GK1, Ji1], without describing explicitly the dynamics of traffic on each lane. Indeed, lane interactions are quite complex, [LD1, Ng1, SC2], therefore limiting a model to the macroscopic impact of multiple lanes is a good option for some applications. Other models describe the interaction between lanes as flow transfers between lanes [Ng1], or represent a multi-lane link as a succession of multi-link nodes ([LK1], based on a supply/demand approach; [Da2, SC2]). The approach of the present paper considers that the essential mechanism of lane change is the formation of a local equilibrium between lanes [Da1, LK1, LK3] based on user utility.

Summarizing: the paper describes GSOM modeling on networks with multilane extension, in conjunction with instantaneous travel time estimation, and consider some impacts of instantaneous travel time information on travel flow. Indeed travel times are the basic components of assignment and path choice.

2   GSOM Models

2.1   GSOM Family: Description

The starting point of macroscopic modelling is the LWR model [LW1, Ri1], which describes major macroscopic aspects of traffic flow such as kinematic waves, fundamental diagram and capacity constraints. The GSOM model combines the LWR model with lagrangian driver attributes; thus its main features of a GSOM model (refer to [LK4]) are the following

- the density ρ of traffic behaves in a way similar to the LWR model and satisfies a conservation law;
- the fundamental diagram depends on density (or spacing), and on driver/car attributes I. In GSOM models, the attributes are lagrangian;
- the attributes are dynamic, possibly multi-dimensional, and the dynamics of the attributes is described by a system of partial differential equations in eulerian (x, t) coordinates, or ordinary differential equationsin lagrangian coordinates.

In a network, the (lagrangian) space coordinate x will usually denote the position ξ_a on an arc $(a) \overset{def}{=} (T(a), H(a))$. Let us also introduce lagrangian coordinates: the vehicle index n and the time t, and let $r \overset{def}{=} 1/\rho$ denote the spacing. Let q and v denote flow and velocity. The GSOM model can be expressed in lagrangian resp. eulerian coordinates as:

$$
\begin{aligned}
\partial_t \rho + \partial_x q &= 0 && \text{Conservation of vehicles} \\
\partial_t I &= \varphi(I, W_t, n) && \text{Dynamics of the driver attribute } I \\
v &= \mathcal{V}(r, I, n) && \text{Driver dependent FD}
\end{aligned}
\tag{1}
$$

$$
\begin{aligned}
\partial_t \rho + \partial_x q &= 0 && \text{Conservation of vehicles} \\
\partial_t \rho I + v \partial_x(\rho v I) &= \rho \varphi(I, x) && \text{Dynamics of the driver attribute } I \\
v &= \mathcal{V}(1/\rho, x) && \text{Driver dependent FD}
\end{aligned}
\tag{2}
$$

Note that the modelling capabilities of the two expressions are not strictly equivalent. $\mathcal{V}(r, I)$ denotes the fundamental diagram in lagrangian coordinates. The process $I(t)$ is (possibly) a stochastic ODE driven by a brownian motion B_t; $W_t \overset{def}{=} \frac{dB_t}{dt}$.

2.2   Application to Traffic with Information

Attributes can be passive ($\varphi = 0$) or active ($\varphi \neq 0$). Examples of passive attributes are:

- OD, path, destination;
- Vehicle type (electrical, gasoline)/Driver type (autonomous, human);
- Mode (taxi, bus, demand responsive, autolib, internet service);
- Availability of information and communication equipment (GPS, mobile);

Examples of active attributes are:

- Driver behavioral attribute (patient/impatient, fast/slow ...);
- Traffic state transition;
- Stochastic components (behaviour, parameter, perturbation);
- Vehicle physical state: engine (cold/hot), battery charge;
- Connectivity;
- V2V, V2S information.

For traffic management it is necessary to address intersection modeling. In the present context intersection modeling is developed based on [LK2,LK4,CL1]. It must be noted that if attributes do not impact the fundamental diagram (ODs, paths) simple point-wise nodes can be used [LK2]. Otherwise internal state node models must be used [LK4,LK2].

3   Travel times

One of the most basic tools which can be used for traffic management in the context of information are travel times. It is necessary to specify definitions. Let us consider a network, and a path between an origin and a destination in the network. We distinguish between three path travel times [KL1]:

- *ETT*: experienced travel time (it is evaluated at the arrival at the destination of the path);
- *PTT*: predictive travel time (it is evaluated at the departure at the beginning of the path);
- *ITT*: instantaneous travel time (it is an index of the state of traffic along the path).

In a context of V2V and V2S information, *ETT*s are readily available at least for a fraction of travellers. For instance they can be provided by communicating vehicles or GPS equipped vehicles. Further, the velocity field of traffic, and possibly the density of traffic, can be obtained from equipped vehicles and used in order to deduce *ITT*s. *PTT* estimation requires more complex procedures: traffic demand prediction at the network boundary supplemented with network traffic modelling, or big data approach of traffic and travel times on the network.

In this section we address the simpler question of estimating instantaneous travel times which are readily obtainable. We consider them preferable to predictive and experienced travel times for the following reason. *ETT* yields an image of the past state of the network. *PTT* yields an image of the future state of the network (including the impact of intervening traffic management) but requires prediction. *ITT* yields an image of the present state of the network.

A simple and intuitive expression on a path, parameterized by ξ between its extremal points g and h, would be the following:

$$ITT(g, h; t) = \int_g^h d\xi / V(\xi, t) \tag{3}$$

with $(x, t) \rightarrow V(x, t)$ the velocity field and $ITT(g, h; t)$ the instantaneous travel time from g to h estimated at time t. This expression (3) is not satisfactory, since the speed of vehicles can eventually be null because of traffic lights or congestion, and the estimate of the ITT will then be infinite. Following [KL1], we propose an alternate formula for the ITT based on the velocity field, in which the ITT from x to h, i.e. on a link with head h, is given by:

$$\left| \begin{array}{l} -V \frac{\partial S}{\partial x} + (1 - \frac{V}{V_{max}}) \frac{\partial S}{\partial t} = 1 \\ \\ S(x, h, t) = 0 \quad (\forall t). \end{array} \right. \tag{4}$$

with $S(x, h, t) = ITT(g, h; t)$. The proposed formula can be adapted easily to the GSOM model by applying $V(x, t) = V_e(\rho(x, t), I(x, t); x)$ and $V_{max} = V_{max}(I(x, t); x)$. (note that $V_e(\rho, I) \stackrel{def}{=} \mathcal{V}(r, I)$).

Let us consider some special cases. If $v = 0$, S increases as time. If v is maximum: then S is equal to the ITT as given by (3). Consider in the plane (x, t) the field $\mathcal{X} \stackrel{def}{=} (-V(x, t), (1 - V(x, t)/V_{max})$, S increases as time along the field lines. Under conditions of periodicity (which imply predictability) $S = PTT$. The travel time S provided by (4) is adequate for estimating the ITT and fully compatible with GSOM modelling.

4   Multilane Flow

This section is based on [LK3, FH1, KL3]. We consider traffic flow modelled by a GSOM model with passive attributes. Specifically $\mathcal{D}$ denotes a set of user classes (destinations, paths, vehicle class such as taxi, or any combination thereof), $\chi^d(x, t)$ denotes the fraction of traffic belonging to class d and $\chi \stackrel{def}{=} (\chi^d)_{d \in \mathcal{D}}$. Let $(a) \stackrel{def}{=} (T(a), H(a))$ be a link, $\mathcal{I}$ be the set of lanes of (a), and I^d the set of lanes accessible to vehicles of class d on (a). x is a short script for the absciss ξ_a on (a). i can refer to a single lane or to a group of lanes. Notations are the following:

- $\rho^d = \chi^d \rho$: partial density of vehicles $d \in \mathcal{D}$;
- $\rho_{max,i}$: the maximum density of lanes i;
- ρ_i^d the density of vehicles d in lanes $i \in \mathcal{I}$.

We denote: $\rho_i^d \stackrel{def}{=} \varphi_i^d \rho$. The lane-assignment splits the density ρ^d between the lanes $i \in Ic$, the result being the densities ρ_i^d, which are the unknowns of the proble and must be determined as functions of the ρ^d. The density ρ_i of lane $i \in \mathcal{I}$ is given by

$$\rho_i \stackrel{def}{=} \sum_{d \in \mathcal{D}} \rho_i^d$$

The speed of traffic on lane i is given by the lane fundamental diagram $V_{e,i}$:

$$v_i = V_{e,i}(\rho_i)$$

The partial densities ρ_i^d are subject to the following constraints:

$$\left|\begin{array}{l} \rho_i^d \geq 0 \quad \forall d \in \mathcal{D} \,, \; \forall i \in \mathcal{I}^d \\ \sum_{i \in \mathcal{I}^d} \rho_i^d = \rho^d \quad \forall d \in \mathcal{D} \\ \sum_{d/i \in \mathcal{I}^d} \rho_i^d \leq \rho_{max,i} \quad \forall i \in \mathcal{I} \end{array}\right. \tag{5}$$

We assume that each driver chooses the lane with highest utility available to him: the problem admits a structure similar to the structure of assignment problems. All or nothing lane assignment has been considered in the case of two types of vehicles - and two lanes by [Da1]. Following [FH1] driver behaviour is better modelled by taking into account the variability of driver behaviour and perception. This variability can be expressed by a stochastic utility

$$U_i^d \stackrel{def}{=} \theta \left(v_i + w_i^d \right) + \ln(\rho_{max,i}) + \eta_i^d$$

Each driver is assumed to chose the lane with highest utility.

If the stochastic component η_i^d is Gumbel distributed, a Logit split of densities results:

$$\max_{\left(\rho_i^d\right)_{d \in \mathcal{D}, i \in \mathcal{I}^d}} \sum_{i \in \mathcal{I}} \left(\int_{i \in \mathcal{I}}^{\sum_{d/i \in \mathcal{I}^d} \rho_i^d} V_{e,i}(s)ds - \frac{1}{\theta} \sum_{\delta/i \in \mathcal{I}^\delta} \rho_{max,i} H\left(\frac{\rho_i^\delta}{\rho_{max,i}}\right) - w_i^\delta \rho_i^\delta \right) \tag{6}$$

with $H(x) \stackrel{def}{=} x \left(ln(x) - 1 \right)$ the neguentropy. This Logit assignment of partial densities in lanes can be expressed as the following fixed point:

$$\left|\begin{array}{l} \varphi_i^d = \rho_i^d/\rho^d = \rho_{max,i}e^{\theta\left(v_i+w_i^d\right)} \Big/ \sum_{\ell \in \mathcal{I}^d} \rho_{max,\ell}e^{\theta\left(v\ell+w_i^\ell\right)} \quad \forall d \in \mathcal{D}, \quad \forall i \in \mathcal{I}^d \\ v_i = V_{e,i}\left(\sum_{d/i \in \mathcal{I}^d} \rho_i^d \right) \quad \forall i \in \mathcal{I} \\ \rho^d = \sum_{i \in \mathcal{I}^d} \rho_i^d \quad \forall d \in \mathcal{D} \end{array}\right. \tag{7}$$

The split densities ρ_i^d and the lane densities ρ_i are smooth functions of $\overline{\rho} \stackrel{def}{=} \left(\rho^d\right)_{d \in \mathcal{D}}$.

Thus the flow for class d users is given by

$$q^d \stackrel{def}{=} \mathcal{Q}^d(\overline{\rho}) = \sum_{i \in \mathcal{I}} \rho_i^d V_{i,e}(\rho_i)$$

and the conservation of vehicles d can be expressed as

$$\frac{\partial \rho^d}{\partial t} + \frac{\partial}{\partial x}\left(\mathcal{Q}^d(\overline{\rho})\right) = 0 \quad \forall d \in \mathcal{D} \tag{8}$$

with a smooth flux function $\mathcal{Q} = \left(\mathcal{Q}^d\right)_{d \in \mathcal{D}}$.

The model (6), (7), (8) simplifies in the homogeneous case ($I^d = \mathcal{I}$ $\forall d \in \mathcal{D}$): it can be shown [KL3] that the split between lanes is independent of the driver class d, and that the total density

$$\rho \stackrel{def}{=} \sum_{d \in \mathcal{D}} \rho_d = \sum_{d \in \mathcal{D}, i \in \mathcal{I}} \rho_i^d$$

satisfies a single conservation law, and (8) reduces a transport equation. Thus the multilane model reduces to a GSOM model with a fundamental diagram which can be obtained as a combination of the lane fundamental diagrams (usually different), even though the traffic flows on each lane propagate a different speeds. Thus the full multi-lane model will only be needed close to intersections, and traffic flow far from intersections will be modelled as GSOM (the homogeneous case) (Fig. 1).

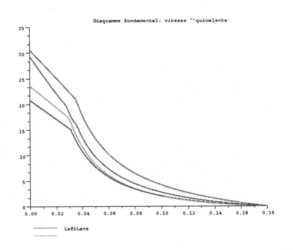

Fig. 1. Speed density relationships per lane (green, turquoise, blue, for resp. left center and right lane), aggregate speed vs total density fundamental diagram (in the homogeneous case i.e. far from intersections). (Color figure online)

5   Dynamic Reactive Assignment

We assume that V2V and V2S yield *ITT*s. Typically *ITT*s are estimated for all links (a) of the network, i.e. $ITT_a(t)$, and inside each link, for each lane (i) i.e. let $S_{a,i}(x,t)$ the *ITT* estimated by (4) with $x = \xi_a$ from x to $H(a)$. Then in (4) v is the speed on lane (i) obtained by solving (7). Let Π_a^d be the shortest path from $H(a)$ to (d) (calculated via Dijkstra). Let Ξ_a^+ be the set of successor links of (a).

The choice of the next arc $(b) \in \Xi_a^+$ by travellers on (a) at location x with destination (d) can be modelled by a Logit model [ML1] yielding the probability $\lambda_{a,b}^d(x,t)$ of choice of (b):

$$\lambda_{a,b}^d(x,t) = \frac{\exp\left(-\tau\left(ITT_{x,H(a)}^d + ITT_b^d(t) + \Pi_b^d\right)\right)}{\sum\limits_{\beta \in \Xi_a^+} \exp\left(-\tau\left(ITT_{x,H(a)}^d + ITT_\beta^d(t) + \Pi_\beta^d\right)\right)} \tag{9}$$

In the above Formula (9) we could replace $ITT_{x,H(a)}^d$ by $S_{a,i}(x,t)$ where (i) denotes the set of preselection lanes for link (b). This allows to adapt the preference coefficients w_j^d. The complete model (2), (7)–(9) (supplemented by a node model) can be considered as a mean field model on the network. The Logit model is relevant here because any paradoxical effects are avoided: the alternatives (next links) are true alternatives and (9) can be interpreted as a nested logit model. Nevertheless it is important that the sensitivity coefficient τ be estimated with care.

Using ITT reactive dynamic assignment induces a network behaviour very far from optimality, a situation much improved by the use of PTT. The following figures illustrate this fact.

Let us consider an example. The network is a standard network with 16 ODs, 4 intersections and 14 paths. The two main paths joining the dominant OD pair are considered. The morning demand peak for this OD is depicted in Fig. 2. The demand exceeds the system capacity when the peak reaches its maximum and induces queues lcated at the intersections. On the two main competing paths, users are provided with either ITT or PTT values for both paths at all times when they enter the system. Thus the drivers can chose between the two paths.

The effective path travel times are given by Fig. 4, under two hypotheses: (i) if users receive the ITT information (left) or (ii) if users receive the PTT information (right). The PTT information improves the effective travel times:

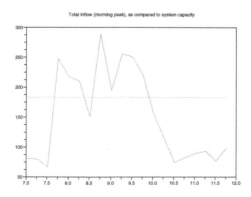

Fig. 2. Total system demand (green) vs system capacity (turquoise) (Color figure online)

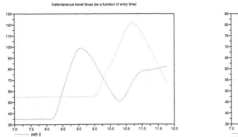
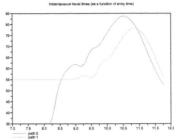

Fig. 3. Instantaneous travel times of paths 1 (turquoise) and 2 (green), based on instantaneous (left) and predictive (right) information (Color figure online)

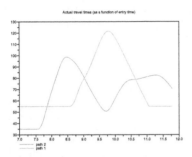
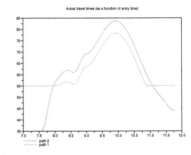

Fig. 4. Effective travel times of paths 1 (turquoise) and 2 (green), based on instantaneous (left) and predictive (right) information (Color figure online)

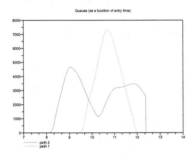
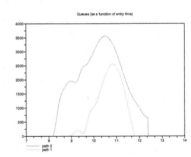

Fig. 5. Total queues on paths 1 (turquoise) and 2 (green), based on instantaneous (left) and predictive (right) information (Color figure online)

for instance the maximum travel path time is reduced to 85 min versus 120 min if users receive the ITT information. Figure 5 depicts the queues in the system which are significantly improved by the use of PTT (by a factor 2). If we consider Fig. 3, and compare it with Fig. 4, it can be noted that the path ITT estimates disagree with the effective travel time when the ITT is used for driver path choice. Actually the profiles are qualitatively similar but there is a time lag of about 1 h. Further when the PTT is used for routing the ITT estimates are quite close to the actual travel times. It can also be noted that the use of *PTT* information improves the coordination of the two paths (and regulates the load in the network): their travel times are close to each other.

6   Conclusion

The modelling chain introduced in this paper is suited for describing, analyzing and managing transportation sytems in conditions of vehicular multi-modality and information. In particular it can be used to assess and implement dynamic reactive dynamic traffic assignment (RDTA) schemes based on information. The use of PTT information improves the RDTA results, as expected.

Other strategies are conceivable and must be assessed. If several operators provide competing information services the situation becomes more complicated, as the competition of these operators can induce negative effects. V2V and V2S information may be put to other uses (local control based on cooperation for instance), the study of which will require further developments of the model.

References

[CF1]   Codeca, L., Frank, R., Engel, T.: Traffic routing in urban environments: the impact of partial information. In: Proceedings of 17th IEEE International Conference on Intelligent Transportation Systems (ITSC 2014), pp. 2511–2516 (2014)

[CL1]   Costeseque, G., Lebacque, J.P., Khelifi, A.: Lagrangian GSOM traffic flow models on junctions. IFAC PapersOnLine **48**(1), 147–152 (2015)

[Da1]   Daganzo, C.F.: A continuum theory of traffic dynamics for motorways with special lanes. Transp. Res. B **31**, 83–102 (1997)

[Da2]   Daganzo, C.F.: A behavioral theory of multi-lane traffic flow. Merges and the onset of congestion. Transp. Res. Part B **36**, 159–169 (2002)

[DB1]   Darwish, T., Bakar, K.A.: Traffic density estimation in vehicular ad hoc networks: a review. Ad Hoc Netw. **24**, 337–351 (2015)

[DW1]   Ding, J.W., Wang, C.F., Meng, F.H., Wu, T.Y.: Real-time vehicle route guidance using vehicle-to-vehicle communication. IET Commun. **4**(7), 870–883 (2010)

[DR1]   Dorronsoro, B., Ruiz, P., Danoy, G., Pign, Y., Bouvry, P.: Evolutionary Algorithms for Mobile Ad hoc Networks. Wiley, Hoboken (2014). ISBN: 978-1-118-34113-1

[FH1]   Farhi, N., Haj-Salem, H., Khoshyaran, M.M., Lebacque, J.P., Salvarani, F., Schnetzler, B., de Vuyst, F.: Logit lane assignment model: first results. Presented at TRB (2013). arXiv preprint arXiv:1302.0142

[GK1]   Greenberg, J.M., Klar, A., Rascle, M.: Congestion on multilane highways. SIAM J. Appl. Math. **63**(3), 813–818 (2003)

[Ji1]   Jin, W.L.: A kinematic wave theory of lane-changing vehicular traffic (2006). eprint arXiv:math/0503036

[KL1]   Khoshyaran, M.M., Lebacque, J.P.: A reactive dynamic assignment scheme. In: Proceedings of the 3rd IMA International Conference on Mathematics in Transport Planning and Control (1998)

[KL2] Khoshyaran, M.M., Lebacque, J.P.: Lagrangian modelling of intersections for the GSOM generic macroscopic traffic flow model. In: Proceedings of the 2008 AATT Conference in Athens (2008)

[KL3] Khoshyaran, M.M., Lebacque, J.P.: Numerical solutions to the logit lane assignment model. Proceedia Soc. Behav. Sci. **54**, 907–916 (2012)

[LD1] Laval, G., Daganzo, C.F.: Multi-lane hybrid traffic flow model: a theory on the impacts of lane-changing manoeuvres. TRB Annual Meeting (2005)

[LK1] Lebacque, J.P., Khoshyaran, M.M.: Macroscopic flow models. In: Patriksson, M., Labbé, M. (eds.) Presented at the 6th Meeting of the EURO Working Group on Transportation 1998. Published in "Transportation Planning: The state of the Art", pp. 119–139. Kluwer Academic Press (2002)

[LK2] Lebacque, J.P., Khoshyaran, M.M.: First order macroscopic traffic flow models: intersection modeling, network modeling. Dynamics and Human Interaction. In: Transportation and Traffic Theory, Flow. Elsevier, Amsterdam (2005)

[LK3] Lebacque, J.P., Khoshyaran, M.M.: A stochastic lane assignment scheme for macroscopic multi-lane traffic flow modelling. TRISTAN VII, Tromsø (2009)

[LK4] Lebacque, J.P., Khoshyaran, M.M.: A variational formulation for higher order macroscopic traffic flow models of the GSOM family. Transp. Res. Part B **57**, 245–265 (2013)

[LM1] Lebacque, J.P., Mammar, S., Haj-Salem, H.: Generic second order traffic flow modeling. In: Allsop, R.E., Bell, M.G.H., Heydecker, B.G. (eds.) Transportation and Traffic Flow Theory 2007 (2007)

[LW1] Lighthill, M.H., Whitham, G.B.: On kinematic waves II: a theory of traffic flow on long crowded roads. Proc. Royal Soc. (Lond.) **A 229**, 317–345 (1955)

[LW2] Li, F., Wang, Y.: Routing in vehicular ad hoc networks: a survey. IEEE Veh. Technol. Mag. **C47**, 12–22 (2007)

[ML1] Ma, T.-Y., Lebacque, J.P.: A cross-entropy based multiagent approach for multimodal activity chain modeling and simulation. Transp. Res. Part C **C28**, 116–129 (2013)

[MB1] Medetov, S., Bakhouya, M., Gaber, J., Zinedine, K., Wack, M., Lorenz, P.: A decentralized approach for information dissemination in vehicular ad hoc networks. J. Netw. Comput. Appl. **46**(154–165), 2014 (2014)

[Ng1] Ngoduy, D.: Macroscopic discontinuity modelling for multiclass multilane traffic flow operations. TRAIL thesis series. Delft University Press (2006)

[Ri1] Richards, P.I.: Shock-waves on the highway. Opns. Res. **4**, 42–51 (1956)

[SC2] Schnetzler, B., Lebacque, J.P., Louis, X.: A multilane junction model. Transportmetrica **8**(4) (2012)

[TG1] Thomin, P., Gibaud, A., Koutcherawy, P.: Deployment of a fully distributed system for improving urban traffic flows: a simulation-based performance analysis. Simul. Modell. Pract. Theory **31**, 22–38 (2013)

[YL1] Yuan, Q., Liu, Z., Li, J., Zhang, J., Yang, F.: A traffic congestion detection and information dissemination scheme for urban expressways using vehicular networks. Transp. Res. Part C **C47**, 114–127 (2014)

Designing Mass-Customized Service Subject to Public Grid-Like Network Constraints

Grzegorz Bocewicz[1(✉)], Robert Wójcik[2], and Zbigniew Banaszak[1]

[1] Department of Computer Science and Management,
Koszalin University of Technology, Śniadeckich 2, 75-453 Koszalin, Poland
bocewicz@ie.tu.koszalin.pl
[2] Faculty of Electronics, Department of Computer Engineering,
Wrocław University of Science and Technology, Wrocław, Poland
robert.wojcik@pwr.wroc.pl

Abstract. The paper introduces the concept of a fractal topology routes network (FTRN) in which different transportation modes interact with each other via distinguished subsets of common shared hubs as to provide a variety of mass customized passenger services. Passenger flows following assumed set of travel destinations transit through different transportation modes. The network of repetitively acting local transportation modes routed along road loops of FTRN structure provides a framework for passengers' origin-destination trip routing and scheduling. In that context, first of all some mass-customized services following presumed scenarios caused by seasonal timetable changes and major sporting events or other special events, e.g. emergency transportation ones, have to be assessed in advance. In general case such problems belong to NP-hard ones. However, the passenger travel schedules can be estimated easily while taking into account both structural features of transportation network, e.g. its regularity, and cyclic behaviour of local transportation modes employed in FTRN. Therefore, the goal is to provide a declarative model allowing one to formulate a constraint satisfaction problem enabling assessment of a city network period as well as development of conditions guaranteeing the right match-up of local transportation line schedules to a given passenger flow itineraries.

Keywords: Public transport · Traffic flow · Grid-like network · Multimodal process · Declarative modeling

1 Introduction

Numerous road network patterns deployed in cities range from the tightly structured fractal network with perpendicular roads in a regular raster pattern to the hierarchical network with sprawling secondary and tertiary roads feeding into arterial roads in a branch like system [6, 10, 12]. Regular structure (e.g. fractal topology networks) supporting mass customized services (e.g. transportation, delivery, supply) are found in different application domains (such as manufacturing, intercity fright transportation supply chains, multimodal passenger transport network combining several unimodal networks (bus, tram, metro, train, etc.) as well as service domains (including

J. Świątek and J.M. Tomczak (eds.), *Advances in Systems Science*, Advances in Intelligent Systems and Computing 539, DOI 10.1007/978-3-319-48944-5_21

passenger/cargo transportation systems, e.g. ferry, ship, airline, automated guided vehicle (AGV), train networks, as well as data and supply media flows, e.g., cloud computing, oil pipeline and overhead power line networks) [3, 4, 11].

The throughput of passengers and/or freight depends on geometrical and operational characteristics of FTRN. In that context the main problems concern routing and scheduling of multimodal processes of passengers flows. Multimodal transportation processes (MTP) involve the movement of objects (passenger/cargo) using different modes of transport in a single, integrated transport chain on a given route [1, 14, 16, 21]. Examples of such processes follow from daily commuting (bus – streetcar – subway), courier services (e.g., DHL), etc. Their transport routes are made up of local segments operated by one mode of transport or involving one type of transport processes, and the objects are moved by suitable local means of transport.

The idea standing behind of the proposed approach resembles Glenday sieve supporting concept of repetitive flexible supply [8] and a concept of fractal organization structure [17].

Main advantage of an approach proposed follows from the fact that the passenger travel schedules can be estimated easily while taking into account cyclic behaviour of both: local transportation modes and the whole transportation network. That is because, since the transportation processes executed by particular lines are usually cyclic, hence the MTP supported by them have also periodic character. In that context it provides attractive alternative to usually considered problems concerning routing and scheduling of passengers flows which are in general NP-hard [13].

The research presented can be seen as a continuation of our former work [3, 4] regarding AGVs fleet scheduling in a regular, mesh-like layout of a Flexible Manufacturing System (FMS) transportation routes. The solution proposed employs concepts of a grid-like network of transportation modes and supporting them fractal organisation of public transport. The goal is to create a fixed schedule of locally acting transportation means as to match-up a given set of passengers' itineraries tracing assumed multimodal process.

The rest of the paper is organized as follows: Sect. 2 provides brief insight into related work. Section 3 introduces to a concept of fractal structure of multimodal transportation network encompassing FTRN topology. Section 4 provides the declarative problem formulation focused on routing and scheduling of MTPs. Computational experiments and directions of further research are presented in Sects. 5 and 6, respectively.

2 Related Work

A material transport system configuration is one of the most important aspects of public transport design and organization. The fundamental questions that need to be answered are what transport technologies are going to be used in a urban transport system (UTS) and how will they be used [1, 2]. The UTS design problem usually focuses on the transportation routes topology and transportation means selection. In turn, the traffic flow organization problem boils down to local transport modes and MTP routing as well as their scheduling [5, 14, 15, 19].

UTSs are the subject of intensive research [14, 19, 20]. In practice, however, most of this research is limited to either identifying the fractal pattern of the analyzed transport network, or to estimating qualitative and quantitative parameters of urban transport system [2, 17]. Their operation depends on the manner fractal patterns are propagated in order to satisfy the needs of urbanized communication infrastructures.

In general, UTSs belong to a class of transport systems responsible for movement of cargo, people, goods, energy, financial capital, data, etc. from a point of origin to a destination. A commonly accepted definition of a transport system describes it as being characterized, regardless of its specific nature and character, by a structure (the component subsystems and relations between them) and a behavior, which determines the responses of the system to changes in (expectations of) the environment [15].

To summarize, UTSs are investigated in a context of their design and operation, i.e. corresponding synthesis and analysis problems. The analysis problem, more specifically the reachability problem, focuses on specific behaviors of a system with a given structure. In turn, the synthesis problem is aimed at searching for a system structure which can guarantee its desired behavior.

In other words, the analysis problem boils down to the routing and scheduling of MTPs carried out in transport networks with a fixed grid-like structure. Formulated in this way, it assumes that network topology, routes, parameters of local means of transport, dispatching rules governing access to shared stations or stops, and initial and terminal points of the alternative routes of MTP are given. What is sought are the variants of routes which guarantee delivery times not exceeding a pre-set deadline. In turn, the synthesis problem, it is assumed that the topology of the transport network, the routes, and parameters of the means of local transport are given, and the unknown are the dispatching rules governing access to shared stations or stops, which guarantee timely completion of an MTP, as scheduled.

The main advantages following from the regular structure of MTP layout comes down to the flexibility and robustness being vital to improve public transport robustness [7, 9, 20]. Moreover, such structures provide a chance to evaluate variants of admissible routings and schedules following assumed itineraries in a polynomial time. Among numerous reports concerning mesh-like or grid-like as well as fractal-like structures of urban transportation networks the following ones should be mentioned [2, 17].

3   Grid-Like Structure Networks

For the purposes of further discussion, it is assumed that a UTS encompasses all possible branches of transport and transport technologies, including road and rail (surface and underground) transport, e.g. buses, streetcars, subway lines. It is worth noting that the various modes of transport (buses, streetcars, commuter rail, subway lines) which form traffic flows in a UTS run to scheduled timetables, moving along a fixed, closed-loop route.

In the context of research submitted in [12] "When certain economic conditions are met, roads are built first around the central parcel, and then gradually cover the parcels on the periphery. The tree-like (nonredundant) structure is the emergent topological characteristic in the first stage; as time progresses, the network not only reaches other parcels farther from the center, but also provides multiple paths for parcels that are already connected." The class of FTRN analyzed here is limited to network structures with a regular, recursive morphology typical of tree or mesh (grid) topologies.

As an example of city district growth and its corresponding fractal representation let us consider FTRN shown in Fig. 1.

The last FTSN shown in Fig. 1 (a) has been obtained after 16 iterations of the following fractal generating algorithm.

Fractal Generating Algorithm

Given a vertex.

Step 1.   Link this vertex with the next one by vertically oriented edge.

Step 2.   For newly obtained vertex repeat Step 1.

Step 3.   Newly obtained vertex links with two vertices: the first one by the edge oriented in the same direction as the preceding edge, and the second one by the edge 90° clockwise oriented in terms of the succeeding edge.

Step 4.   Newly obtained vertices link with vertices by edges oriented in the same direction as preceding them edges.

Step 5.   Newly obtained vertices link with new ones following the rule:

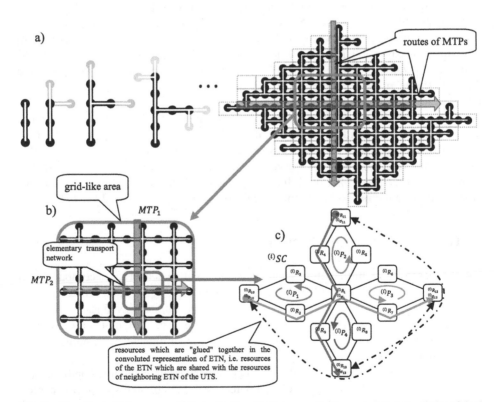

Fig. 1. Fractal model representation of city district growth "snapshots" (a), distinguished grid-like area (b), graph model of elementary transport network (c)

(a) In the case of two edges being 90° clockwise oriented the new vertices are added - first one linked by the edge oriented in the same direction as the preceding one, and the second vertex linked by the edge 90° clockwise oriented in terms of the preceding edge.

(b) In the case of two edges being vertically oriented the new vertices are added - first one linked by the edge oriented in the same direction as the preceding one, and the second vertex linked by the edge 90° anti-clockwise oriented in terms of the preceding edge.

Step 6.   For newly obtained vertex repeat Step 4.

Step 7.   Each of newly obtained vertices link with two new ones following the rule:

(a) In the case a vertex is linked to the last one of two same direction oriented (previously 90° clockwise oriented) subsequently preceding edges the new vertices are added - first one linked by the edge oriented in the same direction as the preceding one, and the second vertex linked by the edge 90° clockwise oriented in terms of the preceding edge.

(b) In the case a vertex is linked to the last one of four same direction oriented (previously 90° clockwise oriented) subsequently preceding edges the new vertices are added - first one linked by the edge oriented in the same direction as the preceding one, and the second vertex linked by the edge 90° anti-clockwise oriented in terms of the preceding edge.

(c) In the case a vertex is linked to the last one of two same vertically oriented subsequently preceding edges the new vertices are added - first one linked by the edge oriented in the same direction as the preceding one, and the second vertex linked by the edge 90° clockwise oriented in terms of the preceding edge.

(d) In the case a vertex is linked to the last one of two same direction oriented (previously 90° anti-clockwise oriented) subsequently preceding edges the new vertices are added - first one linked by the edge oriented in the same direction as the preceding one, and the second vertex linked by the edge 90° clockwise oriented in terms of the preceding edge.

Step 8.   Repeat Step 4.

Note that Fig. 1 (a) shows results of the first, second, third, fourth and sixteen iterations. Besides of above, other kinds of fractal topologies can be considered as well, see Fig. 2.

Let us consider three types of Elementary Transport Network (ETN) creating fractal structures from Figs. 1 (a), 2 (a) and (b) as those shown in Fig. 3 (a). Graph theoretical models of local transport processes executed in these structures (along elementary cyclic paths) are shown in Fig. 3 (b), where vertices represent network resources, i.e. stations, stops and shared route sections and edges represent relationship linking resources. Network resources are denoted by $^{(i)}R_r$ that means the r-th resource in the i-th ETN. Local transport processes are marked with labeled arcs whose orientation

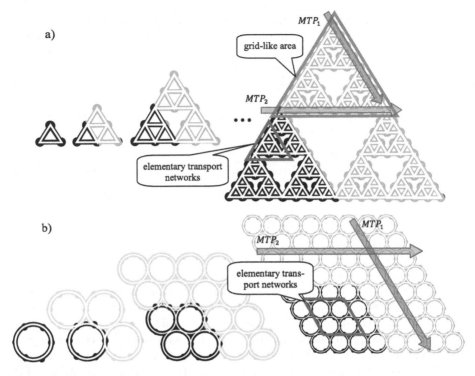

Fig. 2. Examples of FTRNs generated in course of growing fractal structures and composed of repeating so called elementary transport networks: △ - like (a), and ◯ – like (b)

indicates the direction of flow of local traffic (transport modes); for example, the arc labeled $^{(i)}P_j$ means the j-th local transport process, in in the i-th ETN (Fig. 3).

Component local cyclic processes are executed along given routes composed of two kinds of resources: shared or unshared by competing processes. For example, access to the common shared resources $^{(i)}R_1$ in Fig. 3 (a) is synchronized by a mutual exclusion protocol following the given priority dispatching rule $^{(i)}\sigma_1$ determining the order in which local processes can access the shared resource $^{(i)}R_1$, for instance $^{(i)}\sigma_1 = (^{(i)}P_1, ^{(i)}P_2, ^{(i)}P_3, ^{(i)}P_4)$, process $^{(i)}P_1$ is allowed to access the resource $^{(i)}R_1$ as first, then the processes $^{(i)}P_2$, $^{(i)}P_3$, and next process $^{(i)}P_4$, and then once again $^{(i)}P_1$, and so on. MTP routes, showing the sequences of the resources between which objects are moved are represented graphically with bold symbols of nodes and arcs; see Fig. 1 (a).

Let us consider convoluted representations of ETN from Figs. 1 (b), 2 (a) and (b) shown in Fig. 3 (b). As it can be easily noted, convoluted forms arise as a result of gluing together of selected vertices of ETN. Just which vertices are glued together in the convoluted form is determined by the choice of those resources of the ETN which are shared with the resources of neighboring structures of the transport network. For

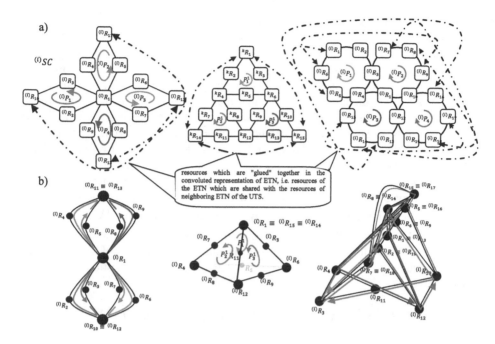

Fig. 3. Graph models of ETNs (a) and their convoluted counterpart representations (b)

example in Fig. 3 (a), a vertex corresponding to resource $^{(i)}R_{10}$ is glued with a vertex corresponding to resource $^{(i)}R_{12}$, because resource $^{(i)}R_{10}$ is a counterpart of $^{(i)}R_{12}$, see Fig. 1.

It can be shown that if the traffic flow in a given convoluted representation of an ETN is free of congestion, i.e. the schedule which specifies it is a cyclic schedule, then the flow of traffic in the entire transport network consisting of ETN structures also has a cyclic nature [3, 4].

4   Declarative Problem Formulation

Consider declarative model of UTS comprising:

- Sets of decision variables describing the structures of
  - local transportation processes, i.e. the type and number of resources and modes of transport they use, as well as the associated travel/dwelling times,
  - the MTP, i.e. the type and number of resources in a chain and the type and number of transport modes used, as well as the associated travel/dwelling times,
- Domains of decision variables,
- Sets of determining constraints:
  - sets of dispatching rules assigned to shared network resources
  - transport schedules determining the periods (takts) and dates of delivery of transported goods.

These assumptions, on the one hand, explicitly constrain the topology of UTS routes to transport networks with fractal structures and, on the other hand, implicitly make the efficiency of potential MTPs (e.g. regarding the possible delivery dates) conditional on the admissible flow of traffic (e.g. congestion-free traffic) operating under local transport processes. This observation implies that research can be limited to certain elementary structures, e.g. ETNs, that make up the whole transport network.

It can be shown that if the traffic flow in a given convoluted representation of an ETN is free of congestion, i.e. the schedule which specifies it is a cyclic schedule, then the flow of traffic in the entire UTS's transport network composed of ETNs also has a cyclic nature. This observation allows one to focus on formulating the following constraint satisfaction problem, the solution to which is a set of dispatching rules that guarantees a congestion-free flow of traffic. In other words, assuming that the behavior of each i-th ETN is represented by a cyclic schedule $^{(i)}X' = \left(^{(i)}X_k \mid k = 1, \ldots, h, \ldots, L_i\right)$, where: $^{(i)}X_h$ is a set of beginning moments of operation of the h-th local process of the i-th ETN and L_i denotes the cardinality of the set of local processes comprising the i-th ETN, the constraint satisfaction problem in question has the following form:

$$PS_i = ((\{^{(i)}X', {}^{(i)}\Theta, {}^{(i)}\alpha\}, \{D_X, D_\Theta, D_\alpha\}), \{C_L, C_M, C_D\}) \tag{1}$$

where:

$^{(i)}X'$, $^{(i)}\Theta$, $^{(i)}\alpha$ – decision variables,

- $^{(i)}X'$ – cyclic schedule of the i-th ETN,
- $^{(i)}\Theta$ – set of dispatching rules determining the order of operations competing for access to the common resources of the i-th ETN,
- $^{(i)}\alpha$ – set of values of periods of local processes occurring in the i-th ETN,

  D_X, D_Θ, D_α – domains of admissible values of discrete decision variables
  C_L, C_M, C_D – finite sets of constraints limiting the values of decision variables

- C_L, C_M – sets of conditions constraining the set of potential behaviors of the i-th ETN [3],
- C_D – a set of sufficient conditions the satisfaction of which guarantees congestion-free (i.e. deadlock-free and collision-free) flow of traffic in a transport network modeled by the i-th ETN [4].

Note that a constraint satisfaction problem: $PSO = ((X, D), C)$ is usually given by [18] a finite set of decision variables $X = \{x_1, x_2, \ldots, x_n\}$, a finite family of finite domains of discrete decision variables $D = \{D_i \mid D_i = \{d_{i,1}, d_{i,2}, \ldots, d_{i,j} \ldots, d_{i,m}\}, i = 1..n\}$, and a finite set of constraints limiting the values of the decision variables $C = \{C_i \mid i = 1..L\}$, where: C_i is a predicate $P[x_k, x_l, \ldots, x_h]$ defined on a subset of set X. What is sought is an admissible solution, i.e. a solution in which the values of all decision variables X satisfy all constraints C.

5 Computational Experiments

As an illustration of the approach, consider three different FTRN as shown in Figs. 1 (a), 2 (a) and (c). Problem (1) considered for the selected ETNs from Fig. 3 (b) was implemented and solved in the constraint programming environment OzMozart (CPU Intel Core 2 Duo 3 GHz RAM 4 GB). When the assumption was made that all operation times in local processes are the same and equal to $t_{i,j} = 1$ u.t. (unit of time), the first acceptable solution was obtained in less than one second. An analysis of cyclic schedules $^{(i)}X'$ allows an easy deduction of parameters regarding whole UTS as well as MTP distinguished in Figs. 1 and 2 while collected in the Table 1. The length of all MTP routes is the same and equal to four sections of the road.

A head time is the amount of time that must elapse between two consecutive (following the same MTP) transport operations completions and a cycle length means the period during which a sequence of a recurring succession of processes to resources is completed. In turn, the running time is the time spent traveling between stops or stations, and dwell time is the time spent stopped at locations to allow passengers to the mode of transport.

Obtained results confirm, that operating characteristics of UTS depend on its structural features. It means the mass-customized services available in a given UTN depend on its topological constraints, especially structural characteristics of FTRN. Consequently, the proposed approach can be employed in course of prototyping of grid-like structures as well as traffic flow organization in regular structure urban districts.

Table 1. Operating characteristics of UTSs following different FTRNs

FTRN/ETN	Dispatching rules	Cycle length [u.t.]	MTP	Total travel time [u.t.]	Dwell time [u.t.]	Headway [u.t]
	$^{(i)}\sigma_1 = \left(^{(i)}P_1, {}^{(i)}P_1, {}^{(i)}P_2, {}^{(i)}P_3\right)$ $^{(i)}\sigma_{10-12} = \left(^{(i)}P_1, {}^{(i)}P_3\right)$ $^{(i)}\sigma_{11-13} = \left(^{(i)}P_2, {}^{(i)}P_4\right)$	8	MTP$_1$	64	36	8
			MTP$_2$	64	36	8
	$^{(i)}\sigma_{1-15-14} = \left(^{(i)}P_1, {}^{(i)}P_3, {}^{(i)}P_{14}\right)$ $^{(i)}\sigma_4 = \left(^{(i)}P_1, {}^{(i)}P_2\right)$ $^{(i)}\sigma_6 = \left(^{(i)}P_3, {}^{(i)}P_1\right)$ $^{(i)}\sigma_{12} = \left(^{(i)}P_2, {}^{(i)}P_3\right)$	6	MTP$_1$	48	28	6
			MTP$_2$	48	28	6
	$^{(i)}\sigma_{1-13} = \left(^{(i)}P_3, {}^{(i)}P_1\right)$ $^{(i)}\sigma_{2-19} = \left(^{(i)}P_4, {}^{(i)}P_1\right)$ $^{(i)}\sigma_{5-16} = \left(^{(i)}P_1, {}^{(i)}P_4\right)$ $^{(i)}\sigma_{6-9} = \left(^{(i)}P_1, {}^{(i)}P_2\right)$ $^{(i)}\sigma_{7-18} = \left(^{(i)}P_2, {}^{(i)}P_4\right)$ $^{(i)}\sigma_{8-14} = \left(^{(i)}P_3, {}^{(i)}P_2\right)$ $^{(i)}\sigma_{9-16} = \left(^{(i)}P_1, {}^{(i)}P_2\right)$ $^{(i)}\sigma_{15-17} = \left(^{(i)}P_3, {}^{(i)}P_4\right)$	6	MTP$_1$	48	8	6
			MTP$_2$	48	8	6

6 Conclusions

The declarative reference model of a UTS presented in this study enables an analysis of the relationships between a given FTRN topology and operating characteristics of the following UTS. The model enables to formulate analysis and synthesis problems in terms of a constraint satisfaction problem framework. The questions following problems of both categories can be stated as: Is it possible to make supplies which meet customer demands in a transport network with a preset structure? Is there a transport network structure that ensures deliveries which meet user expectations?

Main advantage of an approach proposed follows from the fact that the passenger travel schedules can be estimated easily while taking into account cyclic behavior of both: local transportation modes and the whole transportation network. That is because, since the transportation processes executed by particular lines are usually cyclic, hence the MTP supported by them have also periodic character. In that context it provides attractive alternative to usually considered problems concerning routing and scheduling of passengers flows which are in general NP-hard.

The main findings obtained confirm quite intuitive research hypothesis assuming that operation of mass-customized services available in a UTN depend on topological network constraints, especially structural characteristics of FTRN. As consequence, they can be employed in course of decision making aimed at urban infrastructure planning and/or assessment of alternative traffic flow organization.

The issues of planning and/or prototyping of alternative structures and/or behavior of UTS within FTRN presented in this work are part of the broader topic of cyclic scheduling which includes problems occurring in tasks associated with determining timetables, telecommunications transmissions, production planning, etc. In future, while continuing along the line of inquiry related to preventing traffic flow congestion in transport networks, we plan to broaden the scope of our research to include the problems of robust scheduling and the related problem of preventing re-scheduling of timetables in UTSs.

REFERENCES

1. Bahrehdar, S.A., Moghaddam, H.R.G.: A decision support system for urban journey planning in multimodal public transit network. Int. J. Adv. Railway Eng. 2(1), 58–71 (2014)
2. Buhl, J., Gautrais, J., Reeves, N., Sol´e, R.V., Valverde, S., Kuntz, P., Theraulaz, G.: Topological patterns in street networks of self-organized urban settlements. Eur. Phys. J. B 49, 513–522 (2006)
3. Bocewicz, G., Banaszak, Z.: Multimodal processes scheduling in mesh-like network environment. Arch. Control Sci. 25(LXI)(2), 237–261 (2015)
4. Bocewicz, G., Nielsen, I., Banaszak, Z.: Automated guided vehicles fleet match-up scheduling with production flow constraints. Eng. Appl. Artif. Intell. 30, 49–62 (2014)
5. Cárdenas, C.J.: Efficient multi-modal route planning with transfer patterns. Master thesis, Albert-Ludwigs-Universität Freiburg, Freiburg (2013)

6. Courtat, T.: Walk on City Maps - Mathematical and Physical phenomenology of the City, a Geometrical approach. Modeling and Simulation. Université Paris-Diderot - Paris VII (2012). https://tel.archives-ouvertes.fr/tel-00714310
7. Duy, N.P., Currie, G., Young, W.: New method for evaluating public transport congestion relief. In: 33rd 2015 Proceedings of the Conference of Australian Institutes of Transport Research (CAITR), Melbourne, Victoria, Australia (2015)
8. Glenday, I.F., Rick, S.: Lean RFS (Repetitive Flexible Supply): Putting the Pieces Together. Productivity Press, New York (2013). 168 p
9. Haghani, A., Oh, S.C.: Formulation and solution of a multi-commodity, multi-modal network flow model for disaster relief operations. Transp. Res. Part A: Policy Pract. **30**, 231–250 (1996)
10. Kelly, G., McCabe, H.: A survey of procedural techniques for city generation. ITB J. (14), 87–130 (2006). http://www.itb.ie/site/researchinnovation/itbjournal.htm
11. Kim, S.-H.: Postponement for designing mass-customized supply chains: categorization and framework for strategic decision making. Int. J Supply Chain Manag. **3**(1), 1–9 (2014)
12. Levinson, D., Huang, A.: A positive theory of network connectivity. Environ. Plan. B: Plan. Des. **39**(2), 308–325 (2012)
13. Levner, E., Kats, V., Alcaide, D., Pablo, L., Cheng, T.C.E.: Complexity of cyclic scheduling problems: a state-of-the-art survey. Comput. Ind. Eng. **59**(2), 352–361 (2010)
14. Li, J.Q., Zhoua, K., Zhanga, L., Zhang, W.-B.: A multimodal trip planning system with real-time traffic and transit information. J. Intell. Transp. Syst. **16**(2), 60–69 (2012)
15. Liu, L.: Data model and algorithms for multimodal route planning with transportation networks. Ph.D. Dissertation theses, Technischen Universität München (2010)
16. Maneengam, A., Laotaweesub, W., Udomsakdigool, A., Sripathomswat, K.: Applying dynamic programming for solving the multimodal transport problem: a case study of thai multimodal transport operator. In: Proceedings of the Asia Pacific Industrial Engineering & Management Systems Conference, Bangkok, Thailand, pp. 1319–1325 (2012)
17. Sandkuhl, K., Kirikova, M.: Analysing enterprise models from a fractal organisation perspective - potentials and limitations. In: Johannesson, P., Krogstie, J., Opdahl, Andreas, L. (eds.) PoEM 2011. LNBIP, vol. 92, pp. 193–207. Springer, Heidelberg (2011). doi:10. 1007/978-3-642-24849-8_15
18. Sitek, P., Wikarek, J.: A hybrid framework for the modelling and optimisation of decision problems in sustainable supply chain management. Int. J. Prod. Res. **53**(21), 1–18 (2015)
19. Sun, Y., Maoxiang Lang, M., Wang, D.: Optimization models and solution algorithms for freight routing planning problem in the multi-modal transportation networks: a review of the state-of-the-art. Open Civil Eng. J. **9**, 714–723 (2015)
20. Susan, J.P.: Vehicle re-routing strategies for congestion avoidance, New Jersey Institute of Technology, 139 p (2014)
21. Zhang, J., Liao, F., Arentze, T., Timmermans, H.: A multimodal transport network model for advanced traveler information systems. Procedia Soc. Behav. Sci. **20**, 313–322 (2011)

Sensing Feedback for the Control of Multi-joint Prosthetic Hand

Andrzej Wołczowski[(⊠)], Mariusz Głebocki, and Janusz Jakubiak

Faculty of Electronics, Department of Cybernetics and Robotics,
Wroclaw University of Science and Technology, Wroclaw, Poland
{andrzej.wolczowski,janusz.jakubiak}@pwr.edu.pl, mariusz@janmar.com.pl

Abstract. The paper presents a design of a sensory system for a model of a hand prosthesis. The system is based on pressure sensors modules that detect a touch with an object, and measure a contact force. First tests show that the sensors are capable to detecting slippage of an object. Kinematic simulations in a CAD program for typical grasps of reference objects made it possible to determine size, shape and placement of the sensors. The developed sensors were integrated with the hand, and preliminary experiments with grasping of various objects were conducted.

1   Introduction

In a construction of a bioprosthesis of a hand two major tasks must be solved: a design of a human-machine interface to recognize the person's intentions, and a design of an autonomic effector device which will execute those intentions.

The tasks of the interface is to measure biosignals coming from a human body during its activity and to transform them into discrete control commands [20]. The commands correspond to the objectives of the prosthesis user. At the decision level of the prosthesis control system biosignals are transformed to control commands. A variety of input signals are used at the decision level: electromyographic (EMG), mechanomyographic (MMG), electroencephalographic (EEG) [19]. After signal acquisition they are analyzed to reduce dimension of the signals through feature extraction. Existing approaches to signal analysis, based on various domains of signals: frequency – Short-Time Fourier Transform (STFT), time and frequency – Discrete Wavelet Transform (DWT) [3]. Features then go through feature selection and transformation process by means of methods like Principal Component Analysis (PCA) [7] or Tensor Factorizations (TF) [4]. The last stage of the decision process is the control decision – a result of selection of the expected move [10].

The main task of the effector device (which is usually a multi-articulated mechanism) is a physical execution of the recognized intentions. The control algorithms form a motoric level of the prosthesis control system allowing interactions with external objects: touching, grabbing and manipulation. Usually these operations are complex as a single command requires a multi-stage motion process. In advanced devices a sensory system with its own processing unit provides some level of autonomy of motions. Major disadvantages of modern hand

© Springer International Publishing AG 2017
J. Świątek and J.M. Tomczak (eds.), *Advances in Systems Science*, Advances in Intelligent
Systems and Computing 539, DOI 10.1007/978-3-319-48944-5_22

Fig. 1. Hand mechanical model, back side

prostheses are their weight and a lack of feedback from grasped objects. As a result, about half of amputees does not use their prostheses in everyday routine [2]. The first of those drawbacks may be reduced firstly by the use of new, lightweight construction materials and, secondly, by energy-efficient actuators and power sources. The latter problem is particularly nagging as it limits the set of objects that can be safely and efficiently operated with an artificial hand. Within a control systems of hand prostheses, sensory feedback is present at several levels. At the level of posture fitting, it is used to match a prosthesis configuration to the shape and the size of the grasped object [11]. This function may be realized by relatively simple binary contact detectors. The next is a force feedback which allows levelling pressure applied to an object from all contact points and control of the force used to grasp the object. This requires more advanced sensors to measure force between the hand and the object. The third level is related to transmitting feeling of a grasp and other modalities, like type of object surface, its temperature and so on to the user. This level allows increasing the precision and comfort of grasp control. In embodiments of the sensory interface dominate the non-invasive methods: electrical [1,8], and mechanical stimulation [5,12]. Independently the work on the implantation of electrodes transmitting sensory sensations directly to the residual nerves in the amputated limb is undertaken [13,15].

This work presents further development of the prosthesis presented in [18]. The capabilities of the mechanical construction were enhanced by the newly designed sensors that replaced simple binary touch detectors. The sensor modules presented in this paper base on pressure sensors and allow receiving force feedback during object grasping.

2 Hand Model

2.1 Hand Mechanics

The model of the prosthesis is equipped with 4 digits (a thumb and 3 fingers) and 13 degrees of freedom (see Fig. 1). Each finger has 4 joints driven by 3 actuators (third and fourth joints are mechanically coupled). All 4 joints of the thumb are driven independently. Joints are driven with 13 Futaba S3150 servo motors. Except the thumb base (thumb joint 0) that is driven directly by a motor mounted in a palm, the remaining 12 motors are placed in a forearm and drive is transmitted to the joints by Bowden cables with 1:1 ratio. All joints are equipped with pulse encoders to provide feedback with real joints orientations.

2.2 Sensory System

The main task of the hand sensory system is to measure interactions between the cybernetic hand and gripped (and then – held) object. The sensors used in the system should inform about the fact of a contact with an object (touch), measure the forces between the hand and the object and in the future – provide information on the tangential motion of the touched or gripped object (slippage).

Size, shape and placement of the sensors mounted on the hand should allow contiguity of all contact points with the object. While for arbitrary object it is too strong to demand and hard to achieve under the constrains of hand kinematics, we limit the project to the set of objects and types of grasp that seem typical in prosthesis use. Verification of the grasp capabilities was made for a set containing: a glass (cylindrical grasp), a tennis ball (spherical grasp), mug with a handle (hook grasp), small ball/credit card (tip/pinch grasp), a mug with a handle (hook grasp) – see Fig. 2. These objects were used as reference, however the results are valid for other objects of similar shapes and sizes.

Fig. 2. Simulation results – grasps of reference objects: tip/pinch, cylindrical, spherical and hook

Before the real sensors were designed, all elements of the hand and motion of kinematic pairs were modelled in TopSolid software to determine possible location and dimensions of the sensors. Using the reference objects (see Fig. 2) and virtually testing their grasping, we analyzed space available for sensors according to the following criteria:

– sensors do not limit mobility of the joints (reachable angles),
– there is no collision between sensors and between sensors and digits,
– when grasping an object, all relevant sensors are in contact with the object.

The ranges of joints mobility of each finger (Table 1) due to the physical limitations of the hand kinematic structure and the influence of some finger poses on the free space remaining between them (the space where the sensors can be placed) have been tested using the TopSolid program.

Table 1. Motion ranges of the joints

Joint no.	0	1	2	3
Thumb	−10 100	−45 45	0 90	0 90
Fingers	−25 25	0 90	0 90	0 90 (coupled with joint 2)

Fig. 3. A configuration with maximum flexion of the fingers – tightening a fist and hook-type grasp of a mug handle

Figure 3 illustrates one of the tested configuration (maximum flexion of the fingers) strongly limiting the space for placement of the sensors. Following the simulations, shapes and permissible dimensions of the external sensors as well as their possible location on fingers were determined. The resulting design of sensors configuration is presented in Fig. 4. The shapes and sizes of sensors will be discussed in Sect. 3.

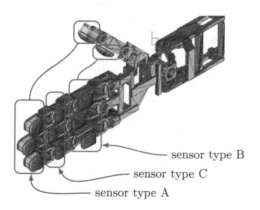

sensor type B

sensor type C

sensor type A

Fig. 4. Sensor shapes and placement

The designed final shapes and placement of sensors are a compromise between sensory system usability and versatility and an acceptable level of its complication. One of the most difficult grasp in respect of the study was the hook grasp of the mug handle. As shown in Fig. 2, only sensors of index, middle and partially thumb finger participate in detection of the contact with the object. This is sufficient to match the grip to the thickness of the mug handle, but only the placement of additional sensors on the sides of the phalanges could ensure the realization of this grip with full feedback.

2.3   Model of Grasping Process

Grasping with a human hand, based on premises presented in [17] and used in [18] may be seen as a 7-stage process. Those stages include: a_0 – rest position, a_1 – grasp preparation, a_2 – grasp closing (until a contact with object is detected), a_3 – grabbing (increasing contact forces until the values resulting from control command are reached), a_4 – maintaining the grasp with force adjustment, a_5 – releasing the grasp, a_6 – transition to the rest position. Possible flow of control between the stages is illustrated in Fig. 5.

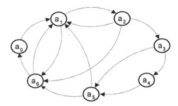

Fig. 5. Graph illustrating state flow of the grasping process

If a prosthesis controller includes an object geometry and receives a force feedback, the stages of the grasping may be formally described as follows. In the description for stage a_i following symbols will be used, with lower index $\cdot_F$ denoting a parameter for all fingers together and $\cdot_{Fjs}$ – for jth phalanx of finger s: $A_F(i)$ – real finger configuration, $A_F^{\max}$ – max degree of opening for the type of grip, $F^\star$ – desired contact force detected by sensor, $V_F(i)$ – finger velocity, and $F_{Fjs}(i)$ – measured (individual) force, $y^\star$ – recognized user's decision, $M(i)$ – measured grasp parameters, K – user's knowledge about grasped object, L – parameters of local control algorithms:

a_0 – fingers move to the rest pose stored in the control algorithm $(A_F(0) \Leftarrow L)$ and stay immobile and passive there $(V_F(0) = 0, F_F(0) = 0)$,

a_1 – type of grip is determined by the control decision based on user's recognized intentions $(y^\star \Leftarrow K)$, the hand is opened to the maximum level suitable for the type of object to be grasped and motion velocity corresponds to the intended arm movement $(A_F(1) \Leftarrow A_F^{\max}(y^\star, L), V_F(1) \Leftarrow M(1))$,

a_2 – interactions of sensors with an object appear in any order, before the contact, motion velocity remains the same as in previous stage $(V_F(2) \Leftarrow M(1))$,

a_3 – motion of each segment is stopped when a signal of touching the object is received $(F_{Fjs}(3) \neq 0)$, the fingers increase the force applied to the object until the expected grasp force is reached $(F_{Fjs}(3) = F^\star \Leftarrow (y^\star, L, M(2)))$

a_4 – contact force in fingers is adjusted to maintain proper grasp $(F_{Fjs}(4) \Leftarrow M(4))$,

a_5 – the fingers are released with velocity depending on the knowledge about the object $(V_F(5) \Leftarrow K)$,

a_6 – the fingers move to the rest position with constant velocity $(A_F(6) \Leftarrow A_F(0), V_F \Leftarrow L)$.

2.4 Kinematics

The scheme of hand kinematics is presented in Fig. 6. Grasping an object requires a control of positions of several touch points on two or more digits at the same time. This assumption requires a definition of kinematics with multiple end-effectors. In a standard robotic approach, this type of kinematics is modelled as a tree-like manipulator using general formalisms like presented in [9,14]. But those methods are general, allowing all joints (both before branching and in all branches) to be actuated. In the case of the analyzed hand, formulas may be simplified due to the observation that all kinematic chains within the hand follow the same pattern: there is an initial branch in a palm, separating the digits, one or two-joint orientating a digit and a digit chain in form of a planar pendulum. The details of kinematic formulation of the hand without sensors was presented in [18], here we enhance the description to introduce multiple touch points.

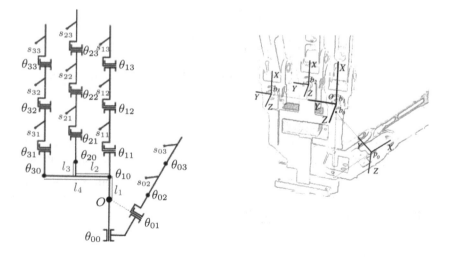

Fig. 6. Kinematic scheme and locations of the origins

The origin of the palm is denoted O and is placed in the center of rotation of the index finger. Digits of the hand are numbered from 0 in the following order: thumb, index, middle and ring finger. A vector of joint angles for a digit i will be denoted $\theta_i = (\theta_{i0}, \theta_{i1}, \theta_{i2}, \theta_{i3})$, where j is a joint number, starting from the joint connecting a digit to the palm, numbered 0. To describe transformations in a kinematic chain of a digit i, we distinguish the following special points in a kinematic chain of each digit:

- b_i is the base of the digit, i.e. the origin of the digit's coordinate frame,
- p_i is the origin of the planar part of the digit,
- s_i^j is a point on a sensor on phalanx j which is the origin of that sensor touch coordinate frame.

Coordinate frame of a contact point is represented as a homogeneous transformation $A_X^Y \in SE(3)$ between coordinate frames X and Y, composed of a 3×3 rotation matrix R_X^Y and a translation vector $t_X^Y \in \mathcal{R}^3$.

With the above notation, the description of each kinematic chain between the palm origin and a sensor touch point may be defined as

$$k_{s_i^j}(\theta_i) = A_O^{s_i^j}(\theta_i) = A_O^{b_i} A_{b_i}^{p_i}(\theta_i) A_{p_i}^{e_i^j}(\theta_i) A_{e_i^j}^{s_i^j}. \tag{1}$$

where the transformations from the palm origin to the base of the finger $A_O^{b_i}$ and from phalanx local frame to a touch point on that phalanx $A_{e_i^j}^{s_i^j}$ are constant.

For fingers $A_O^{b_i}$ is built of identity rotation matrix $R_O^{b_i} = I_3$ and translations vectors $t_O^{b_1} = 0$, $t_O^{b_2} = [l_2 \ l_3 \ 0]^T$, $t_O^{b_3} = [0 \ l_4 \ 0]^T$. In the case of the thumb the transformation is given by

$$R_O^{b_0} = \begin{bmatrix} 0 & 0 & 1 \\ 0 & 1 & 0 \\ -1 & 0 & 0 \end{bmatrix}, \qquad t_O^{b_0} = \begin{bmatrix} -l1 \\ 0 \\ 0 \end{bmatrix}$$

The constant transformation to the sensor in the phalanx frame is defined by

$$R_{e_i^j}^{s_i^j} = I_3 \qquad t_{e_i^j}^{s_i^j} = \begin{bmatrix} -l_{ijx} & l_{ijy} & 0 \end{bmatrix}^T.$$

The remaining two transformations of (1): between a digit base and its planar part $A_{b_i}^{p_i}$ and between the beginning of the planar part of a digit and the phalanx $j - A_{p_i}^{e_i^j}$ – are dependant on joint state vector θ_i. To define them, we use standard description with Denavit-Hartenberg parameters. The parameters for all digits are collected in Table 2.

3   Touch Sensor Construction

The touch sensors are built of three main elements: PCB, base case and a rubber pad (cf. Fig. 7). The layered build of the sensors has some flexibility allowing us to tune the base and pad thickness. Grasp analysis has led to a design of 3 types of sensor modules. Their locations and dimensions are summarized in Table 3.

The main element of the PCB is MPL115A2 – a pressure and temperature sensor. A number of PCBs that can be fitted in cases depends on the segment for

Table 2. Denavit-Hartenberg parameters

joint	thumb θ	d	a	α	
0	θ_{00}	0	0	$\frac{\pi}{4}$	b_0
1	θ_{01}	$-d_1$	L_1	$\frac{\pi}{2}$	p_0
2	θ_{02}	0	L_2	0	
3	θ_{03}	0	L_3	0	e_0^3

joint	fingers 1,2,3 θ	d	a	α	
0	θ_{i0}	0	L_0	$\frac{\pi}{2}$	b_i
1	θ_{i1}	0	L_1	0	p_i
2	θ_{i2}	0	L_2	0	
3	$\theta_{i3}{}^{*}$	0	L_3	0	e_i^3

*finger joints 2 and 3 are coupled, therefore $\theta_{i3} = \theta_{i2}$, for $i = 1, 2, 3$.

a rubber pad —————

PCB —————

base case —————

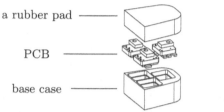

Fig. 7. Sensor elements

which the sensor is dedicated: from 1 for intermediate phalange up to 3 for distal phalanges (Table 3). Such a design may provide more precise information about direction and location of applied force. The base cases were 3D printed. The pad was made of soft (hardness 20 in Shore A scale) polymer rubber VytaFlex. The cast was prepared in vacuum degassing plant to remove remains of air between the pressure sensor and the rubber. Several tests with various shapes of the outer side of the pads has shown that ridged pads have superior grasp properties to those with smooth surfaces and for the best effect the ridges should be irregular, imitating fingerprints. Such shape allows a prosthesis hand not only to obtain a more stable grasp, but it will also allow detection of object slippage (we consider static objects which do not vibrate themselves) [6,16]. Location of sensors PCBs in the cases and the ready sensors are shown in Fig. 8.

The pressure sensors used in the design are factory calibrated for temperature changes and nonlinearity. However, the calibration parameters change after the sensor is covered with a rubber pad. Therefore before the modules can be used,

Table 3. Summary of sensor modules

Type	Phalanx		Dimensions l × w × h	No. of sensors
	Fingers	Thumb		
A	Distal	Distal	19.2 × 18.2 × 14.6	3
B	Proximal	Proximal	10.4 × 15 × 9.6	2
C	Intermediate	—	19.6 × 15 × 14.6	1

Fig. 8. Sensor elements: PCBs and cases (left) and ready sensors (right)

they have to be manually calibrated. The sensor used was heated with pressure and temperature measures, then a linear regression was used to calculate a correction $p = p_{raw} - (aT + b)$. The parameters determined for that sensor were $a = 29.522$, $b = 674.105$. However the process of covering the sensor with rubber is not standard, results for each module vary and each sensor requires separate calibration. Noise after calibration has not exceeded 1 % of the returned value. The calibrated sensor was tested with a load from 0 to 208 g. The readings are almost linear and the sensor detected loads as small as several grams.

The approach adopted assumes the installation of 4 sensor modules with 3 sensors, 3 – with 1 sensor and 4 – with 2 sensors, making a total of 23 sensors. The MPL115A2 sensor has a built-in address for I2C communication interface, therefore a support of 23 sensors require the 23 separate I2C lines. A possible solution is the use of the LTC4316 circuit to change the address of the MPL115A2 sensors for I2C communication interface. This enables direct communication of sensors with the main controller. Setting the address translation system LTC4316 is done using resistors; it be possible to set up to 127 different addresses.

The hardware solution used for communication between the sensors and the main controller enables readings with sensors maximum speed available to them (above 250 Hz). This allows the detection of pressure changes formed in sensors as a result of vibration accompanying the slip of rubber coating on the surface of the object being touched or held. This phenomenon is illustrated in Fig. 9. We see the 4 phases of the sensor operation. The phase 1 (between 7 and 8 s) - an increase in the measured pressure from 1250 hP to max 2000 hP, as a result of touch/shock of finger with object. Phase 2 (from 8 to 10 s), - maintaining pressure at 1800 hp, as a result of constant finger pressure on the object without slipping. Phase 3 (10 to 12 s) the pressure drop as a result of bending the finger rubber pad under the effect of transverse force, without moving. Phase 4 (12–17 s) - jumps/vibration of pressure by slipping of finger over the surface of the object. Phase 5 (17–18 s) - increase pressure to1700 hP, due to arrest the finger movement on the object surface.

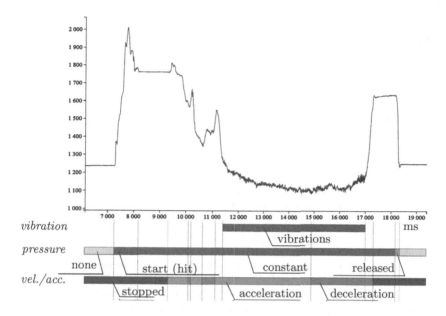

Fig. 9. Sensor response while slippage

4   Conclusion

Limited sensory systems of modern hand prostheses make their users replace touch feeling of a human hand with visual feedback by motion observation. Despite the richness of the information carried in vision channel, the lack of feeling of contact, pressure force or other impressions related to interaction between a prosthesis and a grasped object limits users capabilities of grasping. Comparing to vision systems, touch sensory systems are still not sophisticated enough, as there is no available technology capable of placing on an artificial hand a number of sensors comparable to the number of touch receptors in the human hand.

Despite those limitations, we proposed a solution that at least partially improves touch sensing of a prosthesis. A sensor designed for this purpose was tested and it proved to be capable of detecting and measuring contact forces during a grasp. The hand model was investigated for the best shapes and placement of the sensors to provide grasping capabilities focused on handling typical objects. Kinematics of the hand with sensors was modelled with simplified tree-like structures which can be further used in grasp planning algorithms.

Touch force information provided by the sensors may be used for designing new effective algorithms capable of dexterous grasping and manipulation with typical objects. An additional feature of the designed modules is capability of detecting a slippage of a grasped object. Further research will focus on how to use this information to adjust grasp force in grasp maintaining stage.

The idea of using fingers that are passive in a selected grasp to support an object requires some additional considerations. If the support is meant to be done automatically, the proposed sensors setup must be extended to detect force from a side of a finger. An alternative approach is direct control by the user, however it implies fine-grained intention recognition.

Acknowledgment. The work of the authors was supported by statutory grants.

References

1. Beeker, T.W., During, J., Den Hertog, A.: Artificial touch in a hand-prosthesis. Med. Biol. Eng. **5**(1), 47–49 (1967)
2. Biddiss, E., Chau, T.: Upper-limb prosthetics: critical factors in device abandonment. Am. J. Phys. Med. Rehabil. **86**, 977–987 (2007)
3. Camata, T.V., et al.: Fourier and wavelet spectral analysis of EMG signals in supramaximal constant load dynamic exercise. In: IEEE International Conference Engineering in Medicine and Biology Society (EMBC), pp. 1364–1367 (2010)
4. Cichocki, A., et al.: Nonnegative Matrix and Tensor Factorizations: Applications to Exploratory Multi-way Data Analysis and Blind Source Separation. Wiley, New York (2009)
5. Cipriani, C., et al.: A novel concept for a prosthetic hand with a bidirectional interface: a feasibility study. IEEE Trans. Biomed. Eng. **56**(11), 2739–2743 (2009)
6. Damian, D.D., et al.: Artificial ridged skin for slippage speed detection in prosthetic hand applications. In: IEEE International Conference IROS, Taipei, Taiwan, pp. 904–909 (2010)
7. Güler, N.F., Koçer, S.: Classification of EMG signals using PCA and FFT. J. Med. Syst. **29**(3), 241–250 (2005)
8. Kaczmarek, K.A., et al.: Electrotactile and vibrotactile displays for sensory substitution systems. IEEE Trans. Biomed. Eng. **38**(1), 1–16 (1991)
9. Khalil, W., Dombre, E.: Modeling, Identification and Control of Robot. Butterworth-Heinemann, Oxford (2004)
10. Kurzyński, M., Wołczowski, A.: Hetero- and homogeneous multiclassifier systems based on competence measure applied to the recognition of hand grasping movements. In: Piętka, E., Kawa, J., Wieclawek, W. (eds.) Information Technologies in Biomedicine, Volume 4. AISC, vol. 284, pp. 163–174. Springer, Cham (2014)
11. León, B., Morales, A., Sancho-Bru, J.: Robot grasping foundations. In: León, B., Morales, A., Sancho-Bru, J. (eds.) From Robot to Human Grasping Simulation. Cognitive Systems Monographs, vol. 19, pp. 15–31. Springer, Cham (2014)
12. Pylatiuk, C., Kargov, A., Schulz, S.: Design and evaluation of a low-cost force feedback system for myoelectric prosthetic hands. J. Prosthet. Orthot. **18**, 57–61 (2006)
13. Rossini, P., et al.: Double nerve intraneural interface implant on a human amputee for robotic hand control. Clin. Neurophysiol. **121**(5), 777–783 (2010)
14. Samin, J.C., Fisette, P.: Tree-like multibody structures. In: Samin, J.C., Fisette, P. (eds.) Symbolic Modeling of Multibody Systems. Solid Mechanics and Its Applications, vol. 112, pp. 89–127. Springer, Amsterdam (2003)
15. Tabot, G., et al.: Restoring the sense of touch with a prosthetic hand through a brain interface. Proc. Nat. Acad. Sci. U.S.A. **110**, 18279–18284 (2013)

16. Tegin, J., Wikander, J.: Tactile sensing in intelligent robotic manipulation - a review. Ind. Rob. **32**(1), 64–70 (2005)
17. Wołczowski, A.: Smart hand: the concept of sensor based control. In: Proceedings of MMAR, Miedzyzdroje, pp. 783–790 (2001)
18. Wołczowski, A., Jakubiak, J.: Control of a multi-joint hand prosthesis—an experimental approach. In: Burduk, R., Jackowski, K., Kurzyński, M., Woźniak, M., Żołnierek, A. (eds.) Proceedings of the 9th International Conference on Computer Recognition Systems CORES 2015. AISC, vol. 403, pp. 553–563. Springer, Heidelberg (2016). doi:10.1007/978-3-319-26227-7_52
19. Wołczowski, A., Kurzynski, M.: Control of hand prosthesis using fusion of biosignals and information from prosthesis sensors. In: Borowik, G., Chaczko, Z., Jacak, W., Łuba, T. (eds.) Computational Intelligence and Efficiency in Engineering Systems. SCI, vol. 595, pp. 259–273. Springer, Heidelberg (2015). doi:10.1007/978-3-319-15720-7_19
20. Wołczowski, A., Kurzyński, M.: Human-machine interface in bioprosthesis control using emg signal classification. Expert Syst. **27**(1), 53–70 (2010)

Spatio-temporal Clustering and Forecasting Method for Free-Floating Bike Sharing Systems

Leonardo Caggiani, Michele Ottomanelli, Rosalia Camporeale[⊠], and Mario Binetti

Department of Civil, Environmental and Building Engineering and Chemistry (DICATECh), Politecnico di Bari, Viale Orabona, 4, 70125 Bari, Italy
{leonardo.caggiani,michele.ottomanelli,
rosalia.camporeale,mario.binetti}@poliba.it

Abstract. Free-floating bike sharing systems are an emerging new generation of bike rentals, that eliminates the need for specific stations and allows to leave a bicycle (almost) everywhere in the network. Although free-floating bikes allow much greater spontaneity and flexibility for the user, they need additional operational challenges especially in facing the bike relocation process. Then, we suggest a methodology able to generate spatio-temporal clusters of the usage patterns of the available bikes in every zone of the city, forecast the bicycles use trend (by means of Non-linear Autoregressive Neural Networks) for each cluster, and consequently enhance and simplify the relocation process in the network.

Keywords: Bike sharing systems · Free-floating · Usage patterns · Spatio-temporal clustering · Forecasting · Fleet relocation

1 Introduction

Bicycle sharing systems (BSS) are a new generation of traditional bike rentals [1] that is becoming ever more popular in several modern cities. They provide citizens with an alternative and more sustainable carbon-free mode of transportation (especially suited for short-distance trips), supporting a greener growth of urban environments, and reducing the use of personal car (with all related diseconomies, such as traffic congestion, air pollution and noise).

Through these systems, users are able to pick up a bike from a particular position and return it back closer to their destination. Usually, traditional BSS are station-based; however, there is an emerging new model of bike sharing, called free-floating, that eliminates the need for specific stations, allowing to lock bikes to an ordinary bicycle rack, avoiding the necessity of docking stations and kiosk machines [2]. Although their undeniable qualities, BSS suffer mainly of two problems: the impossibility for the user to find a bike when he/she wants to start his/her journey, and/or the impossibility to leave the bike in the preferred destination due to full stations. If this last drawback can be clearly averted in the free-floating systems thanks to the chance to leave the bicycle

J. Świątek and J.M. Tomczak (eds.), *Advances in Systems Science*, Advances in Intelligent Systems and Computing 539, DOI 10.1007/978-3-319-48944-5_23

(almost) everywhere in the network, there is no guarantee to find a bike near the origin of the desired trip. Consequently, bike sharing providers have to ensure a proper spread of bicycles in the network (and as well free parking slots for returning bikes in the station-based systems) in order to fully satisfy customers. Solving the problem of adequately balance (or relocate) bikes in the city is therefore of paramount importance, and it is strictly connected to a correct prediction of the usage of the system. Forecasting the usage patterns of a BSS is beneficial for both users, able to better plan their trips; and operators, to better manage the system manually relocating bicycles among the zones to improve the customer's experience [22, 23]. Suggest an alternative strategy to reduce the cost of such 'balancing bikes' operation is the final goal of this paper.

We propose to investigate the spatio-temporal correlation pattern of different zones in a city, and integrate it into a forecast model, in order to make easier and more effective the relocation process in BSS.

The remainder of the paper is organized as follows. A literature review about clustering techniques and prediction models is presented; afterwards, the proposed solution methodology is explained in greater detail enclosed with an application to a real case study (London's shared bicycle scheme); conclusions and suggestions for further researches end the discussion.

2 Clustering Techniques

Clustering analysis of a dataset aims at organizing a collection of different patterns into a smaller number of homogeneous groups, without any prior knowledge. A good clustering arrangement is the one able to reach a high correlation among the elements of a single cluster, and a low inter-correlation among distinct clusters.

They have been widely used in literature to explore the activity patterns connected to a bike sharing systems usage, with a diverse range of final purposes. This happens because data collected on such systems are commonly sizeable; therefore, it is difficult try to gain knowledge from them without the help of a method capable of giving a synthetic view of the underlying information.

Typical cluster models include the hierarchical method [3], based on a distance idea, i.e. elements being more related to nearby objects than to the ones further away; or centroid-based methods, where clusters are represented by a central vector. K-means cluster algorithms [4] are among the most well-known of this last category: random procedures are used to generate starting clustering centers, and data objects are assigned to k clusters; based on the initial partitioning, elements are relocated by minimizing their distances within clusters and maximizing their distances among different clusters [5].

Analyzing datasets regarding customers trips, cluster analysis is capable of revealing groups of stations with a similar trend of rental and return activities during the day: Vogel et al. [6] call them "activity clusters", observing that each group of stations shows similar trip purposes represented by the activity. Other studies [7] find that each cluster of bike stations seems to be closely related to the city functions, like employment, leisure, transportation, a determination that can be helpful for a variety of applications, including urban planning and the choice of a business location. This correlation between bike activity stations and external factors has also been modeled, for example applying to it a linear regression [8].

Furthermore, different authors focus on the spatio-temporal correlation among data, considering it a factor of paramount importance in affecting bicycle demands in the system. Froehlich et al. [9] provide a spatio-temporal analysis of Barcelona's shared bicycling system, identifying shared behaviors across stations, finding how these behaviors relate to location, neighborhood and time of the day. More recently, Han et al. [10] correlates the historical usage records of Paris' bike sharing system at both spatial and temporal scale, integrating this analysis into forecasting goals, underlining how this represents a necessary information for predicting bikes demand of each station accurately.

In this paper we want to propose a bi-level method, able to aggregate at first temporally (making use of wavelets and hierarchical clustering) and after that spatially (through k-means clustering) patterns of available bikes in different zones of a city.

3   Bike Share Predictive Modeling

A consistent help towards a better regulation of bike sharing systems lies in forecasting the use of bicycles. Different studies about the prediction of BSS have already been done; roughly, there are 30 predictive methods [11]: dynamic traffic flow distribution, historical mean, regression analysis, time series, Kalman filtering, neural network, fuzzy neural network, non-parametric regression, support vector machine (SVM), and so on.

A short-term forecast can be inferred through methods similar to the ones largely used in road traffic control [12, 13]. Basically, there are linear prediction approaches, where the prediction is based on the main assumption of linearity and stationarity to obtain future trends, such as the historical average model [14]: it is able to forecast future traffic volumes on the basis of the average of past traffic volumes. On the other hand, the non-linear prediction techniques, although with a higher computing complexity, are able to capture the non-linear features of transportation systems, ensuring a more accurate forecasting performance [15].

Neural networks [16] are a flexible method good at recognizing complex, non linear patterns, without a preliminary knowledge about the relationships between input and output variables [17]. Thanks to their characteristics and adaptability, they are among the most significant forecasting techniques in the transportation field.

Among advanced models that have been developed in recent years, there is the SVM, a machine learning technique successfully used in pattern recognition, function regression and function approximation. The underlying reason for using SVM in traffic prediction is its capacity to accurately predict time series data of non-linear, non stationary and not predefined processes [18], outperforming other non-linear techniques. In case of very large datasets, it has been proven that traditional SVM tend to perform worse when trained with the entire data rather than with a set of fine-quality samples [19]. It is also possible to find in literature hybrid methods, like the Wavelet-SVM proposed in [15], that combining two approaches are able to sum the benefits of both techniques and overcome their shortcomings. As a matter of fact, in the field of signal information acquisition and processing, wavelet transformation (a partial transformation in joint time-frequency domain) is rapidly developing [20]. It has a good adaptability for many signals, especially with the non-linear, non-stationary ones, and has shown an excellent behavior in signal de-noising, extraction of weak signals and singularity analysis on signals [21].

4 Proposed Solution Methodology

In this paper, we propose a methodology to correlate the spatio-temporal analysis of a free-floating bike sharing usage with a correct prediction of the trend of available bikes during the day in each cluster. This strategy can be useful in simplifying and enhancing the bikes relocation operations and in the reduction of BSS management costs.

4.1 Spatio-temporal Clustering Method

The preliminary spatio-temporal analysis of the suggested methodology involves the definition of two main categories of clusters: temporal clusters, T_k, $k \in [1...n]$, and spatial clusters S_{kj}, $j \in [1...m_k]$, associated to each temporal one. In the rest of the paper we adopt the following notation.

T_k	Temporal cluster
n	Total number of temporal clusters
$\bar{t}_y \in T_k$	Generic element of the temporal cluster
w_k	Total number of elements of T_k
S_{kj}	Spatial cluster associated to T_k
m_k	Total number of spatial clusters associated to T_k
i	Generic spatial cluster iteration
$\bar{i}$	Total number of spatial cluster iterations
$\bar{s}_{k'j} \in S_{kj}$	Generic element of the spatial cluster
α	Positive integer coefficient
β	Positive integer coefficient
a_{kj}	Total number of elements $\bar{S}_{k'j}$, $k' = k$
b_{kj}	Total number of elements $\bar{S}_{k'j}$, $k' \neq k$

The flowchart in Fig. 1 summarizes the main steps of the process helping in its general understanding. We start with the idea of studying a city with a free floating BSS operating on it. Being this city divided into census districts, the first operation to carry out involves the aggregation of these districts into larger zones.

To build a solid forecasting model, we need to preliminary select a set of days 'uniform', both from the point of view of the type of the day (working days or weekend/holiday days), and the weather conditions. Operated this data selection, it is possible to know the number of available bikes in each moment of the day for every zone (zone temporal pattern). Collecting the bikes usage patterns related to this set of days, we can try to make a plausible prediction of the trend of a 'future' day with similar characteristics.

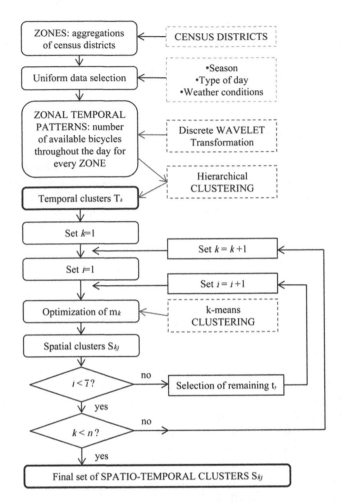

Fig. 1. Spatio-temporal clustering methodology.

Knowing the zone temporal patterns, we can cluster them according to their temporal trends. A discrete wavelet transformation helps in the analysis of signals, leading to the determination of a given number of temporal clusters T_k, $k \in [1 \ldots n]$, applying a hierarchical clustering methodology. In this way each zone belongs to a predetermined T_k.

The next step aims to geographically aggregate groups of zones belonging to the same T_k, creating a certain number of spatial clusters S_{kj}, $j \in [1 \ldots m_k]$, associated to each temporal cluster. We propose to operate this second clusterization using a k-means algorithm. The number of S_{kj} related to each T_k is minimized solving the following optimization problem. For $k \in [1 \ldots n]$:

$$\min m_k \tag{1}$$

s.t.

$$m_k \geq \alpha \tag{2}$$

$$m_k \leq w_k \tag{3}$$

$$\max b_{kj} \leq \beta \tag{4}$$

The objective function (1) aims to minimize the total number m_k of spatial clusters associated to each T_k. This number has to be greater than (or equal to) a positive integer coefficient α (2), and lower than (or equal to) the total number of elements (w_k) belonging to T_k (3). The last constraint (4) forces the maximum among b_{kj} (with $j \in [1\ldots m_k]$) to be smaller or equal to a positive integer coefficient β.

The setting of the thresholds α and β affects the final number of S_{kj}: namely, the greater is α and the lower the value given to β, the higher will be the number of spatial clusters associated to T_k.

At the end of the procedure, we obtain a m_k number of spatial clusters S_{kj} associated to each value of k. The spatial boundary of each cluster S_{kj} is a polygon, having as vertices some of the elements t_y belonging to T_k. If the number of these vertices is equal to or greater than 3, we consider these spatial clusters satisfactory; for the remaining t_y that are not included in the satisfactory spatial clusterization, we have to repeat again the optimization for a number of iterations equal to $\bar{i}$.

The final S_{kj} clusters deduced at the end of the iterations, for each $k \in [1\ldots n]$, are the optimal set of spatio-temporal clusters, starting point to keep on with the next step of the proposed methodology.

4.2 Non-linear Autoregressive Neural Network Prediction

In order to forecast the bicycles use in each S_{kj} belonging to the final optimal spatio-temporal set, we decide to adopt a Nonlinear Autoregressive Neural Network (NARNN). This dynamic neural network can be trained to predict a time series from that series past values.

Given a time series representing the number of available bicycles in a temporal interval for each spatio-temporal cluster S_{kj}, basically we want to find an optimal network structure for the NARNN and train the network to minimize a performance evaluation function, such as the Mean Square Error (MSE) or the Differential Absolute value Error (DAE). The MSE is defined as the average of the squared differences between the forecasted and observed data, while the DAE is the maximum of the difference between real and predicted samples.

There are also two important concepts that it is worthy to explain, as they are used during the analysis of a trained prediction network:

1. the prediction errors should not be correlated in time;
2. the prediction errors should not be correlated with the input sequence.

If the prediction errors were correlated in time, then we would be able to predict the prediction errors and, therefore, improve our original prediction. Also, if the prediction errors were correlated with the input sequence, then we would also be able to use this correlation to predict the errors. In order to test the correlation of the prediction errors in time, we can use the sample autocorrelation function; to test the correlation between the prediction errors and the input sequence, we can use the sample cross-correlation function. Basically, the aim of these functions is allowing us to perform a preliminary skimming of the trained networks. Among the remaining, we have to prefer the one able to minimize to a greater extent the chosen performance evaluation function.

5   Application to London's Shared Bicycling System

London's shared bicycle scheme, called "Barclay's Cycle Hire", was launched by London's public transport authority (Transport for London) on July 30th, 2010. Nowadays, it consists of more than 10000 bikes and over 700 bike docking stations. Indeed, it is difficult to collect data regarding a free-floating BSS applied to a real case study, as it is a relatively new system. Consequently, we pretend that each London's bike sharing station is the ideal centroid of our aggregate zone, and the number of available bicycles in the station corresponds to the number of free-floating bikes spread in the zone at a given moment of the day.

We carry out our analysis selecting 4 Wednesdays of the month of May, 2013: they are working days with similar weather conditions, so suitable for our purposes. We collect data from Transport for London's online interactive map, from which it is possible to acquire information about the location (latitude and longitude) of each station, the number of available bicycles and the weather conditions (temperature, humidity, wind). We scrape data from the website every 2 min, except during moments of intermittent access to the map service and during down times of our own servers. This results in 720 observations of each station's state over any 24 h period.

We apply a wavelet transformation to the zonal temporal patterns of each station (aggregate zone); then, we set $k = 4$. Adopting an ascending hierarchical clustering (Ward's method, with Euclidean distance), we obtain four different temporal clusters T_k, shown in Fig. 2. Each cluster is made by overlapping the zonal temporal patterns of the stations that belong to it; the average is highlighted by the dark red color.

Afterwards, following the solution methodology explained in Sect. 4, adopting a meta-heuristic approach (genetic algorithm) to solve the optimization of the number of spatial clusters to associate to each T_k, we achieve a spatio-temporal clusterization of all London's bike sharing stations (Fig. 3).

Each dot coincides with the ideal centroid of our aggregate zone; the four different colors correspond to a different temporal cluster T_k; every polygon represents a spatio-temporal cluster S_{kj}. We set $\bar{i} = 2$, so we perform two iterations to achieve this clustering configuration. The polygon's edges drawn with the black color delimit the spatio-temporal clusters obtained during the second iteration.

Then, we use the NARNN to forecast the number of bikes of a single spatio-temporal cluster, choosing the one highlighted in grey in Fig. 3, constituted by 13 aggregated zones. We set the 70 % of the average samples to be used for training;

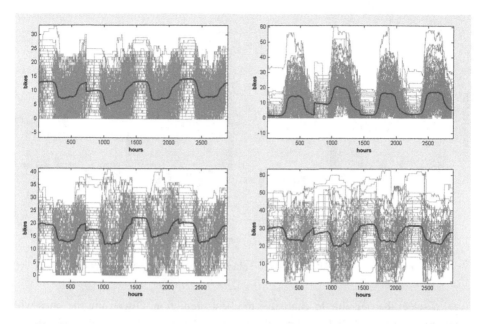

Fig. 2. Temporal clusters T_k of Barclay's Cycle Hire, London. (Color figure online)

Fig. 3. Spatio-temporal clusters S_{kj} of Barclay's Cycle Hire, London. (Color figure online)

the 15 % to be used to validate that the network is generalizing and to stop training before over-fitting; the last 15 % to be used as a completely independent test of network generalization. In order to find an optimal structure for the NARRN, we trained the network 1000 times on the first two days considered in our analysis (first two Wednesdays of May); each trained network allows to forecast the values (number of available bikes) during the third day of the month. Among these 1000 networks, we

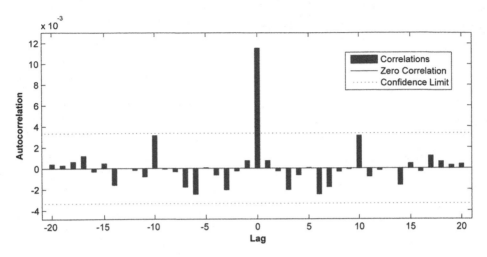

Fig. 4. Autocorrelation function values of the selected neural network.

have to select the one able to guarantee the best prediction. Discarding the ones with prediction errors correlated in time and with the input sequence, we have to chose the one able to minimize the MSE. Figure 4 shows the autocorrelation function values of the best neural network among the trained ones.

Figure 5 shows the real data and the NAR fitting of the chosen neural network. The train set corresponds to the first two days of our analysis; on the basis of the forecast done for the third day, we determine the value of the performance evaluation function.

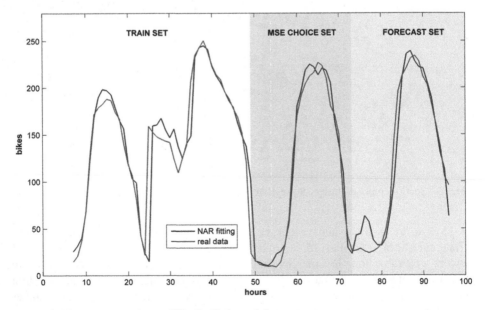

Fig. 5. Train and forecast set.

In this particular case, the selected network reaches the lower value of MSE compared to the remaining ones (MSE = 351.0) The forecast set shows how well the NAR prediction fits with the trend of the real data.

Being able to predict exactly the number of available bikes in each spatio-temporal cluster of the network, it could be possible to enhance the process of bikes relocation, adopting an approximate vehicle routing problem that could be more suitable when referring to a free-floating system.

6   Conclusions and Future Researches

This paper deals with bike sharing systems, in particular it is one of the first in literature addressing to the emerging free-floating ones. Their peculiarity is the spread presence of bicycles in each zone of the city, without the bond to allocate bikes in fixed stations on the network. Although free-floating bikes allow much greater spontaneity and flexibility for the user, they need additional operational challenges especially in facing the bike relocation processes. Then, we propose a methodology able to cluster a city according to its spatio-temporal features, forecast the bicycles trend for each cluster, and consequently simplify the relocation process permitting to solve an approximate vehicle routing problem passing by each cluster (and not each bike) in the network.

The main limitation of the proposed methodology could lie in the presence of zones geographically close but with strongly different temporal patterns of available bikes. There exists the possibility that this might happen, although in very rare situations, as the literature confirms that is quite unusual. This method may be applied also to station-based BSS, or to car sharing systems, that basically behave in an analogue way.

Future researches involve a careful calibration (through a sensitivity analysis) of the number of the temporal cluster (k), and the values of the thresholds (α and β) included into the model of minimization of the number of spatial clusters. Further developments embrace also a full application to a case study, including the prediction of the number of bikes for all the spatio-temporal clusters, accompanied by the selection of the optimal paths to follow during the relocation process.

References

1. Fanaee-T, H., Gama, J.: Event labeling combining ensemble detectors and background knowledge. Progr. Artif. Intell., 1–15. Springer, Heidelberg (2013)
2. Pal, A., Zhang, Y.: Free-floating bike sharing: solving real-life large-scale static rebalancing problems. Technical report, University of South Florida (2015)
3. Zhou, X.: Understanding spatiotemporal patterns of biking behavior by analyzing massive bike sharing data in Chicago. PLoS ONE **10**(10), e0137922 (2015)
4. Xu, H., Ying, J., Wu, H., Lin, F.: Public bicycle traffic flow prediction based on a hybrid model. Appl. Math. Inf. Sci. **7**, 667–674 (2013)
5. Lee, C., Wang, D., Wong, A.: Forecasting utilization in city bike-share program. Technical report, CS 229 2014 Project (2014)

6. Vogel, P., Greiser, T., Mattfeld, D.C.: Understanding bike-sharing systems using data mining: exploring activity patterns. Procedia Soc. Behav. Sci. **20**, 514–523 (2011)
7. Côme, E., Randriamanamihaga, A., Oukhellou, L.: Spatio-temporal Usage Pattern Analysis of the Paris Shared Bicycle Scheme: A Data Mining Approach. Transport Research Arena, Paris (2014)
8. Wang, X., Lindsey, G., Schoner, J.E., Harrison, A.: Modeling bike share station activity: the effects of nearby business and jobs on trips to and from stations. Transp. Res. Rec. **43**, 45 (2012)
9. Froehlich, J., Neumann J., Oliver, N.: Sensing and predicting the pulse of the city through shared bicycling. In: The International Joint Conferences on Artificial Intelligence, pp. 1420–1426 (2009)
10. Han, Y., Côme, E., Oukhellou, L.: Towards bicycle demand prediction of large-scale bicycle sharing system. In: Transportation Research Board 93rd Annual Meeting (2014)
11. Zeng, D., Xu, J., Gu, J., Liu, L., Xu, G.: Short term traffic flow prediction based on online learning SVR. In: Workshop on Power Electronics and Intelligent Transportation System, pp. 616–620 (2008)
12. Vlahogianni, E.I., Karlaftis, M.G., Golias, J.C.: Short-term traffic forecasting: where we are and where we're going. Transp. Res. Part C **43**, 3–19 (2014)
13. Mori, U., Mendiburu, A., Álvarez, M., Lozano, J.A.: A review of travel time estimation and forecasting for advanced traveller information systems. Transportmetrica A Transp. Sci. **11** (2), 119–157 (2015)
14. Smith, B., Demetsky, M.: Traffic flow forecasting: comparison of modeling approaches. J. Transp. Eng. **123**(4), 261–266 (1997)
15. Sun, Y., Leng, B., Guan, W.: A novel wavelet-SVM short-time passenger flow prediction in Beijing subway system. Neurocomputing **166**, 109–121 (2015)
16. Dougherty, M.: A review of neural networks applied to transport. Transp. Res. Part C **3**(4), 247–260 (1995)
17. Wei, Y., Chen, M.: Forecasting the short-term metro passenger flow with empirical model decomposition and neural networks. Transp. Res. Part C **21**(1), 148–162 (2012)
18. Sapankevych, N., Sankar, R.: Time series prediction using support vector machines: a survey. IEEE Intell. Syst. **4**(2), 24–38 (2009)
19. Yu, H., Yang, J., Han, J., Li, X.: Making SVMs scalable to large data sets using hierarchical cluster indexing. Data Min. Knowl. Disc. **11**(3), 295–321 (2005)
20. Zhang, Z., Zhang, P., Yin, Y., Hou, L.: Analysis on urban traffic network states evolution based on grid clustering and wavelet de-noising. In: 11th International IEEE Conference on Intelligent Transportation Systems, Beijing, pp. 1183–1188 (2008)
21. Guan, L., Feng, X.: Research on factors of the signal de-noising effect basing on the wavelet transform and the matlab practice. Autom. Instrum. **6**, 43–46 (2004)
22. Caggiani, L., Ottomanelli, M.: A dynamic simulation based model for optimal fleet repositioning in bike-sharing systems. Procedia Soc. Behav. Sci. **87**, 203–210 (2013)
23. Caggiani, L., Ottomanelli, M.: A modular soft computing based method for vehicles repositioning in bike-sharing systems. Procedia Soc. Behav. Sci. **54**, 675–684 (2012)

Comparison of Algorithms for Constrained Multi-robot Task Allocation

Maciej Hojda[✉]

Faculty of Computer Science and Management, Wroclaw University of Science
and Technology, Wybrzeze Wyspianskiego 27, Wroclaw, Poland
maciej.hojda@pwr.edu.pl

Abstract. This paper introduces and evaluates algorithms for solving
a constrained problem of multi-robot task allocation. The problem is
NP-hard even in its feasibility version and proves difficult to solve in
reasonable time. Tested are dedicated solution algorithms as well as algorithms for the more general multidimensional knapsack/covering problem. Empirical evaluations show that a direct adaptation of knapsack
solving methods is either impossible or leads to significant increase in
the execution time.

Keywords: Multi-robot task allocation · MRTA · Multi-robot systems ·
Constrained task allocation

1   Introduction

Multi-robot systems are varied, dynamic and capable of performing tasks outside
of the scope of capabilities of a single robot. Teams of robots are reported to
successfully tackle problems such as exploration, data gathering, inspection and
monitoring [2, 7, 9, 10]. Their ability to perform well under challenging conditions
is typically attributed not only to sophistication and capabilities of individual
robots but also to advanced coordination methods which improve the efficiency of
the team as a whole. For a synergistic effect to take place in a situation featuring
multiple tasks and a force of heterogeneous robots, it is necessary to apply an
efficient task allocation algorithm. This paper focuses on evaluation of multiple
approaches for solving a problem of task allocation in a multi-task, multi-robot
environment. Considered is a case when robots are capable of working in multiple
modes which differ in efficiency and resource consumption.

Tackled problem of task allocation for a multi-modal multi-robot systems was
first formulated in [5,6] where it was proven that, in general form, even its feasi-
bility version is at least as hard as the $R||C_{\max}$ problem. Due to the expected dif-
ficulty in obtaining feasible solutions in reasonable time the focus was on designing algorithms capable of providing solutions with bounded maximal constraint
violations (i.e. solutions to substitutive problems with less restrictive resource
limits). In this paper, the focus is put on providing solutions without constraint

© Springer International Publishing AG 2017
J. Świątek and J.M. Tomczak (eds.), *Advances in Systems Science*, Advances in Intelligent
Systems and Computing 539, DOI 10.1007/978-3-319-48944-5_24

violations. Tested are dedicated solution algorithms and methods designed for the more general multidimensional knapsack/covering problem [1,8].

The paper is divided into five sections: introduction, problem formulation, presentation of solution algorithms, empirical evaluations and conclusions.

2   Multi-task Allocation Problem

There are J tasks, I executors capable of performing those tasks and K modes of operation. Furthermore, there are sets of (indexes of) executors $\mathbf{I} \triangleq \{1, 2, \ldots, I\}$, tasks $\mathbf{J} \triangleq \{1, 2, \ldots, J\}$ and modes $\mathbf{K} \triangleq \{1, 2, \ldots, K\}$. For every (executor, task, mode) triple we define a single binary decision variable $x_{i,j,k}$ which takes the value of 1 if ith executor performs jth task in kth mode (0 if otherwise). All decision variables are grouped into a matrix $x \triangleq [x_{i,j,k}]_{i \in \mathbf{I}, j \in \mathbf{J}, k \in \mathbf{K}}$. Employment of ith executor for jth task in kth mode increases the completion ratio of the task by a known value $\eta_{i,j,k} \geq 0$ and results in spending of $e_{i,j,k} \geq 0$ of a limited resource of the ith executor. A completion ratio of $E > 0$ has to be met for every task and the resource limit of $F_i > 0$ has to be satisfied for the ith executor (for all executors). Feasibility of a solution x is ensured by constraints given as follows

$$\forall i \in \mathbf{I}, j \in \mathbf{J}, k \in \mathbf{K} \quad x_{i,j,k} \in \{0, 1\}, \tag{1}$$

$$\forall j \in \mathbf{J} \quad \sum_{i \in \mathbf{I}, k \in \mathbf{K}} x_{i,j,k} \eta_{i,j,k} \geq E, \tag{2}$$

$$\forall i \in \mathbf{I} \quad \sum_{j \in \mathbf{J}, k \in \mathbf{K}} x_{i,j,k} e_{i,j,k} \leq F_i, \tag{3}$$

$$\forall i \in \mathbf{I}, j \in \mathbf{J} \quad \sum_{k \in \mathbf{K}} x_{i,j,k} \leq 1. \tag{4}$$

Introduced constraints are responsible for keeping the decision variables binary, completing every task, satisfying the resource limit and ensuring that for every task-executor pair at most one mode of execution is chosen. The main objective is the minimization of the total resource spent on allocation. This quality coefficient is defined as follows

$$Q(x) \triangleq \sum_{i \in \mathbf{I}, j \in \mathbf{J}, k \in \mathbf{K}} x_{i,j,k} e_{i,j,k}. \tag{5}$$

Main problem of this paper – the constrained multi-robot task allocation – is given as follows (definition after [6], with slight modifications regarding presentation).

Problem 1. Multi-task allocation – MTA
Given: $\mathbf{I}, \mathbf{J}, \mathbf{K}; E > 0, F_i > 0,$
$\qquad e_{i,j,k} \geq 0, \eta_{i,j,k} \geq 0 \quad i \in \mathbf{I}, j \in \mathbf{J}, k \in \mathbf{K}.$
Find: x^* where

$$x^* \triangleq \arg \min_{x \in \mathbf{D}_\alpha} Q(x), \tag{6}$$

$$\mathbf{D}_\alpha \triangleq \{x \in \mathbf{X} \triangleq \mathsf{X}_{i \in \mathbf{I}, j \in \mathbf{J}, k \in \mathbf{K}} \mathbf{R} : (1) \wedge (2) \wedge (3) \wedge (4)\}. \tag{7}$$

As a specific example consider a problem of autonomous surveillance with the use of Unmanned Aerial Vehicles with the restriction on operational time (limited by ex. battery capacity). Given are areas of interest subject to inspection. Inspecting of a whole area (at least once) is equivalent to executing a task. Size of the area is the completion ratio of the task. Energy usage depends on the size of the inspected area and the parameters of the executor. The problem is to allocate UAVs to areas while minimizing energy usage. Example of such problems can be found in [3,4]. Solving the problem yields an allocation of tasks to executors and a selection of modes of execution. The following section presents a selection of solution algorithms that can be applied to MTA.

3   Solution Algorithms

Proposed solution algorithms are based on the idea of a functional decomposition which uses the problem of single-task allocation as an intermediary. This problem is defined as follows.

Consider a decision variable $\hat{x} \triangleq [\hat{x}_{i,k}]_{i\in\mathbf{I},k\in\mathbf{K}}$ with binary elements $\hat{x}_{i,k}$ which take the value of 1 if ith executor performs a given task in kth mode (0 if otherwise). For the given task, its completion and resource costs are given by $\hat{\eta}_{i,k} \geq 0$ and $\hat{e}_{i,k} \geq 0$. Feasible solutions are ones satisfying the following constraints

$$\forall i \in \mathbf{I}, k \in \mathbf{K} \quad \hat{x}_{i,k} \in \{0,1\}, \tag{8}$$

$$\sum_{i\in\mathbf{I},k\in\mathbf{K}} \hat{x}_{i,k}\hat{\eta}_{i,k} \geq E, \tag{9}$$

$$\forall i \in \mathbf{I} \quad \sum_{k\in\mathbf{K}} \hat{x}_{i,k}\hat{e}_{i,k} \leq F_i, \tag{10}$$

$$\forall i \in \mathbf{I} \quad \sum_{k\in\mathbf{K}} \hat{x}_{i,k} \leq 1. \tag{11}$$

which ensure that the decision variables take binary values, that the completion ratio $E > 0$ is met, that resource limit $F > 0$ is satisfied and that for every executor at most one mode of execution is chosen. The problem of single-task allocation is defined as follows.

Problem 2. Single task allocation – STA
Given: $\mathbf{I}, \mathbf{K}, E > 0, F_i > 0; \hat{e}_{i,k} \geq 0, \hat{\eta}_{i,k} \geq 0 \ i \in \mathbf{I}, k \in \mathbf{K}$.
Find: $\hat{x}^*$ where

$$\hat{x}^* = \arg\min_{\hat{x}\in\hat{\mathbf{D}}} \hat{Q}(\hat{x}); \quad \hat{Q}(\hat{x}) \triangleq \sum_{i\in\mathbf{I},k\in\mathbf{K}} \hat{x}_{i,k}\hat{e}_{i,k}, \tag{12}$$

$$\hat{\mathbf{D}} = \{\hat{x} \in \mathbf{X} \triangleq \bigtimes_{i\in\mathbf{I},k\in\mathbf{K}}\mathbf{R} : (8) \wedge (9) \wedge (10) \wedge (11)\}. \tag{13}$$

Algorithm 1. A1

1: Set $x^1 := [0]_{i \in \mathbf{I}, k \in \mathbf{K}}$, $j := 0$, $\overline{d}^j := \overline{d}$, $\underline{d}^j := \underline{d}$.
2: Set $j := j + 1$.
3: Set $d^j := \lfloor (\overline{d}^{j-1} + \underline{d}^{j-1})/2 \rfloor$.
4: **if** $RTA(d^j)$ exists and if $\hat{Q}[RA(d^j)] \leq 2d$ **then**
5:     set $x^1 := RA(d^j)$,
6:       set $\overline{d}^j := d^j - 1$, $\underline{d}^j := \underline{d}^{j-1}$,
7:   **else** set $\underline{d}^j := d^j + 1$, $\overline{d}^j := \overline{d}^{j-1}$.
8: **if** $\underline{d}^j \leq \overline{d}^j$ **then**
9:     go to 2.
        return x^1.

The problem is also reducible to $R||C_{\max}$. In [6] a 2-approximate solution algorithm $A1$ was proposed. We provide this algorithm here for the convenience of the reader.

The main idea of the $A1$ algorithm is to perform a binary search on an estimated interval of the optimal quality criterion $\hat{Q}$. At each step, a relaxed version of the STA problem is solved. The algorithm uses an $RTA(d)$ routine for solving the relaxation and an $RA(d)$ rounding routine which resolves all fractional variables of the relaxed solution into binary variables. Subroutine $RTA(d)$ also introduces an additional constraint to the STA problem

$$\hat{e}_{i,k} \leq \min\{F, d\} \Rightarrow \hat{x}_{i,k} = 0. \tag{14}$$

Solution obtained by $RTA(d)$ can contain fractional variables, at most two for every executor $i \in \mathbf{I}$. The rounding subroutine $RA(d)$ for the corresponding $RTA(d)$ solution is as follows. For every $i \in \mathbf{I}$, if there is only one fractional $\hat{x}_{i,k} \in (0,1)$ then let $\hat{x}_{i,k} = 1$. If there are two $k_1, k_2 \in \mathbf{K}$ for which $\hat{x}_{i,k_1}, \hat{x}_{i,k_2} \in (0,1)$, and $\hat{\eta}_{i_1,k_1} \geq \hat{\eta}_{i_1,k_2}$ then set $\hat{x}^d_{i_1,k_1} := 1$, $\hat{x}^d_{i_1,k_2} = 0$.

Solution to the STA problem is used in two algorithms for solving the MTA problem. The first algorithm, $A2f$ performs individual task allocations for each task separately. With every task, the resource limit F_i is deceased by the value already used up. The second algorithm $A2e$, which is a modification of $A2f$, increases the costs of task execution instead of decreasing the resource limit. We note that both $A2f$ and $A2e$ use the routine $A1(j)$ for solving the allocation sub-problem for the jth task.

We compare algorithms A2F and A2e with algorithms for solving the more general multidimensional knapsack/covering problem. We use two state of the art algorithms for comparison, the Adaptive Memory Search (Almha) [1] and the Alternating Control Tree (ACT) [8].

The Almha is an algorithm based on tabu search with move evaluation dependent on both, the quality of the solution and the constraint violation. Importance of constraints, and therefore their influence on move evaluation varies depending on the duration for which they remained violated. The ACT algorithm is an iterative improvement routine which solves, at each step, a substitutive problem.

Algorithm 2. A2f

1: Set $j := 1$.
2: $\hat{x}_j^2 \triangleq [\hat{x}_{i,j,k}^2]_{i\in\mathbf{I},k\in\mathbf{K}} := A1(j)$.
3: If $\hat{x}_j^2$ does not exits then **return** no solution.
4: $F_i := F_i - \sum_{i\in\mathbf{I},k\in\mathbf{K}} \hat{x}_{i,j,k}^2 e_{i,j,k}$.
5: $j := j + 1$.
6: If $j \leq J$ go to 2.
$\quad\quad\quad$ **return** $x^2 \triangleq [x_{i,j,k}^2]_{i\in\mathbf{I},j\in\mathbf{J},k\in\mathbf{K}}$ where $x_{i,j,k}^2 = \hat{x}_{i,j,k}^2$.

Algorithm 3. A2e (modification of A2f)

4: $\forall n \in \mathbf{J}, i \in \mathbf{I}, k \in \mathbf{K} : e_{i,n,k} := e_{i,n,k} + \sum_{l\in\mathbf{K}} \hat{x}_{i,j,l}^2 e_{i,j,l}$.

The substitutive solution is then used in order to introduce additional constraints to the original problem. The structure of the substitutive problem allows for it to be solved with the use of different methods. In the experiments in this paper an exact solver is used. For more details on either algorithm, the reader is referred to the respective articles. Due to the difference in objective of the MTA and the knapsack problem (minimization versus maximization), the objective function was negated for those two methods. In case of ACT, constraint (4) in [8] had to be modified accordingly. To solve linear programming problems and integer programming problems a solver $GLPK$ [11] is used with default settings. To give a better insight into how does the solution change over time, the algorithms were modified to include an iteration limit further denoted as N.

We consider several versions of all the aforementioned algorithms

- $A2f$,
- $A2e$,
- ALx which is the $Almha$ method run for up to x iterations,
- ACx which is the ACT method run for up to x iterations.

They are evaluated and compared in the following section.

4   Empirical Evaluation

We perform three series of experiments. In the first series we tune the $Almha$ algorithm. In the second series we evaluate the algorithms for a set of small sized problems and in the third series we compare them for large sized problems. Due to random nature of some of the data, all the experiments were repeated five times. Presented are the averages over those five repetitions. For all experiments a time limit of 30 minutes was set. Lack of a result, either due to time limit or due to the inability of the algorithm to provide one is marked accordingly (by number of failed repetitions or by "−" if all of them failed). Finally, due to random nature of the data, some of the generated problems were unfeasible for all of the algorithms tested. Those cases were omitted from the results and the experiment was repeated.

Series 1. The first series is performed for the algorithm Almha exclusively. The goal is to fine tune the parameters of the method. The following problem data is used

- $J = 10$,
- $I = 10$,
- $K = 2$,
- $\eta_{i,j,k} \in [0, 10]$,
- $e_{i,j,k} \in [0, 10]$,
- $F_i \in [0, \sum_{i \in \mathbf{I}, j \in \mathbf{J}} \max_{k \in \mathbf{K}} e_{i,j,k}]$,
- $E = 10$.

Intervals represent integers randomly selected from an uniform distribution. Nominal parameters of the algorithm are $base = 10$, $pert = 0$, $\alpha_+ = \beta_+ = 2$, $\alpha_* = \beta_* = 1$, $w_{inc} = 0$.

The first experiment was done to estimate the number of iterations sufficient for further tuning experiments. The quality coefficient used was the $P \triangleq \frac{Q}{Q_*}T$ where Q is the quality obtained by the algorithm, Q^* is the optimal solution and T is the algorithm execution time. Results are presented in Table 1.

Table 1. (Series 1) Quality P for varying N for the Almha algorithm

N	1	5	10	50	100	500	1000
P	-	-	-	27	13	60	108

The number of iterations set for the reminder of tuning was $N = 100$. The results are presented in the Table 4. The quality coefficient used for the presentation of the results is the relative quality $R \triangleq \frac{Q}{Q_*}$.

Table 2. (Series 1) Quality R for varying parameters for the Almha algorithm

$base$	1	3	5	10	30	$\alpha_+.\beta_+$	0	0.1	0.5	1	2	5
R	1.53	2.30	1.4	1.35	1.39	R	2.18	1.33	2.67	1.47	1.41	1.31

$\alpha_*.\beta_*$	1	1.1	1.5	2	5	w_{inc}	0	0.1	0.5	1	2
R	1.24	1.45	1.54	1.55	1.46	R	1.36	1.42	7.38	12.26	13.39

From the results visible in the Table 2 we select parameters $base = 10$, $\alpha_+ = \beta_+ = 5$, $\alpha_* = \beta_* = 1$, $w_{inc} = 0$ for further evaluations of this algorithm.

Series 2. The second series of experiments was performed in order to compare selected algorithms for solving the stated MTA problem on small instances with small energy requirements. The problem data remains as in the first series

Table 3. (Series 2) Quality Q for varying I.

I	A2f	A2e	AL50	AL100	AL500	AL1000	AC1	AC5	AC50	AC500
10	27.6	31.4	159.4	32.4	30	30.2	28	28	26.4	26.4
20	29	32	177	39.6	32	29.8	27.8	27.8	26.8	26.8
30	26	29.8	135.4	29.8	30.2	32.2	25	25	24.4	24.4
40	27.4	31.2	203.6	35.6	37.6	32.6	28	28	26	26
50	29.6	32.4	175.6	79.6	33.6	32.2	28.6	28.6	27.8	27.8

Table 4. (Series 2) Execution time $T[s]$ for varying I.

I	A2f	A2e	AL50	AL100	AL500	AL1000	AC1	AC5	AC50	AC500
10	0.13	0.16	0.92	1.8	8.94	17.94	0.21	0.2	1.97	195.34
20	0.13	0.16	0.91	1.8	8.89	17.8	0.2	0.2	1.95	76.68
30	0.13	0.16	0.91	1.8	8.9	17.78	0.21	0.2	1.96	56.03
40	0.13	0.16	0.92	1.81	9.02	17.87	0.21	0.2	1.98	130.06
50	0.13	0.16	0.91	1.8	8.88	17.77	0.2	0.2	1.98	68.81

Table 5. (Series 2) Quality Q for varying J.

J	A2f	A2e	AL50	AL100	AL500	AL1000	AC1	AC5	AC50	AC500
10	27.6	31.4	159.4	32.4	30	30.2	28	28	26.4	26.4
20	54.8	67.2	730.66 (2)	326.4	90.8	117.8	53.8	53.8	51.8	-
30	78.8	94.4	-	887.2	112.8	111	80	80	77.6	-

Table 6. (Series 2) Execution time $T[s]$ for varying J.

J	A2f	A2e	AL50	AL100	AL500	AL1000	AC1	AC5	AC50	AC500
10	0.13	0.16	0.92	1.8	8.94	17.94	0.21	0.2	1.97	195.34
20	0.26	0.38	2.84 (2)	5.61	28.2	56.27	1.32	1.31	13.17	-
30	0.4	0.65	7.2	14.35	70.86	141.46	4.65	4.67	47.37	-

with the exception of I and J which change as follows $I \in \{10, 20, 30, 40, 50\}$, $J \in \{10, 20, 30\}$. Results are presented in Tables 3, 4, 5 and 6.

Series 3. Final series of experiments compares the algorithms on large instances for higher energy requirements. The problem data remains as in the first series with the exception of I and J which change as follows $I \in \{10, 30, 50, 100, 300, 500\}$, $J \in \{10, 20, 30\}$ and $E = 50$. For the experiments with varying J we set $I = 100$. Results are presented in Tables 7, 8, 9 and 10.

The general conclusion from the results presented in Tables 3, 4, 5 and 6 is that for small sized instances, all of the tested algorithms provide solutions that

Table 7. (Series 3) Quality Q for varying I.

I	A2f	A2e	AL50	AL100	AL500	AL1000	AC1	AC5	AC50
10	250.6	282.6	355 (4)	299.75 (1)	299.75 (1)	309.6	240.6	240.6	237
30	109	129.8	-	883	138.6	142.8	103	103	102.2
50	83.2	91.4	-	-	132.6	113.2	81.2	81.2	80.8
100	64.8	69.8	-	-	-	-	64	64	-
300	52	53	-	-	-	-	-	-	-
500	50	50.2	-	-	-	-	-	-	-

Table 8. (Series 3) Execution time $T[s]$ for varying I.

I	A2f	A2e	AL50	AL100	AL500	AL1000	AC1	AC5	AC50
10	0.16	0.25	0.96 (4)	1.82 (1)	9.13 (1)	17.99	0.14	0.14	2.11
30	0.5	0.67	-	14.19	70.69	141.83	4.19	4.21	42.76
50	0.9	1.1	-	-	271.17	544.92	23.27	23.27	236.28
100	1.99	2.33	-	-	-	-	202.59	203.14	-
300	6.82	7.78	-	-	-	-	-	-	-
500	12.53	14.24	-	-	-	-	-	-	-

Table 9. (Series 3) Quality Q for varying J.

J	A2f	A2e	AL50	AL100	AC1	AC5	AC50
10	65.2	68.6	-	-	64.6	64.6	64.6
20	123.2	136.6	-	-	123 (1)	123 (1)	-
30	190	219.6	-	-	-	-	-

Table 10. (Series 3) Execution time $T[s]$ for varying J.

J	A2f	A2e	AL50	AL100	AC1	AC5	AC50
10	1.90	2.25	673.14	1368.5	204.99	204.62	-
20	3.81	5.19	-	-	1277.28 (1)	1279.28 (1)	-
30	5.79	9.14	-	-	-	-	-

do not significantly differ in quality given sufficient execution time. Increasing the number of executors alone does not significantly decrease the quality nor does it increase the execution time. This is unsurprising, considering that the low energy requirement allows for a small number of executors to be sufficient. Opposite to that is the influence of a changing number of tasks. As the number of tasks grows the quality and the execution time increase as well. Furthermore, a low number of iterations (50) for the Almha algorithm becomes insufficient to handle most instances. More iterations and therefore a significantly longer execution time is required to obtain a solution. Similar results can be noticed for ACT although their extent is smaller.

For problems of increased size and complexity illustrated in Tables 7, 8, 9 and 10 it is clear that the Tabu Search based Almha method does not provide acceptable solutions within the time limit for most of the tested cases. ACT performs better but still fails to provide any solutions for cases with over 100 executors. It is worth noting that for a low number of 10 executors the solutions provided by ACT are, on average, better than those yielded by the A2f and A2e algorithms.

Relatively weak performance of the general ACT method can be attributed to a large number of decision variables which result in a need to solve integer programming problems of proportional size. Even worse efficiency of the Almha algorithm is a result of considering only single variable flip moves at every iteration while maintaining a neighbourhood that is large and thus costly to evaluate. Algorithms A2f and A2e, which are inherently based on decomposition, need to solve significantly less complex problems at a time. Of the two tested dedicated methods, the A2f performed, on average, better than the A2e, providing a better result and finishing the execution faster.

5   Conclusions

This paper contains an evaluation of selected solution algorithms for a problem of task allocation in a multi-robot system with multi-modal task execution. Performed comparison has validated the notion that direct application of existing algorithms for solving the more general knapsack/covering problem does not yield satisfactory results outside of small-scope problems. It is therefore imperative to work on improving the existing solution algorithms.

References

1. Arntzen, H., Hvattum, L., Lokketangen, A.: Adaptive memory search for multi-demand multidimensional knapsack problems. Comput. Oper. Res. **33**, 2508–2525 (2005). Elsevier
2. Correll, N., Martinoli, A.: Multirobot inspection of industrial machinery. IEEE Robot. Autom. Mag. **16**, 103–112 (2009)
3. Maza, I., Ollero, A.: Multiple UAV cooperative searching operation using polygon area decomposition and efficient coverage algorithms. In: Alami, R., Chatila, R., Asama, H. (eds.) Distributed Autonomous Robotic Systems 6, pp. 221–230. Springer, Heidelberg (2007)
4. Melodia, T., Pompili, D., Akyildiz, I.: Handling mobility in wireless sensor and actor networks. IEEE Trans. Mob. Comput. **10**(2), 160–173 (2010)
5. Hojda, M.: Task Allocation with Energy Constraints in Multi-Robot Systems. Aktualne problemy automatyki i robotyki, pp. 796–805. Akademicka Oficyna Wydawnicza EXIT (2014)

6. Hojda, M.: Multi-robot task allocation algorithms in systems with energy constraints. Automatyzacja procesw dyskretnych : teoria i zastosowania, pp. 105–113. Wydawnictwo Pracowni Komputerowej Jacka Skalmierskiego (2014)
7. Hoshino, S., Seki, H., Naka, Y.: Development of a exible and agile multi-robot manufacturing system. In: Proceedings of 17th World Congress of The International Federation of Automatic Control, pp. 15786–15791 (2008)
8. Hvattum, L., Arntzen, H., Lokketangen, A., Glover, F.: Alternating control tree search for knapsack/covering problems. J. Heuristics **16**, 239–258 (2008). Springer
9. Pham, V., Juang, J.: An improved active SLAM algorithm for multi-robot exploration. In: Proceedings of SICE, pp. 1660–1665 (2011)
10. Shah, R., Sumit, R., Sushant, J., Brunette, W.: Data mules: modeling and analysis of a three-tier architecture for sparse sensor networks. Ad Hoc Netw. **1**(2), 215–233 (2003)
11. GNU Linear Programming Kit. https://www.gnu.org/software/glpk/

The Methodology of Multiple Criteria Decision Making/Aiding as a System-Oriented Analysis for Transportation and Logistics

Jacek Żak[✉]

Poznan University of Technology, 3 Piotrowo street, 60-965 Poznan, Poland
jacek.zak@put.poznan.pl

Abstract. The paper presents the principles of **Multiple Criteria Decision Making/Aiding (MCDM/A)** and its application in **transportation and logistics**. The Methodology of MCDM/A is presented as a universal, system-oriented tool of analyzing complex objects, problems and systems, including transportation/logistics decision problems. The article describes basic notions and concepts associated with MCDM/A methodology. It presents the rules, procedures and paradigms of MCDM/A and points out their resemblance to System Analysis. The author presents **the process of Multiple Criteria Decision Making/Aiding** and explains all its phases, bodies involved and generated results. The theoretical background is supported by a **real world case study** focused on the analysis, mathematical modeling and solution procedure of a multiple criteria crew sizing problem for a freight transportation and delivery process. The decision problem is solved in a **two-stage procedure** composed of: generation of a set of Pareto optimal solutions by a heuristic method and their review and selection of the compromise solution by an interactive method with graphical facilities (LBS method). The case study shows practical applicability of the presented approach.

Keywords: Multiple criteria decision making/aiding · System analysis · Transportation and logistics

1 Introduction

The methodology of Multiple Criteria Decision Making/Aiding (MCDM/A) [4, 10, 15, 20, 21, 24] has become very popular in the recent years. It has been applied in many areas, including transportation and logistics [4, 18, 24–26]. Many multiple criteria computational procedures and software tools have been produced to assist the decision makers (DMs) in solving complex, multiple criteria decision situations. At the same time the MCDM/A methodology is still applied intuitively in many areas and in some cases it is not used in an fully appropriate way. This article aims at showing the major concepts and principles of MCDM/A and bring it closer to the reader as a very powerful, universal and system – oriented methodology. The paper has an instructional character and thus, defines basic notions/terms used in MCDM/A and gives them transportation – logistics interpretation. It presents the application of MCDM/A methodology in the fields of transportation and logistics.

© Springer International Publishing AG 2017
J. Świątek and J.M. Tomczak (eds.), *Advances in Systems Science*, Advances in Intelligent Systems and Computing 539, DOI 10.1007/978-3-319-48944-5_25

The author undertakes a challenge to prove that MCDM/A is a system oriented methodology and demonstrates its features that make it similar to System Analysis (SA). The rules and paradigms of MCDM/A are described and compared with core principles of System Analysis (SA). The case study shows the practical applicability of the MCDM/A to analyzing and solving a complex transportation – logistics decision problem.

Despite the variety of proposed definitions many authors would agree that **logistics** is [3, 19]: a field of study, an industry or a function that deals with planning, implementation and control of the efficient flow (forward and reverse) and storage of goods (including raw materials, components and finished products), between their points of origin (fields, mines, sites) and their destinations (points of consumption) in order to satisfy customers' requirements and expectations. Logistics controls the transformation of raw materials into semi-finished products, processing of the latter and their delivery in the form of finished goods to consumers for final consumption and/or utilization [3]. Logistcs involves several major activities such as: material processing (manufacturing), warehousing, material handling and control (loading and unloading, packaging, inspecting), material shipping and transporting, material monitoring and returning to suppliers. Logistics manages and controls three major flows of the supply chain or network, i.e. material flow, information flow and cash flow [3].

Transportation is also [23, 24] a field of study, an industry or an activity that focuses on the movement of people (passengers), animals and/or goods (freight) from one location (origin) to another (destination) by certain means of transport (vehicles) and with the application of a certain infrastructure. Based on the character of the environment in which transportation is carried out it is divided into the following modes: air, rail, road, water, cable, pipeline, space. As far as the carried load is concerned transportation can be generally classified as passenger and freight transportation [23, 24]. Thus, when the freight movement is concerned transportation can be considered as a component of logistics. As opposed to freight transportation, the passenger transportation is beyond traditional understanding of logistics. The most popular category of transportation, both in passengers' and freight movements, is a sub-kind of land/surface transportation, called road transportation.

To carry out their tasks and functions **transportation** and **logistics** require **specific/specialized resources**, such as:

- transportation/logistics infrastructure (warehouses – logistics centers, depots, hubs, transportation terminals, harbors, airports, customs agencies, border crossings, roads, railways, logistics equipment – storage equipment, feeders, belt conveyers);
- human resources – transportation and logistics managers, transportation engineers and planners, dispatchers, drivers, pilots, cabin crew, terminal staff, maintenance staff, warehouse staff;
- moving fleet (vehicles) – planes, container ships, cruisers, trucks, vans, tractors and trailers, coaches, buses, trams, trains, fork lifts, cranes.

These components are properly utilized thanks to the application of certain business processes characteristic for transportation and logistics activities and coordinated through the application of selected organizational rules [3, 24].

Both fields are very **complex, volatile and dynamically changing.** They are strongly influenced by random phenomena and thus transportation and logistics activities are carried out under uncertainty. A strong interaction between transportation/logistics infrastructure, human resources and fleet is observed in transportation and logistics systems. In many cases the performance of transportation/ logistics activities/functions requires that all these components are well coordinated and mutually matched. In many cases both transportation and logistics activities require that the expectations and interests of many stakeholders are satisfied. The major bodies involved and interested in the results of transportation and logistics activities are as follows: customers (business units, individual passengers); operators (transport operators, logistics service providers), local communities (residents), road users, local authorities, governmental institutions, suppliers.

The existence of many stakeholders and the complexity of logistics and transportation requires that analysis of transportation and logistics processes, systems and activities, evaluation of certain solutions for these areas, design and assessment of original transportation/logistics concepts requires **a variety of measures and characteristics.** Transportation and logistics need to be assessed from different perspectives, including, among others: economic, technical, social, safety-oriented and environmental aspects. In many cases the objectives of various stakeholders are in conflict and the existence of trade-offs is transparent. Thus, the search for a balanced, sustainable and compromise solution is necessary.

In both the above mentioned areas there are **many decision problems** that require a solution and/or searching for the most desirable (rational) decision. The typical transportation-logistics problems include: finding a location of an infrastructure (terminals, warehouses), selecting routes and designing transportation corridors, fleet management (composition, routing, replacement), design of logistics centers, inventory management, design and adjustment of the transportation-logistics portfolio, management of transportation – logistics processes. Based on his research J. Żak [24] proves that **these decision problems have a multiple criteria character.**

For all the above mentioned reasons the necessity to apply Multiple Criteria Decision Making/Aiding (MCDM/A) in transportation and logistics becomes obvious and very transparent. MCDM/A is a field of study, originating from Operations Research (OR), which aims at giving the DM some tools/methods in order to enable him/her to advance in solving complex decision problems in which several – often contradictory – points of view must be taken into account [20, 21, 24]. The methodology of Multiple Criteria Decision Making/Aiding (MCDM/A) can help the Decision Makers (DM-s) in analyzing and solving complex transportation – logistics decision situations/problems, in assessing innovative transportation – logistics projects, concepts and solutions, in analyzing trade-offs and balancing conflicting interests associated with the operations of certain transportation-logistics processes and systems, in searching for the most desired, compromise transportation – logistics decisions.

The major objective of this paper is to demonstrate the values of Multiple Criteria Decision Making/Aiding (MCDM/A) for transportation and logistics and its applicability in these fields. MCDM/A is presented as a system – oriented methodology and its practical application is demonstrated in a comprehensive case study. The basic concepts of MCDM/A are defined and its transportation – logistics interpretation is given.

The article refers to a traditional MCDM/A-based paradigm of solving multiple criteria decision problems and compares it with the universal procedure of System Analysis.

The paper is composed of 5 sections. The introduction includes the background of the presented analysis, the definition of major notions and concepts and the research objectives of the paper. In Sect. 2 the MCDM/A is presented as a system – oriented methodology and its major notions and concepts are defined. Section 3 is focused on the presentation of real world transportation – logistics case study in which the application of MCDM/A methods is demonstrated. The system – oriented analysis and the results of computational experiments are shown. Section 4 includes conclusions. The paper is completed by a list of references.

2   Basic Concepts and Notions of Multiple Criteria Decision Making/Aiding (MCDM/A). MCDM/A as a System – Oriented Methodology

As indicated in Sect. 1 Multiple Criteria Decision Making/Aiding (MCDM/A) focuses its efforts on developing methods that can assist the DM and support him/her in solving multiple criteria decision problems. In contrast to the classical techniques of operations research, multiple criteria methods do not yield "objectively best" solutions, because it is impossible to generate such solutions which are the best from all points of view, simultaneously. In these circumstances a notion of a compromise solution [20, 21, 24] that takes into account both the trade-offs between criteria and the DM's preferences, is much more rational.

In each formulation of the multiple criteria decision problem there are always two major components, i.e. a set of actions/variants/solutions A and a consistent family of criteria F. The set of A can be defined directly in the form of a complete list or indirectly in the form of certain rules and formulas that determine feasible actions/variants/solutions, e.g. in the form of constraints [20, 21, 24]. The consistent family of criteria F should be characterized by the following features [15, 21, 24]:

- it should provide a comprehensive and complete evaluation of A,
- each criterion in F should have a specific direction of preferences (minimized – min or maximized – max) to adequately indicate the global preferences and expectations of the DM,
- each criterion in F should not be redundant with other criteria in F; the domain of each criterion in F should be disjoint with the domains of other criteria.

The definition of A and F is a critical component of each multiple criteria analysis.

There are three major categories of the above mentioned multiple criteria decision problems, i.e.:

- choice problems,
- sorting problems,
- ranking problems.

While analyzing the decision situation it is necessary to recognize what features it represents and thus which category of the decision problem it belongs to. Choice problems are focused on determining the best subset of actions/variants/solutions in A according to F. They are usually formulated as multiple criteria mathematical programming (optimization) problems. Sorting problems consist in dividing A into subsets representing specific classes of actions/variants/solutions, according to concrete classification rules. Ranking problems are such decision situations in which the DM tends to rank actions/variants/solutions in A from the best to the worst, according to F. The decision problem considered in this paper in the case study (Sect. 3) is a multiple criteria choice (optimization) problem.

Multiple criteria decision problems require specific procedures, methods and decision-making/aiding tools to be solved. These methods can be classified according to several criteria, including [21, 24]:

- the overall objective of the decision method correlated with the category of the decision problem,
- the moment of the definition of the DM's preferences,
- the manner of the preference aggregation.

Based on the first classification criterion one can distinguish the following categories of MCDM/A methods [21, 24]:

- multiple criteria choice (optimization) methods (e.g. LBS, Steuer Procedure, Topsis),
- multiple criteria sorting (classification) methods (e.g. Electre Tri, 4eMka),
- multiple criteria ranking methods (e.g. Electre III/IV, AHP).

With respect to the second division criterion three categories of methods are identified [21, 24]:

- methods with an a'priori defined preferences (e.g. Electre methods, Promethee I and II, UTA, Mappac, Oreste),
- methods with an a'posteriori defined preferences (e.g. PSA method),
- interactive methods (e.g. GDF, SWT, Steuer Procedure, STEM, VIG, LBS).

According to the third classification criterion one can distinguish [21, 24]:

- the methods of American inspiration, based on the utility function [10] (e.g. AHP, UTA),
- the methods of the European/French origin, based on the outranking relation [15, 16] (e.g. Electre methods, Promethee I and II).

These methods are computer-based mathematical procedures that solve the above mentioned categories of the multiple criteria decision problems. They are usually developed as user-friendly, interactive Decision Support Systems [23] that assist the DM in solving the considered decision problems.

Based on the works of B. Roy [15], J. Figueira et al. [4] and J. Żak [24] one can distinguish five major stages of the solution procedure of the multiple criteria decision problems:

- Investigation of the decision problem considered and its verbal description. Recognition of the category of the decision problem, definition of the major objective of the analysis, the DM and stakeholders.
- Problem structuring and mathematical modeling. Analysis of the parameters characterizing the decision problem, collecting data, definition of the set of variants/feasible solutions A and construction of the consistent family of criteria F. Modeling and aggregating the DM's and stakeholders' preferences (synthesis of the global model of preferences).
- Review, evaluation and selection of the appropriate methods and algorithms to solve the considered decision problem. Matching the specific characteristics of the decision problem with appropriate multiple criteria method.
- Carrying out a series of computational experiments. Solving the decision problem with an application of the global model of preferences and a selected MCDM/A method. Review and analysis of the generated results. Sensitivity analysis – investigation of the stability of generated results.
- Selection of the most desired solution/variant/action considered to be a compromise solution. Possible implementation of this solution in the real world.

The above described solution procedure resembles in many aspects the paradigms proposed by such fields as: system analysis [11, 12] and operations research [6]. In the author's opinion the following features transparently make the multiple criteria decision making/aiding (MCDM/A) process similar procedure to a system oriented analysis:

- It is focused on a comprehensive analysis of a certain object/organization/system/problem, including its internal components and connections with external world. It involves the definition of the major players of the decision process and the complete evaluation of the organization/system/object/problem. Similarly to system analysis in the MCDM/A process the real world is investigated, modelled and structured.
- MCDM/A looks at the object/organization/system/problem at stake from a broad perspective. It applies system approach to problem analysis and solution. It investigates the object/problem from different perspectives and considers its various aspects (technical, economic, social, environmental, political).
- As a result it applies different measures (criteria) to evaluate the object/problem and thus searches for solutions that attempt to resolve conflicts between different components of the organization/system and satisfy its overall objectives.
- It requires a variety of skills and competences and thus a team approach to carry out a complete study of a certain decision problem or the full analysis of a certain object. Mutual interactions between major players of the decision process (DM, analyst and stakeholders) are required to reach the final result.
- MCDM/A uses scientific methods and rigid, systematic way of thinking to investigate the object/problem of concern; as a result it generates a transparent and well-structured picture of the considered situation/object.
- It attempts to find the most desirable, satisfactory solution for the object/problem under consideration; as in system approach it tends to generate the solutions that are

satisfactory for different entities and assessed/perceived from different perspectives (points of view).

Solving a multiple criteria decision problem consists in finding a(n) action/variant/solution x in A (in the decision space for multiple objective optimization problems) which image in the objective (criteria) space z^x generates the most desired values of the objective function $f(x)$ [23, 27]. The image of a solution x in the objective space is a point $\mathbf{z}^x = [z_1^x, \ldots, z_J^x] = \mathbf{f}(\mathbf{x})$, such that $z_j^x = f_j(\mathbf{x})$, for $j = 1, \ldots, J$. Important notions in MCDM/A context are: dominance relation and Pareto-optimal (efficient) solution that refer to optimality in a multiple objective sense. Point $\mathbf{z}^1$ dominates $\mathbf{z}^2$, $\mathbf{z}^1 \succ \mathbf{z}^2$, if $\forall j \, z_j^1 \geq z_j^2$ and $z_j^1 > z_j^2$ for at least one j. Solution $\mathbf{x}^1$ dominates $\mathbf{x}^2$ if the image of $\mathbf{x}^1$ dominates the image of $\mathbf{x}^2$ in the objective space. A solution $\mathbf{x} \in A$ is Pareto-optimal (efficient) if there is no $\mathbf{x}' \in A$ that dominates $\mathbf{x}$. Point being an image of a Pareto-optimal solution in the objective space is called nondominated. The set PO of all Pareto-optimal solutions is called the Pareto-optimal set. The image ND of the Pareto-optimal set in the objective space is called the non-dominated set or Pareto front [20, 21, 24].

While searching for the best compromise solution in multiple criteria decision problems the DM can be interested in knowing what is the range of the criteria values in the Pareto-optimal set. The dispersion of the criteria values is given by two virtual points, called an ideal and a nadir point. The ideal point is represented by coordinates $\left(z_j^* \right)$ characterized by the maximum values of all criteria j $z_j^* = Max f_j(x)$, while the nadir point is represented by coordinates $\left(z_j^{\wedge} \right)$ characterized by the minimum values of all criteria $z_j^{\wedge} = Min \, f_j(x)$.

The MCDM/A methodology [4, 15, 20, 21, 24] clearly identifies major participants of the decision making/aiding process, such as: the decision maker (DM), the analyst and the interveners (stakeholders) and describes their roles in this process. The DM (an individual or a group of individuals) defines the objectives of the decision process, expresses preferences and finally evaluates the generated results. The analyst, who is external to the decision problem, handles the decision making/aiding process and supports the DM during its course. His/her role is to construct a decision model and suggest the most appropriate tools and methods in order to advance the DM in solving the decision problem. The analyst explains to the DM the consequences of certain actions/solutions and finally recommends the most desired one. The interveners, frequently called stakeholders, represent different individuals, organizations and groups that are interested in generating a rational solution to a certain decision problem. In many decision situations they express their own preferences and present concrete stand points that should be taken into account while analyzing the decision problem. Final decision usually has a strong impact on satisfaction of the interveners'/stakehoders' interests.

The above considerations lead us to a universal/generic paradigm of the multiple criteria decision making/aiding process, presented in Fig. 1. As described above it involves a comprehensive analysis of the real world, which is the object of investigations, with a special emphasis on a specific decision problem that arises in the real

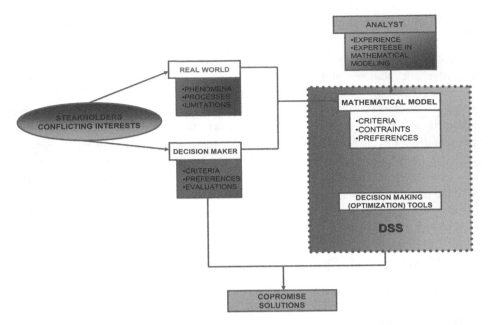

Fig. 1. The generic paradigm of the system – oriented, multiple criteria decision making/aiding process

world. Thus, it represents transparent features that make it very similar to a system analysis based on the careful investigation of internal processes of an object and its relationships with the external environment [11, 12]. The multiple criteria decision making/aiding process is based on the mutual interactions between its major participants. Thus, we observe here the two-way relationships between the decision maker (DM) and stakeholders (interveners), the DM and Analyst as well as the DM and the Decision Support System (DSS) based on the MCDM/A methods. As a result of the multiple criteria decision making/aiding process the compromise solutions are generated.

As presented in Fig. 1 the MCDM/A - based decision process is strongly supported by computer-based systems. These systems have an interactive character and help the DM to generate final, compromise solutions in an iterative manner. They never replace the DM and do not make the final decision instead of him, her or them. Their role is to assist the DM in the whole decision – making process, present a variety of solutions and help him/her/them to investigate those solutions. The computer – based decision support systems help the DM to express and model his/her or their preferences and better understand the considered decision problem.

For each specific transportation/logistics problem the above described generic scheme of the MCDM/A process is customized to the concrete transportation/logistics environment and specific character of the considered decision situation. The described approach has been applied in solving a transportation/logistics decision problem, presented as a case study and analyzed in Sect. 3 of this paper. The customization of this generic/universal paradigm is presented there.

3   Application of MCDM/A in Transportation and Logistics. Case Study on Multiple Objective Optimization of the Crew Size for a Transportation Process [24]

3.1   Major Characteristics of the Considered Decision Problem and the Applied Solution Procedure [24]

The analyzed decision problem consists in optimizing the crew size for a complex transportation process carried out in a large international company CP SA based in Warsaw, Poland. The CP company manufactures and delivers to the market household chemical products, detergents and cosmetics. The transportation process corresponds to the physical distribution of finished goods from manufacturing plants to the retail stores distributed all over the country. It involves the transfer of freight between two manufacturing sites of the CP company (so called "Trunking" = Shipments between warehouses) and the delivery of the final products (finished goods) to customers (retail stores). The "trunking" operations result from different product portfolios of CP company manufacturing plants, located in Warsaw and Wroclaw. Since customers' orders range in scope, transfers of goods between warehouses are required and the consolidation of shipments is carried out. The freight is carried by a fleet of 25 MAN tractors with trailers featured by a capacity of 33 Euro pallets each.

The decision problem has resulted from the analysis of efficiency of the above described goods' transportation and delivery process. This analysis has revealed that the process is characterized by low profitability, resulting from its increasing labor intensity and associated high operating costs. It has been proved that employees' salaries constitute a large portion (almost 70 %) of the generated operational costs. Top management of the company was also concerned about certain customer's complaints regarding service quality and reliability. At the same time the employees complained about monotonous work and difficult working conditions. They realized, however, that under the economic crisis the demand for the company's products substantially decreased. The negotiations between company's top management and representatives of the employees have raised the issue of rationalizing the crew size and labor intensity of certain activities of the delivery process. Both parties agreed that work quality and efficiency should be improved and concrete actions should be undertaken to enhance employees' responsibility for their work. The employees expected to be more empowered and given more flexibility as far as job assignment is concerned. As a consequence of these considerations the goods transportation and delivery process has been thoroughly analyzed to investigate the above mentioned questions and test how the improvement can be reached.

In the above described circumstances the MCDM/A methodology has been applied to analyze the transportation and delivery process and solve the resulting crew sizing problem. The company's top management hired a team of academic consultants (Analysts) to carry out the above described analysis. The universal paradigm of the MCDM/A process, presented in Fig. 1, has been adapted to specific features of the analyzed decision problem. All five phases of this paradigm, described in Sect. 2 have been applied.

3.2   Investigation of the Decision Problem and Its Verbal Description [24]

In the first phase a comprehensive process analysis has been carried out to define the operations of the transportation and delivery process, their scope and certain interactions between them. Parallel and sequential layouts of operations have been discovered, resulting in a complete graphical representation of the analyzed process. The following 15 operations have been defined in the transportation process: PT1 – vehicle preparation and allocation; PT2 – freight loading; PT3 – preparation and printing transport documentation (documents WZ and EP); PT4 – generating and printing the invoice; PT5 – collection of transportation documents (documents WZ and PC); PT6 – parking the loaded vehicle/trailer; PT7 – collection of other transportation documents (complete set); PT8 – freight delivery, driving; PT9 – freight unloading and confirmation of the freight delivery; PT10 – pick-up of raw materials and packing materials; PT11 – delivery of raw materials and packing materials, driving; PT12 – return of the vehicle; PT13 – PT14 – print out of final transportation documents (TL and M) and PT15 – vehicle inspection.

Figure 2 presents the idea of process analysis in the abbreviated form. For scheduling purposes those operations can be further divided into sub-operations and smaller tasks. The process is carried out by a crew of 36 employees and its daily labor costs are equal to 9400 PLN[1]. The analysis let us also define existing bottlenecks in the analyzed process as well as parameters and characteristics of particular operations, such as: capacity, labor intensity, cost, efficiency, etc. The detailed recognition of the transportation process facilitated a precise verbal description of the crew sizing problem. It has been determined which tasks (operations) should be carried out in specific time intervals and what kind of resources (vehicles) are required to fulfill them.

Fig. 2. The general idea and graphical representation of the freight transportation process

In this phase the multiple objective character of the decision problem has been proved and contradictory interests of major stakeholders have been confirmed. The top management of the company shared the role of the DMs with the company's share-holders. The interests of four groups of stakeholders, including: customers, employees, managers and share-holders have been taken into consideration. The

---

[1] PLN – Polish New Zloty - Polish currency; 1 Euro = 4 PLN.

definition of the optimal number of employees has been associated with the assignment and scheduling of duties (operations of the process) to the crew members.

The first phase has been concluded by the interviews with DMs and stakeholders to define their expectations and aspirations. It has been discovered that company's shareholders and managers are interested in cost reduction, company's employees do not want to lose their jobs and expect good working conditions, while customers want to receive high quality product delivered in a reliable manner. It has been proved that job assignment and scheduling may have an impact on the overall costs of the products' transportation and delivery process and overall quality of the service.

3.3   Problem Structuring and Mathematical Modeling [24]

In the second phase the mathematical model of the decision problem has been formulated as a multiple criteria integer programming problem. The components of the model are as follows:

Decision Variables. A binary decision variable has been proposed:

$$x_{ijkl} = \begin{cases} 1 & \text{if employee } k \text{ carries out job } i \text{ at a moment } j \text{ on post } l, \\ 0 & \text{if otherwise,} \end{cases} \tag{1}$$

where: i denominates jobs/tasks ($i = 1, 2, 3,...I$), j defines time intervals $\Delta t = 15$ min ($j = 1, 2, 3,...J$), k is the index of employees ($k = 1, 2, 3, ... K$), l denominates work – stands, vehicles or specific locations where work is performed ($l = 1, 2, 3,...L$). All decision variables constitute a 4 – dimensional, binary decision matrix $X = I \times J \times K \times L$.

Criteria. Based on the defined DMs' and stakeholders' expectations a family of optimization criteria, that satisfies the interests of particular groups of stakeholders, has been defined. Four optimization criteria, including: 1. Number of employees (LP); 2. Efficiency and quality of work (EP); 3. Job dispersion/ differentiation (PZP); 4. Total costs (KRP), have been proposed. Criteria 1 and 4 (minimized) represent the interests of company's management and share-holders. Criterion 2 (maximized) is focused on the customers' satisfaction, while criterion 3 (maximized) guarantees the fulfillment of the employees' aspirations. The consistency of the family of criteria has been investigated. It has been proved that most important aspects of the decision problem are covered by this family of criteria and interests of major stakeholders are represented by its components. All doubts regarding strict correlation between criteria LP and KRP have been dispelled. It has been noticed that cost (KRP) reduction can be achieved not only through the decrease of the number of employees (LP) but also through their reassignment. Even for a fixed crew size the reassignment of employees has resulted in cost reduction. The proposed criteria are characterized below:

1. **Number of employees – LP.** This criterion defines the minimum number of employees required to satisfy the demand for jobs in specific time intervals in the whole time horizon (e.g. day, week, month). The demand for concrete jobs is strongly associated with the transportation schedule. This criterion has a fundamental significance for the crew sizing problem and it is expressed as follows:

$$Min\ LP = \sum_k P_k,$$ (2)

where: P_k is the index that defines whether an employee has a job assignment ($P_k = 1$) or not ($P_k = 0$).

$$P_k = \begin{cases} 1 & \text{if} & \sum_i \sum_j \sum_l x_{ijkl} > 0, & \forall\ k, \\ 0 & \text{if} & \sum_i \sum_j \sum_l x_{ijkl} = 0, & \forall\ k. \end{cases}$$ (3)

2. **Efficiency and quality of work – EP**. This parameter assures the highest quality and efficiency of the goods transportation and delivery process. It is expressed in the following linear form:

$$Max\ EP = \sum_i \sum_j \sum_k \sum_l x_{ijkl} \cdot e_{ijk},$$ (4)

where: e_{ijk} is the efficiency and quality of assigning employee k to job i at a moment (time interval) j; e_{ijk} aggregates such elements as: work quality, employees experience and qualifications, timeliness, reliability and speed of work, cleanness at the work station and appropriate behavior of the employee.

3. **Job dispersion (differentiation) – PZP**. This criterion defines an average number of jobs assigned to an employee. It guarantees that employees perform different categories of tasks in a certain time horizon, which is recommended from a psychological point of view. Based on the modern human resource management concepts [2, 5] the diversity of the duties carried out by a specific employee reduces his/her fatigue and makes the job less monotonous. In some applications this criterion might be also minimized, which would then guarantee the specialization of employees and their focus on concrete categories of jobs. The criterion is expressed in the following way:

$$Max\ PZP = \sum_i \sum_k PP_{ik} \Big/ \sum_k P_k,$$ (5)

where: the numerator in formula (5) is the total number of jobs assigned to particular employees, while the denominator is defined in the same way as in formulas (2) and (3); PP_{ik} is the assignment of employee k to job i and it is calculated as follows:

$$PP_{ik} = \begin{cases} 1 & \text{if} & \sum_j \sum_l x_{ijk} > 0, & \forall\ i, ki \neq k, \\ 0 & \text{if} & \sum_j \sum_l x_{ijk} = 0, & \forall\ i, ki \neq k. \end{cases}$$ (6)

4. **Total costs – KRP**. This criterion minimizes overall costs the freight transportation and delivery process fulfillment. These costs depend on the size and composition of

the crew that carries out particular operations of the process. The criterion is expressed as follows:

$$Min\,KPP = \sum_i \sum_j \sum_k \sum_l x_{ijkl} \cdot k_k, \qquad (7)$$

where: k_k is a unit labor cost for an employee k in time interval Δt; k_k is defined by formula (8).

$$k_k = MP_k \bigg/ \left(\frac{F_k}{\Delta t} \cdot LDP\right) \qquad (8)$$

where: MP_k is a salary of employee k with overheads per time horizon (day, week, month), F_k is a daily number of available standard hours (expressed in minutes) per employee k, LDP is an average number of days in a certain time horizon (day, week, month).

Constraints. The following constraints have been proposed to define the space of feasible solutions:

- Each employee k carries out no more than 1 job at a time interval j,
- The number of jobs carried out at a moment j cannot exceed the number of available work – stations,
- The sum of time intervals assigned to job i must be equal to the labor intensity (in man – hours) of each job i,
- Daily work load for each employee k must be fulfilled within a certain time limit (e.g. shift),
- Some jobs must be assigned to specific employees, based on their qualifications, experience and received certificates.

For this formulation of the decision problem all data has been collected and specific input parameters have been computed.

3.4 Review, Evaluation and Selection of the Appropriate Solution Methods

In this phase a comprehensive review of different multiple criteria procedures has been performed. Strengths and weaknesses of different methods have been defined. It has been concluded that the decision problem has a specific structure and there are no commercial procedures that would allow for its efficient way of solving. It has been checked that the computational complexity of the decision problem is high and thus exact optimization algorithms might fail while being applied to solve it.

In addition, based on the interviews with the DMs it has been noticed that they would like to have a chance to better investigate different solutions and understand the existing trade-offs between criteria. The DMs suggested they would like to explore various regions of the generated solution space and select the most desirable one according to their expressed preferences and aspirations.

Based on these observations and statements the analysts recommended a two phase solution procedures:

– Generation of a good approximation of a large Pareto optimal set by a heuristic method [13]. For this phase a customized procedure has been advised.
– Review and evaluation of solutions based on the preferences and aspirations of the DM and stakeholders. For this phase an interactive approached supported by graphical analysis of solutions has been recommended.

Specialized multiple objective heuristic for the crew sizing problem (Software package PEOPLE). This customized algorithm proposed by J. Zak [24] utilizes specific features of the decision problem. It tends to optimize the number of employees while assigning specific jobs to each of them. The computational procedure can be divided into the following three phases:

1. Generating an initial random solution (nucleus) that satisfies all constraints. The initial solution is generated in the following steps: (1) all qualified employees are sequentially assigned to each job and one of them is selected randomly; (2) time intervals are defined for each job and one of them is selected randomly; (3) those steps are repeated until the initial solution is found; (4) the values of criteria constituting the objective function are calculated for the initial solution.
2. Searching for the improvement of the initial solution. In this phase the algorithm acts as a greedy local search metaheuristic. The procedure makes elementary moves that build the neighborhood of the initial solution, composed of new solutions The elementary move consists in exchanging tasks between two selected employees. The values of criteria for each newly generated solution are calculated and compared with the values of criteria for the initial solution. A newly generated solution is accepted if it dominates an initial solution. Thus, a list of existing Pareto-optimal solutions is constructed. The process is repeated for several iterations. In each of them the newly generated solution is compared with the list of existing Pareto-optimal solutions. If a new solution dominates some of the existing Pareto-optimal solutions they are removed from the list, while the new solution is added to the list.
3. Generating a new initial solution and repeating phase 2. This step is carried out until a stop condition is reached. The stop condition is defined by a number of iterations or by the values of particular criteria, representing the DM's aspirations.

Light Beam Search method. Light Beam Search (LBS) method, proposed by A. Jaszkiewicz and R. Slowinski [8, 9] belongs to efficient, interactive solution procedures for multiple criteria optimization problems. In the iterative process of searching for a compromise solution the computational phases alternate with the decision making phases. In each computational phase, a solution, or a sample of solutions, is selected for examination in the decision phase. The algorithm of the LBS method is composed of the following six steps:

1. Finding the ideal and the nadir points [20, 21] and selecting the reference point that expresses the DM's aspirations.
2. Computing the initial middle point by projecting the reference point onto the non-dominated set. This projection corresponds to finding the Pareto-optimal (efficient) point, which is the best on the current achievement scalarizing function, e.g. Tchebycheff scalarizing function [20] with a reference point used as a benchmark on the set of available points.
3. Defining the DM's preferences and generating the outranking neighborhood of the middle point. The DM's preferences are expressed in terms of indifference (q_j), and optionally, preference (p_j) and veto (v_j) thresholds. In this phase characteristic neighbors (Pareto-optimal solutions) of the middle point are found. Each newly generated solution z is compared with the middle point z^c and tested whether it outranks this point. The outranking relation is used to test the outranking conditions for all available points z.
4. Reviewing, analyzing and evaluating the generated solutions z. In this decision phase the middle point and its neighbors are presented to the DM both numerically and graphically. If the number of neighbors is too large the DM may apply a filtering procedure to generate and review only a sample of solutions. The procedure stops when one of the presented points is satisfactory to the DM.
5. Detailed and extended analysis of the generated solutions z. This step is carried out if the DM is not satisfied with none of the points generated in the previous step. He/she may change then the reference point (redefinition of the aspiration levels for each criterion) or shift the middle point to another point from the neighborhood. Those actions move the neighborhood across the whole set of Pareto-optimal (efficient) solutions and give the DM a possibility to scan it. Those movements resemble the process of illuminating a certain area of the Pareto-optimal set by a focused beam of light from a spotlight in a reference point. Thus, the name of the method is Light Beam Search. The DM may also change his/her preferences (redefinition of thresholds q_j, p_j, v_j).
6. Selecting the most desired solution and accepting it as the best compromise.

Light Beam Search (LBS) method has been used to review and evaluate the generated Pareto optimal solutions. Based on its application the compromise solution has been generated in the iterative and interactive process described above.

3.5 Computational Experiments. Selection of the Compromise Solution [24]

As stated above the solution procedure of the crew sizing problem has been divided into two phases. In the first phase a set of Pareto-optimal solutions has been generated with an application of the above described specialized heuristic procedure, implemented in a software package PEOPLE. The screen of this software package, representing the employees – jobs assignment matrix is presented in Fig. 3. This matrix is one of 2176 Pareto-optimal solutions, generated in the first phase of the computational experiments. The presented portion of the matrix contains the assignment of the operation (PT8) to 11

Time intervals, Δt=15 min - dt(1) - dt(13)

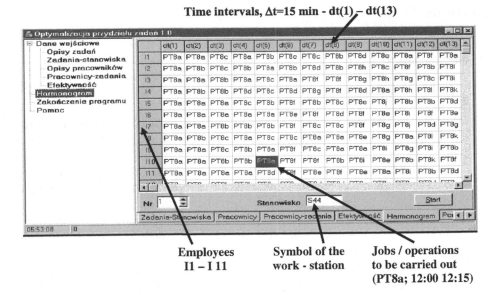

Employees	Symbol of the	Jobs / operations
I1 – I 11	work - station	to be carried out
		(PT8a; 12:00 12:15)

Fig. 3. Employees – jobs assignment matrix. Computational results generated by the heuristic procedure implemented in the software package PEOPLE

employees (I1 to I11) in a time span of 3 h and 15 min [dt(1) to dt(13)]. As one can see each operation can be split into specific sub-operations (a, b, c,…, k).

In the second phase the generated solutions have been reviewed and evaluated with an application of the LBS method. The resultant of this phase is the selection of the most desired, compromise solution. The searching process for the compromise solution has been initiated by the definition of the ideal and the nadir points (see Table 1). The criteria values for these points determine the direction of the search procedure.

The DM defines the reference point that represents his/her aspirations and expectations. Afterwards the computational procedure generates the middle point and its outranking neighborhood. The neighborhood is composed of, so called, characteristic neighbors that are computed based on the definition of the DM's preferences, represented by indifference (q_j), and optionally, preference (p_j) and veto (v_j) thresholds. Figure 4 presents numerical and graphical comparison of the reference and middle points. Graphically, black solid lines represent the values of each criterion (LP – Pracownicy, EP – Efektywnosc, PZP – Zroznicowanie, KRP – Koszty) for the reference point, while shaded bars (rectangles) correspond to those values for the middle point.

Table 1. Extreme values of criteria for the generated set of Pareto-optimal solutions represented by the ideal and the nadir points

| | Criteria | | | |
	LP	EP	PZP	KRP
Ideal point	32	2880.9	1.2	8286.3
Nadir point	39	2746.7	1.0	9124.3

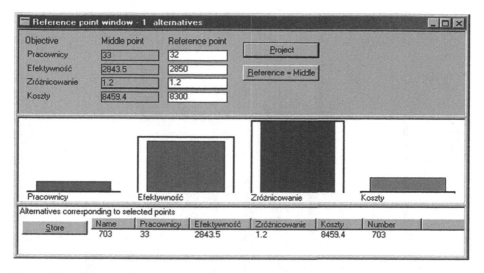

Fig. 4. The comparison between the reference and the middle point and criteria values for the compromise solution 703. Computational results generated by LBS method.

In the analyzed case the DM focuses on the reduction of the number of employees (LP) and high level of efficiency and quality of work (EP). Those criteria constitute major interests of the DM and correspond to the following major objectives of the decision making process: rationalizing of the crew size and reduction of the passengers' complaints regarding the quality and efficiency of the transportation service. In the neighborhood of the middle point the DM searches for the most satisfactory solution. In this interactive decision making process the DM redefines several times the reference point and/or shifts it to another point in the neighborhood. The example of the neighborhood, composed of 26 solutions generated around the middle point 235, is presented in Fig. 5. 13 out of all 26 solutions, denominated by numbers between 194 and 1225, constituting the neighborhood, are presented on the screen. Each solution, e.g.: 194, 496, 830, 1018 or 1225 is featured by its numerical and graphical representation. In the lower part of the screen the colorful bars demonstrate the values of specific criteria and correspond to their numerical equivalents, presented above. Criterion LP – Number of employees (Pracownicy) is presented in red, criterion EP – Efficiency and quality of work (Efektywnosc) is in green, criterion PZP – Job dispersion/differentiation (Zroznicowanie) is in blue, while criterion KRP – Total costs (Koszty) is presented in yellow. For sake of clarity it is important to explain that for solution 827 for instance, the numerical values of criteria are as follows: LP = 34; EP = 2857.9; PZP = 1.1 and KRK = 8788.9.

After several iterations the final compromise has been reached. The compromise solution – number 703 is presented in Fig. 4. It is characterized by the following values of criteria: LP = 33; EP = 2843.5; PZP = 1.2; KRP = 8459.4. Finally accepted compromise solution provides high level of satisfaction on those criteria that represent major interests of the DM. The values of LP and EP for the compromise solution constitute 97 % and 99 % of the most desired values for the respective criteria

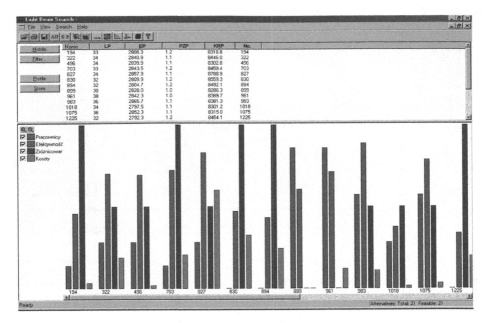

Fig. 5. The neighborhood of the middle point 235 represented in the graphical and numerical form.

represented by the ideal point. At the same time the values of two other criteria (PZP and KRP) for the compromise solution are also very satisfactory and represent 100 % and 98 % of the criteria values for the ideal point, respectively. One can, thus, conclude that compromise solution satisfies the interests of the DMs and major stakeholders to a high degree.

The selected solution has been implemented. Thanks to that the number of the employees has been reduced by 20 % (from 36 to 33) and the total daily costs have been cut by 16 % (from 9400 PLN to 8500 PLN). The values of other two criteria have not been changed. The proposed changes resulted in the total annual savings pf approximately 400 000 PLN (100 000 Euro) and the noticeable improvement of the transportation process profitability from 4 % to 7 %. It is also clear that interests of all stakeholders have not been satisfied.

4   Conclusions

The paper presents the theoretical background and practical application of MCDM/A methodology in transportation and logistics. It demonstrates the universal, generic paradigm of solving multiple criteria decision problems and shows its usefulness for solving complex transportation – logistics problems. The MCDM/A methodology is presented as a system – oriented analysis. In the case study a multiple criteria crew sizing problem is analyzed and solved.

From a methodological point of view the major output of this research is the presentation in a nut shell basic methodological rules of MCDM/A and its characterization as a system oriented analysis. The article demonstrates all phases of the solution procedure, including: investigation of the decision problem and its verbal description; problem structuring and mathematical modeling; selection of the appropriate solution methods; computational experiments; final selection of the most desirable solution. The author claims that the MCDM/A methodology resembles in many aspects the system – oriented analysis and points out major similarities between those two approaches, such as:

– Broad perspective and comprehensive, system approach to problem analysis and solution. Investigation of the object/problem at stake from different perspectives and considering its various aspects (technical, economic, social, environmental, political).
– Interdisciplinary character of the analysis. Necessity to apply a variety of skills and competences and thus a team approach to carry out a complete study of a certain decision problem or the full analysis of a certain object.
– Application of scientific methods and rigid, systematic way of thinking to investigate the object/problem of concern; as a result generation of a transparent and well-structured picture of the considered situation/object.

From a practical point of view the critical component of the presented research is the real world application of MCDM/A methodology in the fields of transportation and logistics. The author proves that MCDM/A rules can be successfully applied to solve complex, multiple criteria decision problems, such as crew sizing problem. The generated results show that the obtained solution improves the current status on two out of four analyzed criteria by 20 % (reduction of the number of employees) and 16 % (reduction of daily costs), respectively without degradation of the outcome on two remaining criteria. This means that the existing practice was not Pareto – optimal and the newly introduced changes have brought the transportation process operations into the Pareto front. This observation proves practical applicability of the proposed approach and shows that MCDM/A can generate substantial improvements in real world that translate into concrete annual savings of 400 000 PLN (100 000 Euro). In effect the efficiency of the transportation process is improved and its profitability increased by 3 percentage points.

References

1. Brans, J., Vincke, P., Mareschal, B.: How to select and how to rank projects: the PROMETHEE method. Eur. J. Oper. Res. **24**, 228–238 (1986)
2. Campbell, G.: Cross-utilization of workers whose capabilities differ. Manage. Sci. **45**(5), 722–732 (1999)
3. Coyle, J.J., Bardi, E.J., Langley Jr., J.: The Management of Business Logistics. West Publishing Company, St. Paul (1996)
4. Figueira, J., Greco, S., Ehrgott, M. (eds.): Multiple Criteria Decision Analysis. State of the Art Surveys. Springer, New York (2005)

5. Gomez-Mejia, L., Balkin, D., Cardy, R.: Managing Human Resources. Prentice Hall, Englewood Cliffs (1995)

6. Hillier, F., Lieberman, G.: Introduction to Operations Research. McGraw-Hill, New York (2005)

7. Jacquet-Lagreze, E., Siskos, J.: Assessing a set of additive utility functions for multicriteria decision-making, the UTA method. Eur. J. Oper. Res. **10**, 151–164 (1982)

8. Jaszkiewicz, A., Słowiński, R.: The "Light Beam Search" approach – an overview of methodology and applications. Eur. J. Oper. Res. **113**(2), 300–314 (1999)

9. Jaszkiewicz, A., Słowiński, R.: The light beam search – outranking based interactive procedure for multiple-objective mathematical programming. In: Pardalos, P.M., Siskos, Y., Zopounidis, C. (eds.) Advances in Multicriteria Analysis, pp. 129–146. Kluwer Academic Publishers, Dordrecht (1995)

10. Keeney, R., Raiffa, H.: Decisions with Multiple Objectives. Preferences and Value Tradeoffs. Cambridge University Press, Cambridge (1993)

11. Klir, G.: Facets of Systems Science. Kluwer/Plenum Press, New York (2001)

12. Klir, G., Elias, D.: Architecture of Systems Problem Solving. Kluwer/Plenum, New York (2003)

13. Michalewicz, Z., Fogel, D.: How to Solve It: Modern Heuristics. Springer, Berlin (2000)

14. Roubens, M.: Preference relations on actions and criteria in multicriteria decision making. Eur. J. Oper. Res. **10**, 51–55 (1982)

15. Roy, B.: Decision-aid and decision making. Eur. J. Oper. Res. **45**, 324–331 (1990)

16. Roy, B.: The outranking approach and the foundations of ELECTRE methods. In: Bana e Costa, C.A. (ed.) Readings in Multiple Criteria Decision Aid, pp. 155–183. Springer, Berlin (1990)

17. Saaty, T.: The Analytic Hierarchy Process: Planning, Priority Setting Resource Allocation. McGraw-Hill, New York (1980)

18. Saaty, T.: Transport planning with multiple criteria: the analytic hierarchy process applications and progress review. J. Adv. Transp. **29**(1), 81–126 (1995)

19. Shapiro, R., Heskett, J.: Logistics Strategy. West Publishing, St. Paul (1985)

20. Steuer, R.: Multiple Criteria Optimization: Theory, Computation and Application. Wiley, New York (1986)

21. Vincke, P.: Multicriteria Decision-Aid. Wiley, Chichester (1992)

22. Żak, J.: Application of operations research techniques to the redesign of the distribution systems. In: Dangelmaier, W., Blecken, A., Delius, R., Klöpfer, S. (eds.) Advanced Manufacturing and Sustainable Logistics. LNBIP, vol. 46, pp. 57–72. Springer, Heidelberg (2010)

23. Żak, J.: Decision support systems in transportation. In: Jain, L., Lim, C. (eds.) Handbook on Decision Making – vol. 1: Techniques and Applications. Intelligent Systems Reference Library, vol. 4, pp. 249–294. Springer, Berlin (2010)

24. Żak, J.: Multiple Criteria Decision Aiding in Road Transportation. Poznan University of Technology Publishers, Poznan (2005) (in Polish)

25. Żak J.: The MCDA methodology applied to solve complex transportation decision problems. In: Proceedings of the 9th Meeting of the EURO Working Group on Transportation: Intermodality, Sustainability and Intelligent Transportation Systems, Bari, 10–13 June 2002, pp. 685–693 (2002)

26. Żak, J.: The methodology of multiple criteria decision making/aiding in public transportation. J. Adv. Transp. **45**(1), 1–20 (2011)

27. Żak, J., Redmer, A., Sawicki, P.: Multiple objective optimization of the fleet sizing problem for road freight transportation. J. Adv. Transp. **45**(4), 321–347 (2011)

A Joint Problem of Track Closure Planning and Train Run Rescheduling with Detours

Maciej Hojda$^{(\boxtimes)}$ and Grzegorz Filcek

Faculty of Computer Science and Management, Wroclaw University of Science and Technology, 27 Wyb. Wyspianskiego Street, 50-370 Wroclaw, Poland
{Maciej.Hojda,Grzegorz.Filcek}@pwr.edu.pl

Abstract. The main goal of this paper is to present an extended model and a solution algorithm for the joint problem of train run rescheduling and track closure planning. Inclusion of the possibility of rerouting increases the complexity, on one hand, but potentially leads to better overall solutions, on the other. Presented and tested is a heuristic solution algorithm working under the assumption of obligatory character of both runs and closures. The algorithm finds both, time shifts of planned runs and execution times of closures while attempting to minimize the total time divergence of the new schedule as compared with the original schedule. This approach fits squarely into a concept of a multi-stage train run and closure selection and scheduling system which is referenced in this work.

Keywords: Train scheduling · Train routing · Track closure planning · Rescheduling · Train timetabling

1 Introduction

In a well-performing train scheduling system it is imperative to expeditiously and efficiently ensure the continuity of operation of the railway lines [3,5]. A prominent cause behind the need for rescheduling and rerouting is the need for execution of planned and unplanned track closures. This problem is well-known, widely described [2,3,7–10,12] and usually formulated as an integer or mixed programming problem [10–12]. However, specific versions of the problem differ significantly, depending on the complexity of the railway structure and on additional constraints taken into account. A contender for the title of the most practical one is a case of multi-track network with constraints on delays and speed-ups of train runs and with limited capacities of tracks and station. Inclusion of the need to compare the new schedule with the original one manifests itself not only with specific constraints but also with a dedicated quality criterion. The latter is the delay (or acceleration) of train arrivals at preselected stations [2,9,10]. Additional complications arise when rail closures are also subject to decision making [1,5].

© Springer International Publishing AG 2017
J. Świątek and J.M. Tomczak (eds.), *Advances in Systems Science*, Advances in Intelligent Systems and Computing 539, DOI 10.1007/978-3-319-48944-5_26

In earlier work [4] the authors proposed a multi-stage train run and closure selection and scheduling system for a complex multi-track network. A key element of that system was the algorithm for obtaining time shifts in schedules of train runs and track closures plans under the assumption of obligatory character of both runs and closures. This element of the system (see Fig. 1) is under additional consideration in this paper. More specifically, authors explore the possibility of including a rerouting subroutine in order to improve the quality and the feasibility of the solutions obtained by the time shift determination algorithm. It is further assumed that trains are given a priority ordering and their inclusion in the schedule depends on it. It is further shown that, in general, the inclusion of a rerouting subroutine enhances the capabilities of the algorithm.

The paper is divided into five sections: the introduction, problem formulation, solution algorithm presentation, empirical evaluation and conclusions.

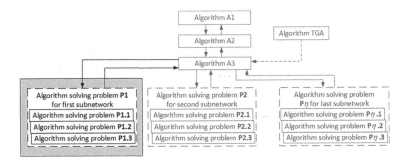

Fig. 1. Multi-level decision making system.

The work consists of five parts: this introduction, formulation of the decision-making problem, presentation of solution algorithm, empirical studies and a summary.

2   Problem Formulation

Notation was replicated after [4] with slight modifications. The rail connection network is described by an undirected graph $\mathbf{G} = <\mathbf{V}, \mathbf{E}>$, where $\mathbf{V} = \{0, 1, 2, \ldots, V\}$ is a set of (indices of) railway stations numbering $V + 1$ and $\mathbf{E} \in \{\{v_1, v_2\} : v_1, v_2 \in \mathbf{V} \wedge v_1 \neq v_2\}$ is a set of connections between stations. Additionally, by $\mathbf{K}_{i,j} = \{1, 2, \ldots, K_{i,j}\}$ we denote a set of track indices on section $i \leftrightarrow j$ where $K_{i,j}$ is their number. The node with the index 0 represents the artificial origin and destination station for each train run and has been introduced for the convenience of writing.

The number of train runs is denoted as P and a set of all train runs (indices) by $\mathbf{P} = \{1, 2, \ldots, P\}$. The order of train runs to be considered (priority vector) is denoted by $\omega = [\omega^1, \omega^2, \ldots, \omega^P]$ ($\omega^p \in \mathbf{P}$). The first element of the original schedule are train run routes $\rho_p : p \in \mathbf{P}$, where $\rho_p = [\rho_p^{i,j,k}]_{i,j \in \mathbf{V}, k \in \mathbf{K}_{i,j}}$ is the route of pth train run and $\rho_p^{i,j,k} = 1$ only if the train is scheduled to move from the node i to j using the kth track ($= 0$ in the other case). The second element of the schedule are the arrival and departure times for pth train run and the ith node, $\underline{\tau}_{p,i}$ and $\overline{\tau}_{p,i}$ respectively (in case when the train is not scheduled to arrive at the node, these values are undefined). We assume that the route of a train run forms a cycle covering all the nodes assigned to it as well as the node 0. In addition, we denote a visiting sequence of stations for the train run p as $\mathbf{\Psi}_p = [\psi_p^0, \psi_p^1, \dots, \psi_p^{\Psi_p}, \psi_p^{\Psi_p+1}]$, where $\psi_p^0 = \psi_p^{\Psi_p+1} = 0$, and $\Psi_p + 2$ is the number of nodes visited (Ψ_p if the excluding artificial node 0 which is visited twice). Let us also denote travel times between stations as $c_{p,i,j}$ and a set of obligatory stations $\overline{\mathbf{V}}_p \subseteq \mathbf{V}$ that the pth run must visit.

The number of track closures are denoted by N and a set of all track closures (indices) by $\mathbf{N} = \{1, 2, \dots, N\}$. Each closure $n \in \mathbf{N}$ is described by a set of two different nodes $\sigma_n = \{\sigma_n^1, \sigma_n^2\}$ defining the section on which the track closure is planned. By T_n we denote the duration of the closure n. By $\underline{t}_n$ and $\overline{t}_n$ we denote, respectively, the closure's earliest and latest start moment.

The decision variables which represent the schedule are new routes ρ_p', time shifts of train departure times form stations $\Delta\overline{\tau}_{p,i} : p \in \mathbf{P}, i \in \mathbf{V}$ and time shifts of the start moments of the planned track closures $\Delta t_n : n \in \mathbf{N}$. In addition, we assume that all train runs and closures must be executed (they must start and finish) in the given time interval $[0, T]$. All variables and constants with the interpretation of time are real numbers that belong to this interval (unless otherwise noted). Similarly to $\mathbf{\Psi}_p$ we denote as $\mathbf{\Psi}'_p = [\psi'_p{}^0, \psi'_p{}^1, \dots, \psi'_p{}^{\Psi'_p}, \psi'_p{}^{\Psi'_p+1}]$ the new route with Ψ'_p nodes.

The decision variables have to satisfy the following constraints:
– the start time of each track closure is included in the following interval

$$\forall n \in \mathbf{N} \quad \underline{t}_n \leq \underline{t}_n + \Delta t_n \leq \overline{t}_n \tag{1}$$

– the new route of each train run must include obligatory nodes (stations)

$$\forall p \in \mathbf{P}, v \in \overline{\mathbf{V}}_p \quad \sum_{i \in \mathbf{V}, k \in \mathbf{K}_{i,v}} \rho_p'{}^{i,v,k} + \rho_p'{}^{v,i,k} > 0 \tag{2}$$

and those stations must exist

$$\forall p \in \mathbf{P} \quad \{v \in \overline{\mathbf{V}}_p : i \in \{1, 2, \dots, \Psi'_p + 1\} \wedge v = \psi'_p{}^i\} \neq \emptyset, \tag{3}$$

– the new route must be consistent (have no loops, and exactly one cycle going through node 0)

$$\forall p \in \mathbf{P}, v \in \mathbf{V} \quad \sum_{i \in \mathbf{V}, k \in \mathbf{K}_{v,i}} \rho_p'{}^{i,v,k} = \sum_{i \in \mathbf{V}, k \in \mathbf{K}_{v,i}} \rho_p'{}^{v,i,k} = 1 \tag{4}$$

$$\forall p \in \mathbf{P} \quad \sum_{i \in \mathbf{V}, k \in \mathbf{K}_{0,i}} \rho_p'{}^{i,0,k} = 1 \tag{5}$$

$$\forall p \in \mathbf{P}, \mathbf{S} \subset \mathbf{V}, \mathbf{S} \neq \emptyset, \mathbf{S} \neq \mathbf{V} \quad \sum_{i \in \mathbf{S}, j \in \mathbf{S}, k \in \mathbf{K}_{i,j}} \rho_p'{}^{i,j,k} \leq |\mathbf{S}| - 1, \tag{6}$$

– the departure and arrival times of each train run are set in the proper order and belong to the given time interval

$$\forall p \in \mathbf{P}, i \in \{1, 2, \ldots, \Psi'_p + 1\}$$
$$0 \leq \overline{\tau}_{p,\psi'^{i-1}_p} + \Delta\overline{\tau}_{p,\psi'^{i-1}_p} \leq \overline{\tau}_{p,\psi'^{i-1}_p} + \Delta\overline{\tau}_{p,\psi'^{i-1}_p} + c_{p,\psi'^{i-1}_p,\psi'^{i}_p} \leq T, \quad (7)$$

– the dwell times on stations for each train run are limited

$$\forall p \in \mathbf{P}, i \in \{1, 2, \ldots, \Psi_p\}$$
$$\tau_{\min,p,\psi'^{i}_p} \leq \overline{\tau}_{p,\psi'^{i}_p} + \Delta\overline{\tau}_{p,\psi'^{i}_p} - (\overline{\tau}_{p,\psi'^{i-1}_p} + \Delta\overline{\tau}_{p,\psi'^{i-1}_p} + c_{p,\psi'^{i-1}_p,\psi'^{i}_p}) \leq \tau_{\max,p,\psi'^{i}_p} \quad (8)$$

where $\tau_{\min,p,i}$ is minimal, and $\tau_{\max,p,i}$ is the maximal dwell time of the pth train on the ith station,
– delays and speed ups of departures and arrivals are limited

$$\forall p \in \mathbf{P}, i \in \{1, \ldots, \Psi'_p\} \cap \overline{\mathbf{V}}_p \quad \underline{T}_{\mathrm{DEP}}(p, \psi'^{i}_p) \leq \Delta\overline{\tau}_{p,\psi'^{i}_p} \leq \overline{T}_{\mathrm{DEP}}(p, \psi'^{i}_p) \quad (9)$$

$$\forall p \in \mathbf{P}, i \in \{1, \ldots, \Psi'_p\} \cap \overline{\mathbf{V}}_p$$
$$\underline{T}_{\mathrm{ARR}}(p, \psi'^{i}_p) \leq \overline{\tau}_{p,\psi'^{i-1}_p} + \Delta\overline{\tau}_{p,\psi'^{i-1}_p} + c_{p,\psi'^{i-1}_p,\psi'^{i}_p} \leq \overline{T}_{\mathrm{ARR}}(p, \psi'^{i}_p) \quad (10)$$

where $\underline{T}_{\mathrm{DEP}}(p, \psi'^{i}_p), \overline{T}_{\mathrm{DEP}}(p, \psi'^{i}_p), \underline{T}_{\mathrm{ARR}}(p, \psi'^{i}_p), \overline{T}_{\mathrm{ARR}}(p, \psi'^{i}_p)$ are, respectively, minimal and maximal time shifts of departure, and minimal and maximal time shifts of the arrival of train p on station i. In general, those values may not be limited to the given interval $[0, T]$ to signalize the lack of such a constraint (constraint inactive),
– the total travel time extension of the train run is limited

$$\forall p \in \mathbf{P} \quad \Delta\overline{\tau}_{p,\psi'^{\Psi'_p}_p} - \Delta\overline{\tau}_{p,0} \leq \tilde{\tau}_{\max,p} \quad (11)$$

where $\tilde{\tau}_{\max,p}$ is the maximal elongation of the travel time,
– the capacity of the train station is limited

$$\forall p \in \mathbf{P}, i \in \mathbf{V}\backslash\{0\}$$
$$\sum_{j \in \mathbf{V}, k \in \mathbf{K}_{i,j}} \max\{\rho_p'^{i,j,k}, \rho_p'^{j,i,k}\} \sum_{q \in \tilde{\mathbf{P}}_{p,i}, l \in \mathbf{V}, m \in \mathbf{K}_{i,j}} \max\{\rho_p'^{i,l,m}, \rho_p'^{l,i,m}\} \leq \overline{c}_i \quad (12)$$

where $\overline{c}_i$ denotes the capacity of the ith station, and $\tilde{\mathbf{P}}_{p,i}$ is a set of train runs with trains dwelling on the ith station in the same moment as train from the pth train run,
– trains from different train runs do not collide with each other on any section of the route

$$\forall p, q \in \mathbf{P}, p \neq q, i, j \in \mathbf{V}, k \in \mathbf{K}_{i,j}$$
$$\min\{\rho_p'^{i,j,k}(\overline{\tau}_{p,i} + \Delta\overline{\tau}_{p,i} + c_{p,i,j}) + \rho_p'^{j,i,k}(\overline{\tau}_{p,j} + \Delta\overline{\tau}_{p,j} + c_{p,j,i}),$$
$$\rho_q'^{i,j,k}(\overline{\tau}_{q,i} + \Delta\overline{\tau}_{q,i} + c_{q,i,j}) + \rho_q'^{j,i,k}(\overline{\tau}_{q,j} + \Delta\overline{\tau}_{q,j} + c_{q,j,i})\} \leq$$
$$\leq \max\{\rho_p'^{i,j,k}(\overline{\tau}_{p,i} + \Delta\overline{\tau}_{p,i}) + \rho_p'^{j,i,k}(\overline{\tau}_{p,j} + \Delta\overline{\tau}_{p,j}),$$
$$\rho_q'^{i,j,k}(\overline{\tau}_{q,i} + \Delta\overline{\tau}_{q,i}) + \rho_q'^{j,i,k}(\overline{\tau}_{q,j} + \Delta\overline{\tau}_{q,j})\}, \quad (13)$$

– trains from all train runs do not collide with track closures

$$\forall p \in \mathbf{P}, n \in \mathbf{N}, i, j \in \mathbf{V}, k \in \mathbf{K}_{i,j}, i, j \in \sigma_n$$
$$\min\{\rho_p'^{i,j,k}(\overline{\tau}_{p,i} + \Delta\overline{\tau}_{p,i} + c_{p,i,j}), \underline{t}_n + \Delta t_n + T_n\} \leq$$
$$\leq \max\{\rho_p'^{i,j,k}(\overline{\tau}_{p,i} + \Delta\overline{\tau}_{p,i}), \underline{t}_n + \Delta t_n\}, \quad (14)$$

– track closures do not collide with each other

$$\forall n, m \in \mathbf{N}, n \neq m, \sigma_n = \sigma_m$$
$$\min\{\underline{t}_n + \Delta t_n + T_n, \underline{t}_m + \Delta t_m + T_m\} \leq \max\{\underline{t}_n + \Delta t_n, \underline{t}_m + \Delta t_m\}. \quad (15)$$

The quality index is the sum of all time shifts of departures in the new train run schedule in relation to the original one. Summing is done over all obligatory stations.

$$Q = \sum_{p \in \mathbf{P}} \sum_{i \in \overline{\mathbf{V}}_p} |\Delta\overline{\tau}_{p,i}|. \quad (16)$$

The main problem, further abbreviated as $P3r$, is formulated as follows:

For the given $\mathbf{G}$, $K_{i,j}$, $\mathbf{P}$, ω, ρ_p, Ψ_p, $\mathbf{\Psi}_p$, $c_{p,i,j}$, $\overline{V}_p$, N, σ_n, T_n, $\underline{t}_n$, $\overline{t}_n$, T, $\tau_{\min,p,i}$, $\tau_{\max,p,i}$, $\underline{T}_{\text{DEP}}(p,i)$, $\overline{T}_{\text{DEP}}(p,i)$, $\underline{T}_{\text{ARR}}(p,i)$, $\overline{T}_{\text{ARR}}(p,i)$, $\tilde{\tau}_{\max,p}$, $\overline{c}_i$, $\hat{\mathbf{P}}_{p,i}$ $(n \in \mathbf{N}, p \in \mathbf{P}, i \in \mathbf{V})$ find values $\rho_p'^\star$, $\Delta\overline{\tau}_{p,i}^\star$, and $\Delta t_n^\star$ $(n \in \mathbf{N}, p \in \mathbf{P}, i \in \mathbf{V})$ of decision variables minimizing (16) with respect to constraints (1)–(15), i.e.

$$(\rho_1'^\star, \rho_2'^\star, \dots, \rho_P'^\star; \Delta\overline{\tau}_{1,1}^\star, \Delta\overline{\tau}_{1,2}^\star, \dots, \Delta\overline{\tau}_{1,V}^\star, \Delta\overline{\tau}_{2,1}^\star, \Delta\overline{\tau}_{2,2}^\star, \dots, \Delta\overline{\tau}_{P,V}^\star;$$
$$\Delta t_1^\star, \Delta t_2^\star, \dots, \Delta t_N^\star) = \arg\min Q. \quad (17)$$

In the following section, a solution heuristic solution algorithm with polynomial execution time is presented.

3 Solution Algorithm

The solution algorithm is based on the methodology provided in [4] and in more detail in [6] where algorithms for solving a problem of finding time shifts for a fixed set of trains with fixed routes were devised. We denote such an algorithm as $A3$ and the problem itself as $P3$. We denote as RA the algorithm used for rerouting. This algorithm is described in a separate section. We present the algorithm $AP3r$ (see Algorithm 1) for solving $P3r$.

Algorithm 1. AP3r

1: **For** $p \in \{1, 2, \dots P\}$ **do**
2:     For a set of runs up to pth (from the series ω) solve $P3$ using $A3$.
3:     **If** $A3$ returns no solution **then** attempt rerouting using RA.
4:     **If** RA returns a solution **then** update ρ_p', $\underline{T}_{p,\cdot}$, $\overline{\tau}_{p,\cdot}$ according to the solution found
                 and solve $P3$ using $A3$.
5:     **If** RA or $A3$ returns no solution **then** end the algorithm, **return** no feasible solution.
6: **return** the solution ρ' and $\Delta\overline{\tau}$ obtained by the last successful execution of $A3$.

If the algorithm fails to provide a solution then at least one of the train runs could not be accommodated. The algorithm uses RP for solving a rerouting problem which is defined and solved in the next section.

3.1 Rerouting Problem

Rerouting is defined as a decision making problem denoted as RP. The main idea of the problem is to find new routes (and new arrival/departure times for those routes) under constraints of the original problem $P3r$. Decision making is done under fixed schedules for all but one train run. More specifically, in addition to the $P3r$ given also are: closure starting time shifts $\Delta t_n : n \in \mathbf{N}$, departure time shifts $\Delta \bar{\tau}_{p,i} : p \in \mathbf{P}, i \in \mathbf{V}$, current routes $\rho_p^{i,j,k}, (i,j,k) \in \mathbf{E}, p \in \mathbf{P}$ and a single run $q \in \mathbf{P}$ selected for rerouting. Goal is to find a new route $\rho_q'^{i,j,k}, (i,j,k) \in \mathbf{E}$ and time shifts $\Delta \bar{\tau}_{q,i}, i \in \tilde{\mathbf{V}}$ under constraints as in $P3r$.

To solve the RP the following algorithm RA is applied (see Algorithm 2). The algorithm uses a subroutine NS for finding a shortest feasible path between two nodes.

Algorithm 2. RA

1: Using $\overline{\mathbf{V}}_p$ and ρ_p obtain a series of nodes obligatory for the run $\Theta = (\theta^1, \theta^2, \ldots, \theta^M)$
    where $M = |\overline{\mathbf{V}}_p|$.

2: Let $\alpha := 1, G' := (\mathbf{V}', \mathbf{E}') := (\mathbf{V}, \mathbf{E}), \rho_q' := [0]_{i,j \in \mathbf{V}, k \in \mathbf{K}_{i,j}}, \Delta \bar{\tau}_q := [0]_{i \in \mathbf{V}}$.

3: **For** $\alpha \in \{1, 2, \ldots M - 1\}$ **do**

4:    For the modified graph G', use NS to select the path between nodes $\theta^\alpha \to \theta^{\alpha+1}$.
       Use parameters $\underline{\theta} = \theta^\alpha, \bar{\theta} = \theta^{\alpha+1}$.
       Obtain a route $\rho_o = [\rho_o^{i,j,k}]_{i,j \in \mathbf{V}, k \in \mathbf{K}}$ and time shifts $\Delta \bar{\tau}^\circ = [\Delta \bar{\tau}_i^\circ]_{i \in \mathbf{V}}$.

5:    **If** NS failed to return a solution **then** end the algorithm and **return** no solution.

6:    From $\mathbf{E}'$ delete edges adjacent to ρ_o (excluding the last node).

7:    To the route ρ_q' and times $\Delta \bar{\tau}_q$ add ρ_o and times $\Delta \bar{\tau}^\circ$.

8: Add artificial node 0 to the route ρ_q'.

9: **return** $\rho_q', \Delta \bar{\tau}_q$.

Internal subroutine NS is used for finding a path between two nodes (see Algorithm 3).

4 Algorithm Evaluation

This section is divided into two parts. First one illustrates the idea of the algorithms presented in the previous section with the use of a simple example. Second one contains empirical evaluations of the algorithms applied to a section of a real railway network or the Polish Railway.

Algorithm 3. NS

1: For every station $i \in \mathbf{V'}$, calculate time intervals $\Xi_i = \{(\underline{\xi}_i^j, \overline{\xi}_i^j)_j\}$ for which train capacity constraints are satisfied with excess of one.

2: For each track $(i, j, k) \in \mathbf{E'}$, calculate intervals $\Upsilon_{i,j,k} = \{(\underline{\upsilon}_{i,j,k}^n, \overline{\upsilon}_{i,j,k}^n)_n\}$ for which it is empty.

3: For every station $i \in \mathbf{V'}$ set arrival times $\underline{\tau}_i^\circ = \infty$, verification coefficient $\lambda_i = 0$. Set starting point $\theta = \underline{\theta}$, $\underline{\tau}_\theta^\circ = \underline{\tau}_{q,\theta}$, and source node $\pi_i = 0$.

4: Find feasible departure interval $(\underline{\varsigma}, \overline{\varsigma})$ from the current station θ.

5: **For each** $(\theta, j, k) \in \mathbf{E'}$ adjacent to θ **do**

6:     From Ξ_i and $(\underline{\varsigma}, \overline{\varsigma})$ calculate feasible departure times $\Xi_\circ = \{(\underline{\xi}_\circ^j, \overline{\xi}_\circ^j)_j\}$ from the current node.

7:     From $\Xi_\circ$ and $\Upsilon_{\theta,j,k}$ calculate feasible track entering times $\Upsilon_{\circ,j,k} = \{(\underline{\upsilon}_{\circ,j,k}^n, \overline{\upsilon}_{\circ,j,k}^n)_n\}$.

8:     Let $\Upsilon_{\bullet,j,k}$ be $\Upsilon_{\circ,j,k}$ shifted by $c_{q,\theta,j}$.

9:     From $\Upsilon_{\bullet,j,k}$ and Ξ_j calculate feasible entering times $\Xi_\bullet = \{(\underline{\xi}_\bullet^n, \overline{\xi}_\bullet^n)_n\}$ for the destination station j.

10:     If $\Xi_\bullet \neq \emptyset$ and $\min_n \underline{\xi}_\bullet^n < \underline{\tau}_i^\circ$ then set $\underline{\tau}_i^\circ := \min_n \underline{\xi}_\bullet^n$ and $\pi_i = \theta$.

11: Mark the current node as verified $\lambda_\theta = 1$ and move to the next one $\theta = \arg\min_{i \in \mathbf{V'}:\lambda_i=0} \underline{\tau}_i^\circ$.

12: **If** θ exists and $\underline{\tau}_\theta^\circ < \infty$ and $\theta \neq \overline{\theta}$ **then** go to 4.

13: **If** $\underline{\tau}_{\overline{\theta}}^\circ < \infty$ **then return** solution $\rho_\circ$ corresponding to track π_i.

14: **If** $\underline{\tau}_{\overline{\theta}}^\circ = \infty$ **then return** no solution.

4.1 Numerical Example

Consider a railway network with V=5 stations and single track connections $\{1,2\}$, $\{2,3\}$, $\{3,4\}$, $\{1,2\}$, $\{2,4\}$ with travel times $c_{p,i,j}$ the same for every train and given as follows 25, 25, 25, 60, 90. Assume that every non-zero node is connected to the zero node (travel times equal to 0).

Given are two runs, $P = 2$. For $p = 1$, path is $0-1-2-3-4-0$, arrival times $\underline{\tau}_{1,i}$ are 120, 30, 55, 85, 115, departure times $\overline{\tau}_{1,i}$ are 30, 30, 60, 90, 120. Minimal standing times are equal to 5 with the exception of nodes 0, 1 with minimal standing time equal to 0. For $p = 2$, path is $0-2-1-0$, arrival times 30, 25, 0, 0, 0, departure times 0, 30, 0, 0, 0. Minimal standing times are equal to 5 with the exception of nodes 0, 2 with minimal standing time equal to 0. Furthermore $\overline{\mathbf{V}}_1 = \{1, 4\}$, $\overline{\mathbf{V}}_2 = \{1, 2\}$. There are no constraints on minimal or maximal arrival/departure times and no constraints on total route stretching. Global time interval is $[0, 160]$.

First we consider one closure with parameters $\underline{t}_1 = 0$, $\overline{t}_1 = 0$, $T_1 = 30$ that takes place on the connection 1, 2, 1. The algorithm $A3$ finds a solution with time shifts $\Delta\overline{\tau}_{1,i} = [0, 25, 25, 25, 25]$, $\Delta\overline{\tau}_{1,i} = [0, 30, 30, 0, 0]$ and objective $Q = 110$. No rerouting is done.

Changing closure parameters to $T_1 = 50$ yields a different solution with time shifts $\Delta\overline{\tau}_{1,i} = [0, 20, 20, 20, 20]$, $\Delta\overline{\tau}_{1,i} = [0, 65, 0, 30, 0]$ and objective $Q = 105$. Second run was rerouted into $0 - 2 - 3 - 1 - 0$.

4.2 Empirical Evaluation

Separately from the example in the previous subsection, we consider a set of up to 20 train runs for a given real railway schedule on a selected subsection of the Polish Railways (Polskie Linie Kolejowe). We compare two cases: algorithm $AP3r$ and its modification, denoted as $AP3$, that does not perform rerouting. We consider two single closure cases (closures taken separately, $N = 1$ for both cases) with closures on a fixed track and parameters: for closure $n = 1$, $\underline{t} = \overline{t} = 120$, $T = 1200$ (in minutes), for closure $n = 2$, $\underline{t} = \overline{t} = 180$, $T = 1080$. There are no set limits on delays and speed ups of departure, arrival and maximum dwell times: $\underline{T}_{\text{DEP}}(p, i) = \underline{T}_{\text{ARR}}(p, i) = -T$, $\overline{T}_{\text{DEP}}(p, i) = \overline{T}_{\text{ARR}}(p, i) = T$, $\tau_{\max,p,i} = T$, and the minimum dwell time is set to $\tau_{\min,p,i} = 0$. Similarly, we set $\tilde{\tau}_{\max,p} = T$. All considered train runs go directly through the track on which closure is performed. The travel times for that track is either 2 or 3 [minutes], depending on the run. We consider a time horizon of one full day $[0, 1440]$ and execute both algorithms for varying $P \in \{1, 2, \ldots 20\}$ and present the results in Tables 1 and 2. Execution times are denoted as T^{AP3r} and T^{AP3} while obtained quality criteria as Q^{AP3r} and Q^{AP3}.

Table 1. Execution time and quality criterion for $n = 1$

P	1	2	3	4	5	6	7	8	9	10
Q^{AP3}	192	450	932	1267	1658	2186	3306	4303	-	-
Q^{AP3r}	192	450	932	1267	1658	2186	3306	4303	4303	4303
T^{AP3}	0.08	0.24	0.72	1.76	3.37	6.14	10.34	16.89	26.42	25.41
T^{AP3r}	0.06	0.23	0.83	1.78	3.46	5.99	10.32	16.47	33.31	51.15
P	11	12	13	14	15	16	17	18	19	20
Q^{AP3}	-	-	-	-	-	-	-	-	-	-
Q^{AP3r}	4520	4819	5204	5204	5797	5797	5797	5797	5797	5797
T^{AP3}	24.80	24.63	25.72	26	24.38	26.17	25.85	24.91	25.29	24.41
T^{AP3r}	60.91	76.56	98.75	149.30	180.26	254.03	342.31	427.37	533.69	641.56

Since $AP3r$ and $AP3$ give identical solutions as long as $AP3$ can find a feasible solution, we expect that the quality of the solution obtained with both algorithms will be the same for some cases. Indeed, we can see that this is the case for $P \leq 8$ and $n = 1$ (Table 1) as well as $P \leq 15$ and $n = 2$ (Table 2). Likewise, execution times are very close (minor differences can be explained by measurement inaccuracies). It is however clearly visible that the execution time of $AP3r$ grows significantly when it has to contend with rerouting. One can observe nearly doubling of the execution time for $n = 1$ between cases for $P = 8$ and $P = 9$. This is also visible, to a lesser extent, for $n = 2$. This significant time extension may limit the applicability of the proposed subroutine in cases when long execution time is prohibitive.

Table 2. Execution time and quality criterion for $n = 2$

P	1	2	3	4	5	6	7	8	9	10
Q^{AP3}	132	330	692	967	1298	1766	2766	3643	4416	5086
Q^{AP3r}	132	330	692	967	1298	1766	2766	3643	4416	5086
T^{AP3}	0.08	0.25	0.73	1.79	3.4	6.26	10.47	16.64	27.04	38.83
T^{AP3r}	0.06	0.23	0.85	1.8	3.54	6.08	10.24	16.89	26.26	38.55
P	11	12	13	14	15	16	17	18	19	20
Q^{AP3}	5207	5446	5771	6619	7160	-	-	-	-	-
Q^{AP3r}	5207	5446	5771	6619	7160	7160	7160	7160	7160	7160.0
T^{AP3}	55.44	79.58	107.70	146.15	194.6	250.55	253.39	252.49	252.61	250.76
T^{AP3r}	55.94	77.04	106.51	145.52	191.6	307.78	425.51	563.23	710.38	872.56

5   Conclusions

Main contribution of the paper, which is the rerouting algorithm, was shown to significantly improve the ability of finding a feasible solution in a track closure and run rescheduling planning system. Provided empirical evaluation suggests that the execution time might be significant and that improvement on the speed of execution is a necessary step in future works. It is therefore clear that the main the main objective of future works is to improve the speed of the solution algorithm used for finding time shifts as this algorithm is still the bottleneck of the whole rerouting system. This goal was partially accomplished in [6]. It should also be worthwhile to explore the possibility of applying a rerouting subroutine not only for dealing with otherwise unfeasible cases (as was done in this work) but also for improving the overall quality of the solution.

References

1. Albrecht, A.R., Panton, D.M., Lee, D.H.: Rescheduling rail networks with maintenance disruptions using problem space search. Comput. Oper. Res. **40**(3), 703–712 (2013)
2. Brannlund, U., Lindberg, P.O., Nou, A., Nilsson, J.-E.: Railway timetabling using Lagrangian relaxation. Transp. Sci. **32**(4), 358–369 (1998)
3. Cacchiani, V., Huisman, D., Kidd, M., Kroon, L., Toth, P., Veelentruf, L., Wagenaar, J.: An overview of recovery models and algorithms for real-time railway rescheduling. Transp. Res. Part B Methodol. **63**, 15–37 (2014)
4. Filcek, G., Gąsior, D., Hojda, M., Józefczyk, J.: Joint train rescheduling and track closures planning: model and solution algorithm. In: Borzemski, L., Grzech, A., Świątek, J., Wilimowska, Z. (eds.) Information Systems Architecture and Technology: Proceedings of 36th International Conference on Information Systems Architecture and Technology – ISAT 2015 – Part I. AISC, vol. 429, pp. 215–225. Springer, Heidelberg (2016). doi:10.1007/978-3-319-28555-9_19

5. Fokkert, J., Hertog, D., Berg, F., Verhoeven, J.: The Netherlands schedules track maintenance to improve track workers safety. Interfaces **37**(2), 133–142 (2007)
6. Hojda, M.: Optimization of time shifts in train timetables and track closure schedules. In: Proceedings of Krajowa Konferencja Automatyzacji Procesw Dyskretnych 2016, Poland (in print, 2016)
7. Lid, T.: Survey of railway maintenance activities from a planning perspective and literature review concerning the use of mathematical algorithms for solving such planning and scheduling problems, Technical report, Linkpings universitet (2014)
8. Meng, L., Zhou, X.: Simultaneous train rerouting, rescheduling on an N-track network, : A model reformulation with network-based cumulative flow variables. Transp. Res. Part B: Methodological **67**, 208–234 (2014)
9. Niu, H., Zhou, X.: Optimizing urban rail timetable under time-dependent demand and oversaturated conditions. Trans. Res. Part C: Emerg. Technol. **36**, 212–230 (2013)
10. Schobel, A.: Capacity constraints in delay management. J. Public Transp. **1**, 135–154 (2009)
11. Tornquist, J., Persson, J.: Train traffic deviation handling using tabu search and simulated annealing. In: Proceedings of 38th Hawaii International Conference on System Sciences, p. 73a. IEEE (2005)
12. Tornquist, J., Persson, J.: N-tracked railway traffic rescheduling during disturbances. Transp. Res. Part B **41**, 342–362 (2007)

Uncertain Systems

Nonlinear Optimal Control Using Finite-Horizon State Dependent Riccati Equation Combined with Genetic Algorithm

Ahmed Khamis[1($\boxtimes$)], Dawid Zydek[2], and Henry Selvaraj[2]

[1] Guidance and Control Department, Military Technical College, Cairo, Egypt
khamahme@isu.edu
[2] Department of Electrical and Computer Engineering,
University of Nevada, Las Vegas, NV, USA
{Dawid.Zydek,Henry.Selvaraj}@unlv.edu

Abstract. Precise model of DC motor is a nonlinear model. Accurate nonlinear control of DC motors is required. A novel online technique for finite-horizon nonlinear tracking problem is offered in this paper. The idea of the proposed technique is the change of variables that converts the nonlinear differential Riccati equation to a linear Lyapunov differential equation. Genetic algorithm is used as a method to calculate the optimal weighting matrices. Unlike the linear techniques that are used for linearized systems, the proposed technique is effective for wide range of operating points. Simulation results are given to demonstrate the effectiveness of the offered technique.

Keywords: Nonlinear tracking · Genetic algorithm · Finite-horizon state dependent riccati equation

1 Introduction

The need to improve performance in control systems requires more and more accurate modeling [16]. Even if a model is an accurate representation of the real system over a wide range of operating points, it is most often nonlinear [2]. The State Dependent Riccati Equation (SDRE) has become a very effective technique for the systematic design of nonlinear controllers, and it is a very effective algorithm for designing the nonlinear feedback control by taking into consideration the nonlinearities in the system states [4].

A major advantage offered by the SDRE to the control designer is the chance to make tradeoffs between the control effort and the state errors, that tradeoffs can be done by precise tuning of the state dependent coefficients (SDC) matrices. Also, as the SDRE depends only on the current state, the SDRE calculations can be carried out online, in which case the SDRE is defined along the state trajectory. The SDRE calculations can be solved by several methods. One of the approaches that passes the processing and other possible data to the ground infrastructure of computational units is discussed in [6,18,28].

© Springer International Publishing AG 2017
J. Świątek and J.M. Tomczak (eds.), *Advances in Systems Science*, Advances in Intelligent Systems and Computing 539, DOI 10.1007/978-3-319-48944-5_27

Motivated by the high capability of the algebraic SDRE for regulation and tracking of infinite-horizon nonlinear systems [3,7], this paper presents a permanent magnet DC motor position control based on nonlinear motor system dynamics via differential SDRE combined with genetic algorithms (GA). The differential SDRE, strictly speaking could be called state dependent differential Riccati equation (SDDRE), is a technique for finite-horizon optimal control of nonlinear systems. The SDDRE is based on the substitution algorithm [24], that transforms the differential Riccati equation (DRE) to a linear differential Lyapunov equation (DLE) [23]. A full methodology using GA is done as a method to calculate the optimal weighting matrices, which offer good performances and robustness in the closed loop system. At each time step, the coefficients of the linear differential Lyapunov equation are to be calculated and held from current time to the next time step [9]. Then, during online implementation the resulting Lyapunov equation is to be solved in a closed form at each step. The use of Lyapunov-type equations in solving optimal problems is given in [27]. While the nonlinear, optimal, finite-horizon regulation technique was presented in [9], this paper presents the application of a nonlinear, optimal, finite-horizon regulation and tracking technique developed by [12] to a nonlinear model of a DC motor.

The reminder of this paper is organized as follows: the nonlinear mathematical model of the DC motor is discussed in Sect. 2. A brief idea about genetic algorithm is presented in Sect. 3. Section 4 presents the nonlinear finite-horizon tracking technique via SDRE combined with genetic algorithm. Simulation results are discussed in Sect. 5. Finally, conclusions of this paper are given in Sect. 6.

2   DC Motor Nonlinear Mathematical Modeling

In this section, the modeling approach adopted to identify a nonlinear mathematical model for the motor is demonstrated [15]. The DC motor used in this paper is carbon-brush permanent magnet 12 v DC motor. The mathematical model is shown in Fig. 1, where R is the armature resistance, L is the armature inductance, v is the voltage applied to the motor, i is the current through the motor, e is the back emf voltage, J is the moment of inertia of the load, B is the viscous friction coefficient, τ is the torque generated by the motor, θ is the angular position of the motor, and ω is the angular velocity of the motor.

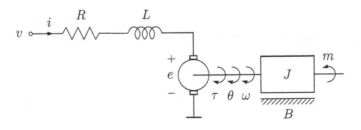

Fig. 1. Permanent magnet DC motor system model [15]

The dynamic equations for the DC motor are:

$$w(t) = \frac{d\theta(t)}{dt}, \tag{1}$$

$$v(t) = L\frac{di(t)}{dt} + Ri(t) + k_w w(t), \tag{2}$$

$$k_i i(t) = J\frac{dw(t)}{dt} + Bw(t) + Csgn(\omega), \tag{3}$$

where k_w is the back emf constant, k_i is the torque constant of the motor, and C is the motor static friction, and the signum function $sgn(\omega)$ is defined as

$$sgn(\omega) = \begin{cases} -1 \text{ for } \omega < 0, \\ 0 \text{ for } \omega = 0, \\ 1 \text{ for } \omega > 0, \end{cases} \tag{4}$$

or it can be written in this form:

$$sgn(\omega) = \frac{|\omega|}{\omega}. \tag{5}$$

The system nonlinear state equations can be written in the form:

$$\dot{x}_1 = x_2, \tag{6}$$

$$\dot{x}_2 = \left(\frac{-B}{J} - \frac{-C}{J|x_2|}\right)x_2 + \frac{k_i}{J}x_3, \tag{7}$$

$$\dot{x}_3 = -\frac{k_w}{L}x_2 - \frac{R}{L}x_3 + \frac{1}{L}u, \tag{8}$$

$$y = x_1, \tag{9}$$

where: $\theta = x_1$, $\dot{\theta} = x_2$, $i = x_3$, $v = u$.

3   Genetic Algorithm

Genetic algorithm (GA) is a stochastic heuristic global random search and optimization technique that simulate the metaphor of natural biological evolution [11]. GA works on a population of potential solutions, applying the principle of survival of the fittest to produce successively better approximations to a solution. At each generation of a GA, a new set of approximations is generated by the method of selecting individuals according to their level of fitness in the problem domain and reproducing them using operators borrowed from natural genetics [21]. This process leads to the evolution of populations of individuals that are more appropriate to their environment than individuals from which they were generated. GA has been shown to be an effective strategy in off-line design of control systems by a number of practitioners [1,17].

GA is not limited by difficult mathematical model, and it is flexible enough for almost all types of design problems. Also, GA can be combined with already developed techniques [20]. There are several approaches to select parents from the old population, and there are different GA techniques that can be used for different selection methods. In the optimization problem, an encoding process is used to represent the solution in a string form [21]. In most cases the binary encoding method is used. The most important operator in GA is crossover. There are many types of crossovers in GA. Another important operator in GA is mutation. The genetic algorithm toolbox uses MATLAB matrix function to build a set of versatile tools for implementing a wide range of GA methods. The GA toolbox is a group of routines that implement the most important function in GA [19]. Figure 2 shows a simple GA flow chart.

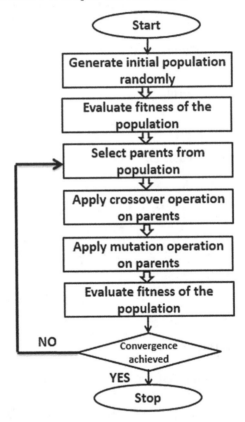

Fig. 2. Simple genetic algorithm flow chart

4   Nonlinear Tracking Using Finite-Horizon Differential SDRE Combined with GA

Nonlinear systems finite-horizon optimal control is an inspiring problem in the control field because of the difficulty of time-dependency of the Hamilton- Jacobi-Bellman (HJB) differential equation [9]. In finite-horizon optimal nonlinear control problem, the DRE cannot be solved in real time since the DRE arising in the optimal control can only be solved backward in time from its known final value. To solve this problem, an approximate analytical method is used [10] to transform the original nonlinear Ricatti equation to a linear DLE that can be solved in closed format at each time step [14].

4.1   Problem Formulation

The nonlinear system considered in this paper is assumed to be in the form:

$$\dot{\mathbf{x}}(t) = \mathbf{f}(\mathbf{x}) + \mathbf{g}(\mathbf{x})\mathbf{u}(t), \tag{10}$$
$$\mathbf{y}(t) = \mathbf{h}(\mathbf{x}). \tag{11}$$

This nonlinear system can be expressed in a state-dependent like-linear form as:

$$\dot{\mathbf{x}}(t) = \mathbf{A}(\mathbf{x})\mathbf{x}(t) + \mathbf{B}(\mathbf{x})\mathbf{u}(t), \tag{12}$$
$$\mathbf{y}(t) = \mathbf{C}(\mathbf{x})\mathbf{x}(t), \tag{13}$$

where $\mathbf{f}(\mathbf{x}) = \mathbf{A}(\mathbf{x})\mathbf{x}(t)$, $\mathbf{B}(\mathbf{x}) = \mathbf{g}(\mathbf{x})$, $\mathbf{h}(\mathbf{x}) = \mathbf{C}(\mathbf{x})\mathbf{x}(t)$.

Let $\mathbf{z}(t)$ be the desired output. The goal is to find a state feedback control law that minimizes the cost function given by [22]:

$$\mathbf{J}(\mathbf{x}, \mathbf{u}) = \frac{1}{2}\mathbf{e}'(t_f)\mathbf{Fe}(t_f) + \frac{1}{2}\int_{t_0}^{t_f} [\mathbf{e}'(t)\mathbf{Q}(\mathbf{x})\mathbf{e}(t) + \mathbf{u}'(\mathbf{x})\mathbf{R}(\mathbf{x})\mathbf{u}(\mathbf{x})]\, dt, \tag{14}$$

where $\mathbf{Q}(\mathbf{x})$ and $\mathbf{F}(\mathbf{x})$ are symmetric positive semi-definite matrices, and $\mathbf{R}(\mathbf{x})$ is a symmetric positive definite matrix. Moreover, $\mathbf{x}'\mathbf{Q}(\mathbf{x})\mathbf{x}$ is a measure of state accuracy and $\mathbf{u}'(\mathbf{x})\mathbf{R}(\mathbf{x})\mathbf{u}(\mathbf{x})$ is a measure of control effort, and the error $\mathbf{e}$ is the difference between the reference output and the actual output, $\mathbf{e}(t) = \mathbf{z}(t) - \mathbf{y}(t)$ [13].

4.2   Solution for Finite-Horizon Differential SDRE Tracking

To minimize the above cost function (14), a feedback control law can be given as

$$\mathbf{u}(\mathbf{x}) = -\mathbf{R}^{-1}(\mathbf{x})\mathbf{B}'(\mathbf{x})[\mathbf{P}(\mathbf{x})\mathbf{x}(t) - \mathbf{g}(\mathbf{x})], \tag{15}$$

where $\mathbf{P}(\mathbf{x})$ is a symmetric, positive-definite solution of the Differential State Dependent Riccati Equation, strictly speaking it could be called State Dependent Differential Riccati Equation (SDDRE), of the form

$$-\dot{\mathbf{P}}(\mathbf{x}) = \mathbf{P}(\mathbf{x})\mathbf{A}(\mathbf{x}) + \mathbf{A}'(\mathbf{x})\mathbf{P}(\mathbf{x}) - \mathbf{P}(\mathbf{x})\mathbf{B}(\mathbf{x})\mathbf{R}^{-1}\mathbf{B}'(\mathbf{x})\mathbf{P}(\mathbf{x}) + \mathbf{C}'(\mathbf{x})\mathbf{Q}(\mathbf{x})\mathbf{C}(\mathbf{x}), \tag{16}$$

with the final condition

$$\mathbf{P}(\mathbf{x}, t_f) = \mathbf{C}'(t_f)\mathbf{FC}(t_f). \tag{17}$$

The resulting SDRE-controlled trajectory becomes the solution of the state-dependent closed-loop dynamics

$$\dot{\mathbf{x}}(t) = [\mathbf{A}(\mathbf{x}) - \mathbf{B}(\mathbf{x})\mathbf{R}^{-1}(\mathbf{x})\mathbf{B}'(\mathbf{x})\mathbf{P}(\mathbf{x})]\mathbf{x}(t) + \mathbf{B}(\mathbf{x})\mathbf{R}^{-1}(\mathbf{x})\mathbf{B}'(\mathbf{x})\mathbf{g}(\mathbf{x}), \tag{18}$$

where $\mathbf{g}(\mathbf{x})$ is a solution of the state-dependent non-homogeneous vector differential equation

$$\dot{\mathbf{g}}(\mathbf{x}) = -[\mathbf{A}(\mathbf{x}) - \mathbf{B}(\mathbf{x})\mathbf{R}^{-1}(\mathbf{x})\mathbf{B}'(\mathbf{x})\mathbf{P}(\mathbf{x})]'\mathbf{g}(\mathbf{x}) - \mathbf{C}'(\mathbf{x})\mathbf{Q}(\mathbf{x})\mathbf{z}(\mathbf{x}), \quad (19)$$

with the final condition

$$\mathbf{g}(\mathbf{x}, t_f) = \mathbf{C}'(t_f)\mathbf{F}\mathbf{z}(t_f). \quad (20)$$

As all the variables are function of $(\mathbf{x}, t)$, the value of the states cannot be calculated ahead of present time step. Consequently, the state dependent coefficients cannot be calculated to solve (16) with the final condition (17) by backward integration from t_f to t_0. To overcome this problem, an approximate analytical approach is used [10, 23, 24]. which converts the original nonlinear Ricatti equation to a linear differential Lyapunov equation. At each time step, the Lyapunov equation can be solved in closed form. The detailed steps of the finite horizon SDRE can be found in [15].

4.3 Genetic Algorithm Combined with the Finite-Horizon Differential SDRE

As a result of the complication of the problem and the large number of possible weighting matrices, Q and R, a GA is used to reduce the number of repetitions. These algorithms have been successfully applied to optimal control problems [5, 25, 26]. GA is a meta-heuristic optimization algorithm that finds solutions using Darwin's theory of natural selection. The different steps of the GA are [8]:

Initial Population: A random population is required to start the evolution process. For an optimal control problem, a population (P_p) of N individuals is produced in the form of coefficients of the weighting matrices (Q, R). Moreover, this population should respect the positive definite principle for the matrix R, and the semi-positive definite principle for the matrix Q.

Evaluation of an Individual: To quantify the adaptation degree of an individual, a fitness function evaluates the performance and the robustness of the controller achieved with the weighting matrices (Q, R) corresponding to the individual. This helps the population to be arranged from the best-fitted individual to the worst one.

Crossover: To generate new individuals, an operator arbitrarily chooses two individuals from the existing population and crosses their genes. To improve the diversity and achieve additional results, two different types of crossover methods are utilized. The first is the uniform crossover. It generates a random binary mask that decides if two genes can be crossed. The second method splits the parents into two or three sections, and each section is crossed to get two individuals.

Mutation: A mutation is defined as varying of the structure of the chromosome. A mutation is generated in an individual when an algorithm randomly selects two genes and permutes them.

Elitist Selection: Elitist selection is based on Darwin's theory. Darwin's theory says that the process of natural selection encourages the most fitted individuals. Only the best individuals will survive and preserve their genetic codes from one generation to another. This technique lets a faster convergence to the detriment of diversity. To avoid this problem, the crossover is done considering all the population, but with more chances given to the better-fit. Thus, even the less fitted individuals can contribute to generate the new generation.

Figure 3 summarizes the overview of the process of finite-horizon SDDRE tracking technique combined with GA.

5   Simulation Results

The system nonlinear state equations of the DC motor (6–8) can be rewritten in state dependent form:

$$
\begin{bmatrix} \dot{x}_1 \\ \dot{x}_2 \\ \dot{x}_3 \end{bmatrix} = \begin{bmatrix} 0 & 1 & 0 \\ 0 & \left(\frac{-B}{J} - \frac{-C}{J|x_2|} \right) & \frac{k_i}{J} \\ 0 & -\frac{k_w}{L} & -\frac{R}{L} \end{bmatrix} \begin{bmatrix} x_1 \\ x_2 \\ x_3 \end{bmatrix} + \begin{bmatrix} 0 \\ 0 \\ \frac{1}{L} \end{bmatrix} u.
\tag{21}
$$

where: $\theta = x_1$, $\dot{\theta} = x_2$, $i = x_3$, $v = u$.

Let the reference angle be

$$
z(t) = 90^o,
\tag{22}
$$

Simulation with the GA controller method shows that the best weighting matrices are

$$
Q = diag(150, 34, 10), R = 2.53.
\tag{23}
$$

The simulations are performed for final time of 10 seconds and the resulting output trajectories is shown in Fig. 4, where the dash-dot line denotes the *reference* angle trajectory, and the solid line denotes the *actual* trajectory. The optimal control is shown in Fig. 5.

Comparing these trajectories in Fig. 4, it is clear that the proposed algorithm after combining the GA with the SDDRE gives very good results as the actual optimal angle is making an accurate tracking to the reference angle with average error of 0.02 %.

From these results, it can be seen that the developed methodology is able to solve the DC motor nonlinear tracking problem with accurate tracking and good robustness.

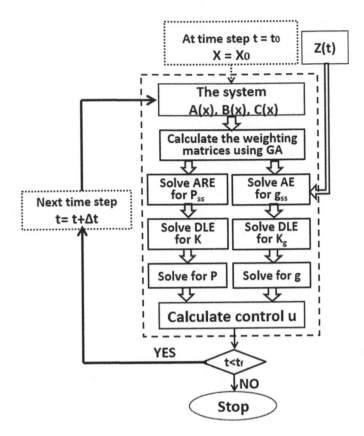

Fig. 3. Overview of the process of finite-horizon differential SDRE combined with GA

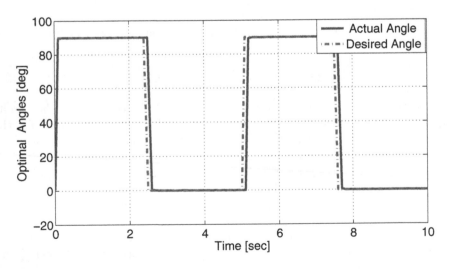

Fig. 4. Optimal angle tracking for the DC motor

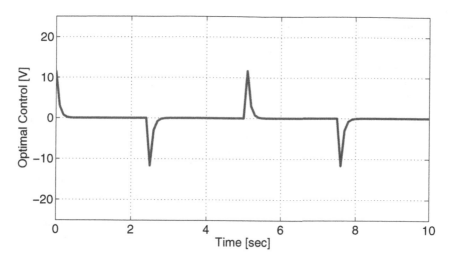

Fig. 5. Optimal control voltage for the DC motor

6   Conclusions

The paper presented a new finite-horizon tracking technique for nonlinear systems. This technique is based on using the genetic algorithm to calculate the weighting matrices in the finite horizon SDRE technique. Change of variables converts the nonlinear differential Riccati equation to a linear Lyapunov equation. The Lyapunov equation is solved in a closed form at the given time step. Simulation results for DC motor are included to demonstrate the effectiveness of the developed technique.

References

1. Bramlette, M.F., Cusic, R.: A comparative evaluation of search methods applied to parametric design of aircraft. In: Proceedings of the Third Iinternational Conference on Genetic Algorithms, pp. 213–218. Morgan Kaufmann Publishers Inc. (1989)
2. Çimen, T.: Recent advances in nonlinear optimal feedback control design. In: Proceedings of the 9th WSEAS International Conference on Applied Mathematics, Istanbul, Turkey, pp. 460–465, May 2006
3. Çimen, T.: Development and validation of a mathematical model for control of constrained nonlinear oil tanker motion. Math. Comput. Model. Dyn. Syst. **15**(1), 17–49 (2009)
4. Çimen, T.: Survey of state-dependent Riccati equation in nonlinear optimal feedback control synthesis. AIAA J. Guidance Control Dyn. **35**(4), 1025–1047 (2012)
5. Chen, B.S., Cheng, Y.M.: A structure-specified H8 optimal control design for practical applications: a genetic approach. IEEE Trans. Control Syst. Technol. **6**(6), 707–718 (1998)
6. Chmaj, G., Walkowiak, K.: A P2P computing system for overlay networks. Future Gener. Comput. Syst. **29**(1), 242–249 (2013)

7. Cloutier, J.: State-dependent riccati equation techniques: an overview. In: Proceedings of the American Control Conference, vol. 2, pp. 932–936 (1997)
8. Ghazi, G., Botez, R.: New robust control analysis methodology for lynx helicopter and cessna citation X aircraft using guardian maps, genetic algorithms and LQR theories combinations. In: AHS 70th Annual Forum, Montreal, Quebec, Canada (2014)
9. Heydari, A., Balakrishnan, S.: Approximate closed-form solutions to finite-horizon optimal control of nonlinear systems. In: 2012 American Control Conference (ACC), pp. 2657–2662. IEEE, June 2012
10. Heydari, A., Balakrishnan, S.: Path planning using a novel finite-horizon suboptimal controller. J. Guidance Control Dyn. **36**(4), 1–5 (2013)
11. Holland, J.H.: Adaptation in Natural and Artificial Systems: An Introductory Analysis with Applications to Biology, Control, and Artificial Intelligence. University of Michigan Press, Ann Arbor (1975)
12. Khamis, A.: Advanced Tracking Strategies for Linear and Nonlinear Control Systems: Theory and Applications. Ph.D. thesis, Idaho State University, Pocatello, Idaho, USA (2014)
13. Khamis, A., Kamel, A., Naidu, D.: Nonlinear optimal tracking for missile gimbaled seeker using finite-horizon state dependent Riccati equation. In: 2014 IEEE 4th Annual International Conference on Cyber Technology in Automation, Control, and Intelligent Systems (CYBER), pp. 88–93, June 2014
14. Khamis, A., Naidu, D., Zydek, D.: Nonlinear optimal control with incomplete state information using state dependent Riccati equation (SDRE). In: Selvaraj, H., Zydek, D., Chmaj, G. (eds.) Progress in Systems Engineering. AISC, vol. 366, pp. 27–33. Springer, Heidelberg (2015)
15. Khamis, A., Naidu, D., Zydek, D.: Nonlinear position control of DC motor using finite-horizon state dependent Riccati equation. In: Selvaraj, H., Zydek, D., Chmaj, G. (eds.) Progress in Systems Engineering, vol. 366, pp. 35–39. Springer, Heidelberg (2015)
16. Khamis, A., Zydek, D., Borowik, G., Naidu, D.S.: Control system design based on modern embedded systems. In: Moreno-Díaz, R., Pichler, F., Quesada-Arencibia, A. (eds.) EUROCAST 2013. LNCS, vol. 8112, pp. 491–498. Springer, Heidelberg (2013). doi:10.1007/978-3-642-53862-9_62
17. Krishnakumar, K., Goldberg, D.E.: Control system optimization using genetic algorithms. J. Guidance Control Dyn. **15**(3), 735–740 (1992)
18. Liu, B., Zydek, D., Selvaraj, H., Gewali, L.: Accelerating high performance computing applications: using cpus, gpus, hybrid cpu/gpu, and fpgas. In: 2012 13th International Conference on Parallel and Distributed Computing, Applications and Technologies (PDCAT), pp. 337–342. IEEE (2012)
19. Magaji, N., Mustafa, M.: Optimal location and signal selection of UPFC device for damping oscillation. Int. J. Electr. Power Energy Syst. **33**(4), 1031–1042 (2011)
20. Magaji, N., Mustafa, M., et al.: Power system damping using GA based fuzzy controlled SVC device. In: TENCON 2009–2009 IEEE Region 10 Conference, pp. 1–7. IEEE (2009)
21. Magaji, N., Hamza, M.F., Dan-Isa, A.: Comparison of GA and LQR tuning of static VAR compensator for damping oscillations. Int. J. Adv. Eng. Technol. **2**(1), 594–601 (2012)
22. Naidu, D.: Optimal Control Systems. CRC Press, Boca Raton (2003)
23. Nazarzadeh, J., Razzaghi, M., Nikravesh, K.: Solution of the matrix Riccati equation for the linear quadratic control problems. Math. Comput. Model. **27**(7), 51–55 (1998)

24. Nguyen, T., Gajic, Z.: Solving the matrix differential riccati equation: a Lyapunov equation approach. IEEE Trans. Automatic Control **55**(1), 191–194 (2010)
25. Patrón, R.F., Kessaci, A., Botez, R.M., Labour, D.: Flight trajectories optimization under the influence of winds using genetic algorithms. In: AIAA Guidance, Navigation, and Control (GNC) Conference, p. 4620 (2013)
26. Patrón, R.F., Botez, R., Labour, D.: Speed and altitude optimization on the FMS CMA-9000 for the Sukhoi superjet 100 using genetic algorithms. In: Integration, and Operations, AIAA Aviation Technology (2013)
27. Vandenberghe, L., Balakrishnan, V., Wallin, R., Hansson, A., Roh, T.: Interior-point methods for semidefinite programming problems derived from the KYP lemma. In: Henrion, D., Garulli, A. (eds.) Positive Polynomials in Control. LNCS, vol. 312, pp. 195–238. Springer, Heidelberg (2005)
28. Zydek, D., Chmaj, G., Chiu, S.: Modeling computational limitations in H-phy and overlay-NoC architectures. J. Supercomputing **70**(1), 1–20 (2013)

Logistic Network Dependability Evaluation – Business Continuity Oriented Approach

Lech Bukowski[1](✉) and Jerzy Feliks[2]

[1] University of Dąbrowa Górnicza, Dąbrowa Górnicza, Poland
lbukowski@wsb.edu.pl
[2] AGH University of Science and Technology, Krakow, Poland
jfeliks@zarz.agh.edu.pl

Abstract. The aim of this paper is to propose a concept of logistic networks dependability evaluation, based on the continuity oriented approach. Because the modern logistic networks are more and more reliable, the number of failures and disruptions is significantly decreased. Therefore the well known probabilistic methods are in this case not more acceptable. We propose an expert system, based on the fuzzy sets theory, to evaluate the level of dependability for different configurations of the logistic networks and to simulate different possible disruptions scenarios.

Keywords: Logistic networks · Dependability · Business continuity

1 Introduction

Global logistic networks, are large scale systems, consisting of many subsystems, which often constitute autonomous organizations. From the point of view of systems science they can be considered as so-called systems of systems, which are a subject of system of systems engineering (SoSE) [6, 7, 11, 19, 20]. This methodology was successfully used in military applications in the last two decades, but nowadays it is increasingly being applied to non-defence related problems, such as architectural design of problems in transportation, healthcare, global communication networks, search and rescue, space exploration and many other domains. SoSE deals with planning, analyzing, organizing, and integrating the capabilities of existing and new systems into a system of systems with the capability greater than the sum of the capabilities of the constituent parts [11]. System of systems may deliver capabilities by combining multiple collaborative and autonomous-yet-interacting systems.

SoSE differs from the classical engineering of a single system and should include the following factors [11]:

- Larger scope and greater complexity of integration efforts,
- Collaborative and dynamic engineering,
- Engineering under the condition of uncertainty,
- Continuing architectural reconfiguration,
- Simultaneous modelling and simulation of emergent systems behaviour,
- Rigorous interface design and management.

© Springer International Publishing AG 2017
J. Świątek and J.M. Tomczak (eds.), *Advances in Systems Science*, Advances in Intelligent Systems and Computing, 539, DOI 10.1007/978-3-319-48944-5_28

Referenced to papers [5], there can be presented the following description of the object of this paper. *A logistic network is a set of systems that results when independent systems are integrated into a larger system that delivers unique capabilities.* The main characteristics of a logistic network are:

- multidimensional complexity – nonlinear and heterogeneous structures (e.g. networks), spatial scope of at least a large scale, dynamic behaviour, and going beyond a single scientific discipline (interdisciplinary or transdisciplinary approach),
- operational and managerial independence of its elements (subsystems) – the subsystems must be able to usefully operate independently and maintain a continuing operational existence independent of the main system,
- emergent behaviour – the logistic network performs functions and carries out purposes that do not reside in any subsystem,
- evolutionary development - development and existence of logistic network is evolutionary with functions and purposes added, removed, and modified with experience.

The typical logistic network is composed of three main subsystems: infrastructure implementing the required functions (e.g. the network organization), protection subsystem (safety and security) and meta-system of management (the flow of information and communication within the logistic network). Example of such logistic network can be e.g. global supply chains, characterized by the following qualities: a distributed structure, a high degree of autonomy of the individual components, a multi-state dynamics, emergent properties and the ability to keep up with rapidly changing target (e.g. demand-driven supply chain).

2 The General Concept of Dependability - Continuity Oriented Approach

The logistic network can be described with the teleological model [12] by the function P and the function G (goal seeking process approach). These functions are taken for the primary measures of system performance (e.g. operating quality) and correspond to well known in the area of applied sciences - the effectiveness (function P) and the efficiency (function G) [6]. Efficiency is the basic economic characteristic and is expressed in practice by such indicators as: profitability, productivity, etc. In contrast, the effectiveness is mainly used in the field of engineering sciences as a measure of the realization of the intended purpose. In the case of operation systems, the basic effectiveness measure is characterized by the degree of fulfilment of the required functions over time, also known as operating quality. It is the way the author understands the *general concept of dependability as a collective term describing the time-related operating quality of a system.* For such systems as logistic networks "the time-related operating quality" can be represented by the term "continuity". We understand this term as follows: *Continuity is a system capability to deliver products (services) at acceptable predefined performance level under the real work conditions (e.g. despite disruptive events).*

The continuity oriented approach is closely related to the idea of resilient enterprise [14] as well as the concept of disruption tolerant operation [2], and based on the methodology of service engineering [13]. The model proposed in the work is based on a typical course of a service delivery process, interrupted by an occurrence of a disruptive event leading to a disruption of this *process continuity*. An example of such a course as a function of time is represented in Fig. 1 (based on [6]), in which there can be distinguished three fundamental phases:

- a phase of normal work, interrupted only by slight disturbances and rare cases of more serious hazards, recognized sufficiently early and "parried" (thwarted), thanks to a properly functioning protection subsystem (safety and security),
- a phase of process continuity disruption, happening directly after an occurrence of a hazard, which has not been effectively blocked by the protection subsystem. During this phase, the system's survivability is its essential property, which is measured by the disruption performance level (DPL) and the post-disruption performance level (PPL), as well as the time t_S
- a phase of system's recovery to an acceptable performance level (APL) after the process continuity disruption, characterized by the APL and the time t_R.

After the removal of the negative effects of the process continuity disruption, a process of the improvement can begin, consisting in an adaptation of the system to new working conditions and a possible increase in the level of performance.

External sources of risk generate threats and hazards which encounter safety and security barriers. As long as these barriers fulfil their roles, the system realizes the required functions in a continuous way. If the barriers turn out to be insufficient to protect a given system, the system must defend itself with its internal resistance

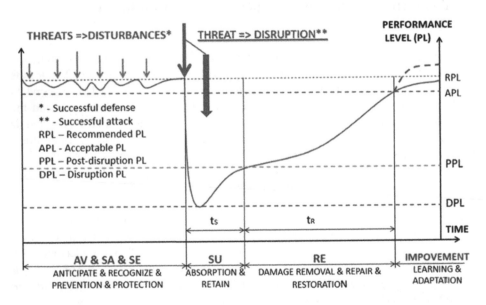

Fig. 1. Typical course of a service delivery process with a continuity disruption

(robustness). In a case in which the resistance is too weak in relation to the working exposure (to absorb disruption impact and to withstand disruption shock), the system goes into the "recovery" state, which encompasses the stages of damage removing, as well as repair and restoration process. If this stage also ends in failure, a total interruption of the process continuity will eventually occur, resulting in the need of reengineering (redesign of structure and processes).

3 The Evaluation of Logistic Networks Dependability

Based on the process continuity oriented approach, the general concept of dependability will be proposed. This concept of dependability includes the constituent properties, that can be represented in the form of a "dependability tree" [1], consisting of three levels. Dependability is divided into five main attributes [6]:

- Availability (AV) – ability to be in state to perform the required functions under given work conditions, is described by:
  - reliability (REL) – ability to perform the required functions, without failure, for a given time interval, under given work conditions;
  - maintainability (MAI) – ability to be retained in, or restored to a state to perform as required, under given conditions of use and maintenance;
  - maintenance support performance (MSP) – effectiveness of an organization in respect to maintenance support;
- Safety (SA) – ability to operate, normally or abnormally, without danger of causing human injury or death and without damage to the system's environment; it consists of:
  - absence of critical damages (ACD);
  - protection of the environment against the effects of any potential critical damages (PRO);
- Security (SE) – ability to prevent an unauthorized access to, or handling of system state, can be described by the concurrent existence of:
  - confidentiality (CON) – unavailability by non-enabled persons;
  - integrity (INT) – impossibility of introducing changes into the system by non-enabled persons;
  - accessibility for enabled users only (ACC);
- Survivability (SU) – capability of a system to fulfil its mission in a timely manner and in the presence of disruptive events; it is described by:
  - detectability (DET) – early threats recognition, supervision and monitoring;
  - robustness (ROB) – resistance and redundancy;
  - adaptability (ADA) – flexibility, agility, fault tolerance;
- Recoverability (RE) – capacity of a system to recover from a failure (restoration) in the acceptable time and costs limits, is divided into:
  - susceptibility to repair (SUS),
  - availability of repair resources (ARR).

Based on this structure of properties the dependability of a system of systems for a given time interval (t_1, t_2), can be described by the model:

$$D = \{w_1 * AV(t_1, \ t_2), w_2 * SA(t_1, \ t_2), w_3 * SE(t_1, \ t_2), w_4 * SU(t_1, \ t_2), w_5 * RE(t_1, \ t_2)\} \quad (1)$$

where: $\sum_{i=1}^{5} w_i = 1$, and $0 \le w_i \le 1$

This model can be interpreted as follows:

A logistic network dependability is the collective term that describes its ability to continuously and safely fulfil the required functions in normal and abnormal operating conditions.

This dependability attributes and some examples of its metrics are shown in Table 1.

Table 1. Dependability attributes and its metrics [6, 10]

Attributes	Examples of metrics
Availability	mean availability – A_m (t_1, t_2); mean down time – MDT;
Reliability	mean failure rate – λ_m (t_1, t_2); mean time to failure – MTTF; mean time between failures – MTBF; reliability function – R (t_1, t_2)
Maintainability	mean repair time MRT; mean time to restoration – MTTR
Maintenance support performance	mean administrative delay – MAD; mean logistic delay – MLD
Safety	probability that a system will be fully functioning or will be failed in a manner that does cause no harm in the time period (t_1, t_2)
Security	probability that a protection subsystem will be able to prevent an unauthorized access to the system in the time period (t_1, t_2)
Survivability	probability that a system will be able to fulfil its mission in the presence of disruptive events in the time period (t_1, t_2); mean time to "bounce back" – MTTBB
Recovery	probability that a system will be able to recover from a failure, in the acceptable time and costs limits; mean time to recover from failure – MTTRFF

The metrics for the various attributes of dependability summarized in Table 1 are probabilistic, therefore, the primary method of determining them is the mathematical statistics. However, at the stage of new logistic network designing, there is no data that would allow to estimate these indicators. In this case, the only effective method is to use the expertise. An example of an expert system using fuzzy logic to evaluate the dependability for the logistic networks of an international manufacturing company is shown in the Sect. 4.

4   The Application Example – A Fuzzy Logic Based Expert System

The application example of the model (1) with the use of fuzzy sets theory [16–18] is based on the idea published in [3, 4]. Fuzzy sets were used because they allow the implementation and analysis of imprecise data. Fuzzy logic oriented systems are a kind of expert systems built on a knowledge base that contains inference algorithms in the form of a rule base. Expert knowledge about availability, safety, security, survivability and recoverability of a system of systems can be expressed in the form of "if … then" rules. The knowledge encoded in a rule base is derived from human experience and intuition, as well as on the basis of theoretical understanding of the properties of the studied object. What distinguishes fuzzy inference in terms of a concept from conventional inference is the lack of analytical description. The approximate inference mechanism transforms knowledge from the rule base into a non-fuzzy form. The non-fuzzy form of any result is obtained by the defuzzification process. Defuzzification transforms the membership degrees of fuzzy sets into a real value. The software WinFACT® [15] was used for building a system in order to evaluate the dependability level. For the design and analysis of fuzzy systems WinFACT offers a comfortable Fuzzy Shell (module FLOP). This component enables the graphical definition of all fuzzy sets and the rule base by some simple mouse clicks. The rules of the fuzzy systems can be defined by means of a table- or matrix-editor as well as by a comfortable text editor. At any time a switching between the two operating modes is possible. If desirable each rule can be combined with a rule specific weighting factor. For the fuzzy inference and the defuzzification process various mechanism are at disposal. Fuzzy model can directly be integrated into the block-oriented simulation module BORIS. By this way even high complex hybrid systems can easily be simulated and analyzed. During the simulation process a so-called "fuzzy debugger" appears, which allows the detailed analysis of the fuzzy model function, the actually active rules and more [15]. The appropriate selection of rules as well as the shape and range of the membership function is verified with a help of a rules viewer and simulation method. The rules viewer displays a roadmap for the whole fuzzy inference process.

An example of the application of this idea at the design stage for the logistic networks of an international manufacturing company is shown in the Fig. 2. The evaluation model of system of systems dependability has a hierarchical structure. At first the Availability (AV), Safety (SA), Security (SE), Survivability (SU) and Recoverability (RE), are evaluated separately, and then the information about them can be represented by the membership function of fuzzy system. Trapezium membership functions are associated with the numerical value of AV and linguistic variable of SA, SE, SU and RE. Each of these parameters was divided into three categories - Low, Moderate and High. Value ranges for the different categories was determined by analogy to the analysis of the correlation coefficient ($0 <$ Low $< 0, 3; 0, 3 <$ Moderate $< 0, 7; 0, 7 <$ High $< 1, 0$). The level of Dependability (D) is represented by the output of this system and depends on the value of inputs and the knowledge which is implemented in the rules base within a fuzzy system.

Figure 3 shows the screenshot of the safety (SA) fuzzy model. The upper-left part of the figure shows the general structure of the model, i.e. inputs and outputs of the system. The inputs of the system are two linguistic variables: ACD (absence of critical damages) and PRO (protection of the environment against the effects of any potential critical damages). The membership functions for this variables shows the right part of the Fig. 3. In the lower-left part of the figure are shown the shapes and ranges of the membership functions for the output variable SA (safety).

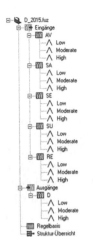

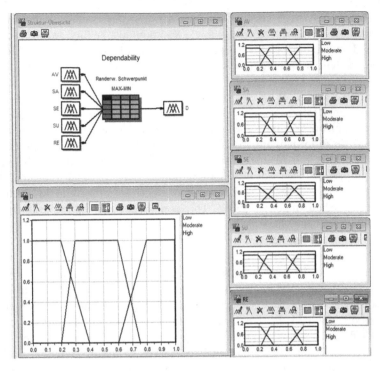

Fig. 2. Screenshot of the dependability fuzzy model

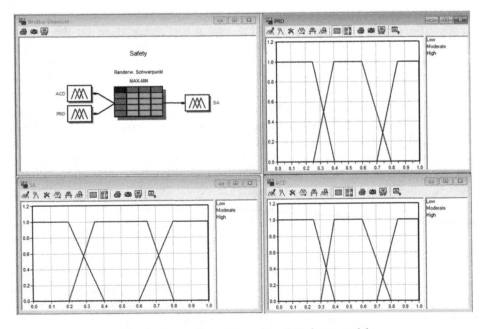

Fig. 3. Screenshot of the safety (SA) fuzzy model

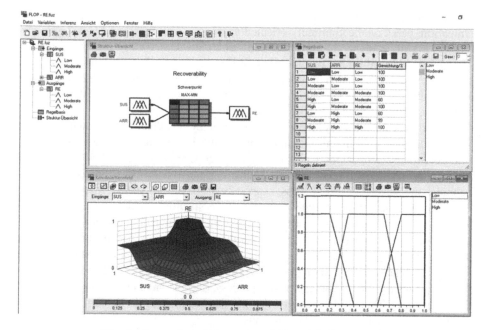

Fig. 4. Screenshot of the recoverability (RE) fuzzy model

On the Fig. 4 are shown: the structure, the base rules and the result of the evaluation of variable recoverability (RE). Assuming that the values of the input variables, namely susceptibility to repair (SUS), availability of repair resources (ARR) and the output variable RE are divided into three ranges (low, moderate and high) complete rule base that allows for the implementation of expertise consists of nine rules. In the process of designing the various components of the system, to different rules they have been assigned different coefficients of confidence (expressed in %). The use of the different coefficients of certainty for the individual rules, many different kinds of defuzzification methods and diverse shapes of membership functions allow us to model the uncertainty in decision-making, the pessimistic, optimistic or moderate tendencies of decision maker or an uncertainty in the estimation of the input data.

5 Conclusion

Logistic networks, especially the global ones, are large scale system of systems characterized by: a distributed structure, a high degree of autonomy of the individual components, a multi-state dynamics, emergent properties and the ability to keep up with rapidly changing target. To reduce total costs, more and more parts of the logistic networks are outsourced to specialized service providers. As a result of this tendency, the business continuity of global organizations is strongly related to the providers service quality and dependability.

The aim of this paper is to propose a concept of logistic networks dependability evaluation, based on the continuity oriented approach. This approach is closely related to the idea of resilient enterprise as well as to the concept of disruption-tolerant operation, and based on the methodology of service engineering. Because the modern logistic networks are more and more reliable, the number of failures and disruptions is significantly decreased. Therefore the well known probabilistic methods, based on statistics, are in this case not more acceptable.

We propose an expert system, based on the fuzzy sets theory, to evaluate the level of dependability for different configurations of the logistic networks and to simulate different possible disruptions scenarios. This system is an universal, "shell" type model, that can be applied to verifying and validating the reliability of various types of technical and sociotechnical systems, especially at the design stage. Adapting this tool to the needs of a particular type of system or a specific practical case requires the estimation of numerical values corresponding to each parameter class. In the case of the application of triangular or trapezoidal membership functions for the linguistic variables, we may assume that the measure of uncertainty in quantitative estimates is the angle of inclination of the triangles or trapezoids sides (a right angle corresponds to the lack of uncertainty in the estimation process, and the smaller the angle, the larger the uncertainty). This model was successfully used to investigate the logistic networks of an international manufacturing company.

References

1. Avizienis, A., et al.: Basic concepts and taxonomy of dependable and secure computing. IEEE Trans. Dependable Secure Comput. **1**(1), 11–33 (2004)
2. Bukowski, L.: Managing disruption risks in the global supply networks – a trans-disciplinary approach. In: Proceedings of International Conference on Industrial Logistics, Croatia, pp. 101–106 (2014)
3. Bukowski, L., Feliks, J.: Application of fuzzy sets in evaluation of failure likelihood. In: Proceedings of the 18-th International Conference on Systems Engineering, Las Vegas, pp. 170–175. IEEE CS (2005)
4. Bukowski, L., Feliks, J.: Vector conception of technical system's dependability. In: Proceedings of the 19-th International Conference on Systems Engineering, Las Vegas, pp. 492–497. IEEE CS (2008)
5. Bukowski, L., Feliks, J.: A unified model of systems dependability and process continuity for complex supply chain. In: Safety and Reliability: Methodology and Applications, pp. 2395–2403. Taylor & Francis Group, A Balkema Book, London (2014)
6. Bukowski, L.: System of systems dependability - theoretical models and applications examples. Reliab. Eng. Syst. Saf. **151**, 76–92 (2016)
7. Gideon, J.M., Dagli, C.H., Miller, A.: Taxonomy of systems-of-systems. In: Proceedings CSER, 23–25 March, Hoboken, NJ, USA, p. 363 (2005)
8. Herrera, I.A., Hovden, J.: The Leading indicators applied to maintenance in the framework of resilience engineering: a conceptual approach. In: 3rd Resilience Engineering Symposium, Antibes- Juan Les Pins, France (2008)
9. Hollnagel, E., Woods, D.W., Leveson, N. (eds.): Resilience Engineering: Concepts and Precepts. Ashgate, Abingdon (2006)

10. IEC, International Standards on Dependability. The World's Online Electrotechnical Vocabulary (2015). www.electropedia.org/
11. Luzeaux, D., et al.: Complex System and Systems of Systems Engineering. ISTE Ltd. and Wiley, Hoboken (2011)
12. Mesarovic, M.D.: Foundation for Mathematical Theory of General Systems. Wiley, New York (1964)
13. Salvendy, G., Karwowski, W.: Introduction to Service Engineering. Wiley, Hoboken (2010)
14. Sheffi, Y.: The Resilient Enterprise: Overcoming Vulnerability for Competitive Advantage. MIT Press, Cambridge (2007)
15. WinFACT, Ingenieurbüro Dr. Kahlert (2016). http://www.kahlert.com/web/english/e_home. php
16. Zadeh, L.A.: Fuzzy sets. Inf. Control **8**, 338–353 (1965)
17. Zadeh, L.A.: Fuzzy sets as a basis for a theory of possibility. Fuzzy Sets Syst. **100**, 3–28 (1978)
18. Zadeh, L.A.: Toward a generalized theory of uncertainty (GTU) – an outline. Inf. Sci. **172**, 1721–1740 (2005)
19. Zio, E.: From complexity science to reliability efficiency: a new way of looking at complex network systems and critical infrastructures. Int. J. Crit. Infrastruct. **3**(3/4), 488–508 (2007)
20. Zio, E.: Reliability engineering: old problems and new challenges. Reliab. Eng. Syst. Saf. **94**, 125–141 (2009)

Robust Finite-Time Control of an Uncertain Aeroelastic System Using Leading- and Trailing-Edge Flaps

Prince Ghorawat[1], Keum W. Lee[2], Sahjendra N. Singh[3(✉)],
and Grzegorz Chmaj[4]

[1] Department of Electrical and Computer Engineering,
University of Nevada, Las Vegas, Las Vegas, NV 89154, USA
ghorawat@unlv.nevada.edu

[2] Department of Electronic Engineering, Catholic Kwandong University Gangneung,
Member of AIAA, Gangneung, Gangwon 25601, Republic of Korea
kwlee@cku.ac.kr

[3] Department of Electrical and Computer Engineering,
University of Nevada, Las Vegas, Las Vegas, NV 89154, USA
sahjendra.singh@unlv.edu

[4] Department of Electrical and Computer Engineering,
University of Nevada, Las Vegas, Las Vegas, NV 89154, USA
grzegorz.chmaj@unlv.edu

Abstract. This paper presents a finite-time sliding mode control system for the stabilization of a MIMO nonlinear aeroelastic system using leading-and trailing-edge control surfaces. The selected two-degree-of-freedom aeroelastic model includes uncertain parameters and gust loads. This wing model exhibits limit cycle oscillations (LCOs) beyond a critical free-stream velocity. The objective is to design a control law for stabilization of the LCOs. A control law is derived for the trajectory tracking of the plunge displacement and pitch angle trajectories. The control law includes a finite-time stabilizing control signal for the system without uncertainties and a discontinuous control signal to nullify the effect of uncertain functions. In the closed-loop system, finite-time stabilization of the state vector to the origin is achieved. Simulation results show that the controller accomplishes suppression of LCOs, despite uncertainties in the system parameters and gust loads.

Keywords: Aeroelastic system · Limit cycle oscillation · Robust finite-time control · Sliding mode control

1 Introduction

Aeroelastic phenomena, such as flutter and limit cycle oscillations (LCOs), severely limit the performance of advanced flexible and combat aircraft. For this reason, considerable effort has been made by researchers in the past to analyze dynamical behavior of nonlinear aeroelastic systems and design control

© Springer International Publishing AG 2017
J. Świątek and J.M. Tomczak (eds.), *Advances in Systems Science*, Advances in Intelligent
Systems and Computing 539, DOI 10.1007/978-3-319-48944-5_29

systems to suppress flutter and LCOs [1–4]. Researchers at the NASA Langley Research Center developed a benchmark active control technology (BACT) wind-tunnel model, explored its instability phenomena, designed controllers for flutter suppression, and validated their performance by wind-tunnel tests [5,6]. Also at the Texas A&M University, a two-degree-of-freedom (2DOF) experimental apparatus has been constructed for the study of the aeroelastic phenomena [7–10]. For this model, a variety of robust, and adaptive control systems have been designed, using a single trailing-edge control surface [7,9,11–15], as well as leading- and trailing-edge control surfaces [10,16–19]. Control of aeroelastic systems using neural network has also been considered [15]. Subsequently, noncertainty-equivalent adaptive laws, based on immersion and invariance theory, have been also derived for flutter suppression [19,20]. A special feature of immersion and invariance based control systems is that once the estimated parameters converge to the actual values at certain instant, then they remain frozen thereafter for all time. An $\mathcal{L}_1$ adaptive controller has been also derived in [21] for this model which can achieve desirable performance bounds by using higher adaptation gains. A robust continuous control system for suppression of LCOs has been also proposed [22].

Researchers have considered finite-time control of nonlinear systems [23,24]. Also for a class of uncertain nonlinear multivariable systems, a higher-order sliding mode control law for the finite-time control has been developed [25]. This controller has a simple structure compared to adaptive systems, and therefore is attractive from the viewpoint of simplicity in implementation. Recently a finite-time stabilizing controller for the suppression of LCOs has been designed [26]. But the aeroelastic model considered for adaptive, robust and finite-time control design, published in literature, are based on simplified models, which ignore sinusoidal functions of pitch angle and nonlinearity in pitch rate of the complete model derived by Seta et al. [8]. Certainly it is of interest to design controllers for the suppression of LCOs of the complete nonlinear aeroelastic model derived in [8]. Although in a recent paper, a differential game based controller for the nonlinear model of [8] has been designed [27], it requires the exact knowledge of system parameters. This is an unrealistic assumption. As such it is of interest to develop a finite-time stabilizing robust control law for the control of the complete aeroelastic model of Seta et al. [8] in the presence of parametric uncertainties and gust loads.

The contribution of this paper lies in the design of robust finite-time stabilizing control law for a 2DOF plunge-pitch aeroelastic model, using leading-and trailing-edge control surfaces. The complete aeroelastic model of [8] with uncertain parameters, including nonlinear functions of pitch angle and pitch rate, and gust load, is considered for design. The objective is to control the plunge and pitch angle trajectories to the origin. A a second-order sliding mode control system for the finite-time stabilization of this nonlinear uncertain aeroelastic system is developed. The control law includes a nominal finite-time stabilizing continuous control signal designed for the model without uncertainties, and a discontinuous control signal for the compensation of uncertain functions and gust load. In the closed-loop system, finite-time control of the complete state vector (pitch angle and plunge displacement and their first derivatives) to the origin is accomplished. Simulation results are presented which show that the controller suppresses the LCOs of the system, despite parameter uncertainties and gust loads.

2   Nonlinear Aeroelastic Model

The model of the multi-input aeroelastic system equipped with leading-and trailing-edge control surfaces is shown in Fig. 1. Its mathematical model has been derived by Seta et al. [8]. The second-order differential equations governing the evolution of the pitch angle (α) and the plunge displacement (h) are given by

$$I_{EA}\ddot{\alpha} + [m_w x_\alpha b cos(\alpha) - m_c r_c b sin(\alpha)]\ddot{h} + c_\alpha \dot{\alpha} + k_\alpha(\alpha)\alpha = M + M_g$$

$$m_t \ddot{h} + [m_w x_\alpha b cos(\alpha) - m_c r_c b sin(\alpha)]\ddot{\alpha} + c_h \dot{h} +$$

$$[-m_w x_\alpha b sin(\alpha) - m_c r_c b cos(\alpha)]\dot{\alpha}^2 + k_h(h)h = -L - L_g \tag{1}$$

where m_W is the mass of the wing section; m_t is the total mass; m_c is the mass of the cam; r_c is nondimensional distance between elastic axis and cam center; b is the semichord of the wing; I_{EA} is the moment of inertia; x_α is the nondimensionalized distance of the center of mass from the elastic axis; c_α and c_h are the pitch and plunge damping coefficients, respectively; and L and M are the aerodynamic lift and moment. It is pointed out that this model, including nonlinear sinusoidal functions of α and nonlinear function of $\dot{\alpha}$, differs from aeroelastic models considered in literature in the past for adaptive and finite-time control system design. In this study, for simplicity, a quasi-steady form of the aerodynamic force and moment given by

$$L = \rho U^2 b C_{l_\alpha} s_p \left[\alpha + (\dot{h}/U) + \left(\frac{1}{2} - a\right)b(\dot{\alpha}/U)\right] + \rho U^2 b C_{l_\beta} s_p \beta + \rho U^2 b C_{l_\gamma} s_p \gamma$$

$$M = \rho U^2 b^2 C_{m\alpha-eff} s_p \left[\alpha + (\dot{h}/U) + \left(\frac{1}{2} - a\right)b(\dot{\alpha}/U)\right]$$

$$+ \rho U^2 b^2 C_{m\beta-eff} s_p \beta + \rho U^2 b^2 C_{m\gamma-eff} s_p \gamma \tag{2}$$

is considered, where a is the nondimensionalized distance from the midchord to the elastic axis, U is the free-stream velocity, s_p is the span, and β and γ are the trailing-edge and leading-edge flap deflections, respectively. The lift and moment derivatives due to α and control surface deflections are C_{l_α}, C_{l_β}, C_{l_γ}, and $C_{m\alpha-eff}$, $C_{m\beta-eff}$, $C_{m\gamma-eff}$, respectively, where

$$C_{m\alpha-eff} = \left(\frac{1}{2} + a\right)C_{l_\alpha} + 2C_{m\alpha}$$

$$C_{m\beta-eff} = \left(\frac{1}{2} + a\right)C_{l_\beta} + 2C_{m\beta}$$

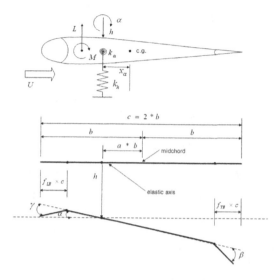

Fig. 1. Aeroelastic system with leading-and trailing-edge flaps

$$Cm_{\gamma-eff} = \left(\frac{1}{2} + a\right) C_{l_\gamma} + 2Cm_\gamma$$

and $C_{m_\alpha} = 0$ for a symmetric airfoil.

The aeroelastic system (1) also includes the lift L_g and moment M_g caused by gust loads. It is assumed that L_g and M_g are of the form [18]

$$L_g = \rho U^2 b s_p C_{l\alpha} w_G(\tau)/U = \rho U b s_p C_{l\alpha} w_G(\tau)$$

$$M_g = (0.5 - a) b L_g \qquad (3)$$

where $w_G(\tau)$ denotes the disturbance velocity, while τ is a dimensionless time variable, defined as $\tau = Ut/b$.

For the purpose of illustration, the functions $k_\alpha(\alpha)$ and $k_h(h)$ are chosen as

$$k_\alpha(\alpha) = 6.861422(1 + 1.1437925\alpha + 96.669627\alpha^2 + 9.513399\alpha^3 - 727.664120\alpha^4)$$

$$k_h(h) = 2844.4 \qquad (4)$$

Solving (1) for $\ddot{\alpha}$ and $\ddot{h}$ gives

$$\begin{bmatrix} \ddot{\alpha} \\ \ddot{h} \end{bmatrix} = \begin{bmatrix} F_{10}(\alpha, h, \dot{\alpha}, \dot{h}, t) \\ F_{20}(\alpha, h, \dot{\alpha}, \dot{h}, t) \end{bmatrix} + [B(\alpha)]u \qquad (5)$$

where $F_0 = [F_{10}, F_{20}]^T \in R^2$ is a nonlinear vector function of indicated arguments, $u = (\beta, \gamma)^T \in R^2$ is the control input vector, and $B(\alpha) \in R^{2\times2}$ is the nonlinear input matrix. The argument t in F_{k0} denotes its dependence on the disturbance input (gust load).

Suppose that $(\alpha_r(t), h_r(t))^T \in R^2$ is a bounded reference trajectory converging to zero. Define

$$z_1 = (z_{11}, z_{21})^T = (\tilde{\alpha}, \dot{\tilde{\alpha}})^T \in R^2; z_2 = (z_{12}, z_{22})^T = (\tilde{h}, \dot{\tilde{h}})^T \in R^2 \qquad (6)$$

where $\tilde{\alpha} = \alpha - \alpha_r$ and $\tilde{h} = h - h_r$ are the pitch and plunge trajectory error, respectively. Also define $z = (z_1^T, z_2^T)^T \in R^4$. Using (5) and (6) gives

$$\dot{z}_{11} = z_{21}; \dot{z}_{12} = z_{22}$$

$$[\dot{z}_{21}, \dot{z}_{22}]^T = F_0(z,t) - [\ddot{\alpha}_r, \ddot{h}_r]^T + B(\alpha)u \doteq F(z,t) + B(\alpha)u \qquad (7)$$

where $F(z,t) = F_0(z) - [\ddot{\alpha}_r, \ddot{h}_r]^T$. The argument t in F also denotes its dependence on the reference trajectory. The vector function $F(z,t)$ and B are assumed to be unknown. We are interested in designing a robust state variable feedback control law for finite-time convergence of $z(t)$ to the origin. Because the reference trajectory converges to zero, this will ensure convergence of the plunge displacement and pitch angle trajectories to zero. Also if the reference trajectory is zero, then controller will achieve finite-time convergence of $z(t)$ to zero.

3   Robust Finite-Time Stabilizing Control Law

In this section, a finite-time stabilizing sliding mode control law for the uncertain system (7) is derived. The control law will consist of a nominal control signal and a discontinuous signal for nullifying the effect of uncertainties in the model.

For the purpose of design, the unknown matrix and the function in Eq. (7) are decomposed into a known and an unknown parts as follows:

$$F(z,t) = F^*(z) + \Delta F(z,t)$$

$$B(\alpha) = B^*(\alpha) + \Delta B(\alpha) \qquad (8)$$

where the quantities with the superscript $*$ denote nominal values; and ΔF and ΔB denote uncertain portions of F and B, respectively. We are interested in the evolution of the trajectories of the system in a compact set $\Omega \in R^4$ in which the input matrix B and B^* remain nonsingular. Then the system dynamics (7) can be expressed as

$$\dot{z}_{11} = z_{21}; \dot{z}_{12} = z_{22}$$

$$[\dot{z}_{21}, \dot{z}_{22}]^T = F^*(z) + \Delta F(z,t) + (B^*(\alpha) + \Delta B(\alpha))u \qquad (9)$$

In view of (9), one selects a control law of the form

$$u = (B^*(\alpha))^{-1}[-F^*(z) + u_n + u_d] \qquad (10)$$

where u_n is a nominal control signal and u_d is a discontinuous signal, (u_n and u_d are yet to be determined). Substituting (10) in (9) gives

$$[\dot{z}_{21}, \dot{z}_{22}]^T = u_n + [I_{2\times2} + \Delta B(B^*)^{-1}]u_d + \Delta F + \Delta B(B^*)^{-1}(u_n - F^*) \qquad (11)$$

For a system described by a chain of integrators, a control input which yields a homogeneous system has been derived by Bhat and Bernstein [23, 24]. For the aeroelastic system, the continuous control signal $u_n = (u_{n1}, u_{n2})^T$ takes the form

$$u_{n1}(z_1) = -p_{11}|z_{11}|^{\nu_{11}} sgn(z_{11}) - p_{21}|z_{21}|^{\nu_{21}} sgn(z_{21})$$

$$u_{n2}(z_2) = -p_{12}|z_{12}|^{\nu_{12}} sgn(z_{12}) - p_{22}|z_{22}|^{\nu_{22}} sgn(z_{22}) \tag{12}$$

where p_{ij} are selected so that the roots of

$$\lambda^2 + p_{2i}\lambda + p_{1i} = 0 \tag{13}$$

are stable ($i = 1, 2$), and the exponents ν_{ij} are chosen to satisfy $\nu_{1i} = \nu_{2i}/(2-\nu_{2i})$ with $\nu_{2i} = \nu_i \in (1 - \epsilon_i, 1)$ and $\epsilon_i \in (0, 1)$. For the system without uncertainties and with control input $u_d = 0$, in view of (11), the closed-loop system yields two decoupled systems and the α dynamics ($i = 1$) and h dynamics ($i = 2$) are described by

$$\dot{z}_{1i} = z_{2i}$$

$$\dot{z}_{2i} = u_{ni} = -p_{1i}|z_{1i}|^{\nu_{11}} sgn(z_{1i})) - p_{2i}|z_{2i}|^{\nu_{21}} sgn(z_{2i}) \tag{14}$$

It is easily verified that the ith-subsystem (13) is homogeneous of negative degree $\mu_i = (\nu_i - 1)/\nu_i$, with dilation $(\nu_{1i}^{-1}, \nu_{2i}^{-1})$. Based on a positive definite radially unbounded Lyapunov function, it has been proven in [23, 24] that there exists $\epsilon_i \in (0, 1)$ such that, for every $\nu_i \in (1-\epsilon, 1)$, the origin $z_i = 0$ of (14) is finite-time stable.

Now for robust control of the uncertain system, the design of a discontinuous control signal u_d based on a higher-order sliding mode design technique of [25] is considered. First we make the following assumption.

Assumption 1: In the admissible region Ω, there exist functions $\gamma_1(z, t)$ and a $\gamma_0 \in [0, 1)$ such that the following inequalities hold :

$$||\Delta F(z, t) + \Delta B(\alpha)(B^*(\alpha))^{-1}(u_n - F^*(z))||_\infty \leq \gamma_1(z, t) \tag{15}$$

$$||\Delta B(\alpha)(B^*(\alpha))^{-1}||_\infty \leq \gamma_0 < 1 \tag{16}$$

For the design, consider a sliding surface $s(z, z_a) = 0$ of the form

$$s(z) = [z_{21}, z_{22}]^T - z_a \tag{17}$$

$$\dot{z}_a = u_n \tag{18}$$

Differentiating s along the solution of (11) and (18) gives

$$\dot{s} = [I_{2 \times 2} + \Delta B(\alpha)(B^*(\alpha))^{-1}]u_d + \Delta F(z, t) + \Delta B(\alpha)(B^*(\alpha))^{-1}(u_n - F^*(z)) \tag{19}$$

For the derivation of u_d, consider a quadratic Lyapunov function

$$W(s) = (s^T s)/2 \tag{20}$$

Differentiating W and using (19) gives

$$\dot{W} = s^T[(I_{2\times2}+\Delta B(\alpha)(B^*(\alpha))^{-1})u_d+\Delta F(z,t)+\Delta B(\alpha)(B^*(\alpha))^{-1}(u_n-F^*(z))]$$
(21)

In view of the inequalities (15) and (16), it easily follows that $\dot{W}$ satisfies

$$\dot{W} \leq s^T[I_{2\times2} + \Delta B(\alpha)(B^*(\alpha))^{-1}]u_d + ||s||_1.||\Delta F(z,t) + \Delta B(\alpha)(B^*(\alpha))^{-1}(u_n - F^*(z))||_\infty$$

$$\leq s^T(I_{2\times2} + \Delta B(\alpha)(B^*(\alpha))^{-1})u_d + \gamma_1(z,t)||s||_1 \qquad (22)$$

where $||.||_1$ and $||.||_\infty$ denote 1 norm and ∞ norm of a vector. For making $s = 0$ attractive, one selects u_d as

$$u_d = -G(z,t)sign(s) \qquad (23)$$

where

$$G(z,t) \geq (1 - \gamma_0)^{-1}[\gamma_1 + \eta] \qquad (24)$$

with $\eta > 0$. Substituting (23) in (22) gives

$$\dot{W} \leq -\eta||s||_1 \leq -\eta||s||_2 \leq -\sqrt{2}\eta\sqrt{W} \qquad (25)$$

This implies that s converges to zero in a finite time and remains zero thereafter. In view of (19), on the sliding manifold $s = 0$, the equivalent control u_{deq} satisfies

$$[I_{2\times2} + \Delta B(\alpha)(B^*(\alpha))^{-1}]u_{deq} + \Delta F(z,t) + \Delta B(\alpha)(B^*(\alpha))^{-1}(u_n - F^*(z)) = 0$$
(26)

Then setting $u_d = u_{deq}$ and using (26) in Eq. (11) gives

$$\dot{z}_{1k} = z_{2k}; [\dot{z}_{12}, \dot{z}_{22}]^T = u_n \qquad (27)$$

Therefore, in view of (14), $z(t)$ converges to zero in a finite time for any $z(0)$. This completes stability analysis for the closed-loop system.

4   Simulation Results

This section presents the results of simulation. The system parameters of [8] are listed in Table 1 in the appendix. Although stability in the closed-loop system is ensured for any uncertainties satisfying (15), here a simplified control law is implemented by setting the nominal vector function F^* to zero. But simulation is done for the choice of $\Delta B(\alpha) = 0$ as well as $\Delta B(\alpha) \neq 0$ for robustness test. The gains in the control law u_n are $p_{11} = 12, p_{21} = 35, p_{12} = 12, p_{22} = 35$, and the fractional exponents are $\nu_{11} = 1/2, \nu_{21} = 2/3, \nu_{12} = 1/2$, and $\nu_{22} = 2/3$. The value of G in the discontinuous control law is $G = 0.05$. For simulation, the sign function in the discontinuous control law is replaced by the saturation function (sat(.)) to avoid control chattering, where $sat(p) = p/\epsilon$ if $|p| < \epsilon$, and

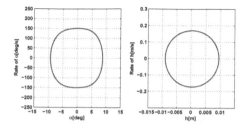

Fig. 2. Limit cycle in the open-loop system: $U = 13.8$ m/s, $a = -0.4$.

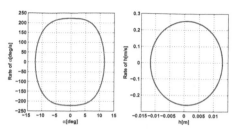

Fig. 3. Limit cycle in the open-loop system, higher U: $U = 18$ m/s, $a = -0.4$.

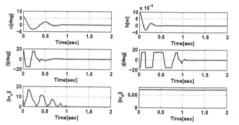

Fig. 4. Robust control: $U = 13.8$ m/s, $a = -0.4$, $W_G = 0$.

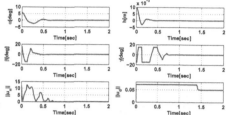

Fig. 5. Robust control, higher U: $U = 18$ m/s, $a = -0.4$, $W_G = 0$

$\text{sat}(p) = \text{sign}(p)$, if $|p| \geq \epsilon$, where $\epsilon = 0.001$. The initial conditions are set to $h(0) = 0.01$ [m], $\alpha(0) = 5.729$ [deg], and $\dot{\alpha}(0) = \dot{h}(0) = 0$.

First stability properties of the uncontrolled wind gust-free system is examined. The open-loop responses for two free-stream velocities are shown in Figs. 2 and 3. It is observed that the open-loop system exhibits limit cycle oscillation in the steady-state. As expected, one observes that the size of LCO increases with the free-stream velocity. Apparently, it is essential to suppress these undesirable oscillations in the plunge and pitch motion.

Now the responses of the closed-loop system, including the control law (10), (12) and (23), in the absence of gust load, for $U = 13.28$ [m/s] and $a = -0.4$ are obtained. It is assumed that F^* and $\Delta B(\alpha)$ are zero. This implies that the vector function $F(z,t)$ is completely unknown. But for robustness test, the model dynamics with actual parameters are simulated. For a realistic simulation, control surface deflections are allowed to saturate at 15 [deg]. It is observed in Fig. 4 that the pitch angle and the plunge displacement converge to zero. The response-time is of the order of less than one second. The control input saturates in the initial transient phase.

Now, performance of the controller at a higher value of free-stream velocity, $U = 18$ [m/s] is evaluated. The controller used to obtain Fig. 4 is retained. Selected responses are shown in Fig. 5. It is seen that, despite larger free-stream velocity, the plunge and pitch trajectories converges to zero. It is seen that convergence time is shorter due to increased effectiveness of flaps at higher free-stream velocity.

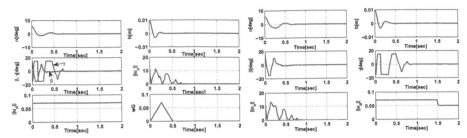

Fig. 6. Robust control, system with triangular gust load: $U = 18$ m/s, $a = -0.4$, $w_0 = 0.07$

Fig. 7. Robust control, uncertainty in input matrix B: $U = 18$ m/s, $a = -0.4$, $W_G = 0$, $\Delta B = -0.25B$

Now the effect of gust load on the performance of the controller is examined. For this, simulation is done in the presence of a triangular gust with $U = 18$ [m/s] and $a = -0.4$. The triangular w_G shown in Fig. 6 is

$$w_G(\tau) = 2w_0 \frac{\tau}{\tau_G} \left(H(\tau) - H\left(\tau - \frac{\tau_G}{2}\right)\right) + 2w_0 \left(\frac{\tau}{\tau_G} - 1\right)\left(H(\tau - \tau_G) - H\left(\tau - \frac{\tau_G}{2}\right)\right)$$

where $H(.)$ denotes the unit step function, $w_0 = 0.07$, $\tau_G = Ut_G/b$ and $t_G = 0.5$ s. The controller used for previous cases is implemented without any modification. Selected responses show that controller suppresses the LCOs, despite the action of chosen wind gust.

In the previous cases, it has been assumed that ΔB is zero. Now to examine the sensitivity of the control law, simulation is done assuming that $B^* = 1.25B$, but the gust load is zero. Thus it is assumed that there is uncertainty in the input matrix B such that ΔB is -0.25 B. Selected responses in Fig. 7 show that the LCOs are suppressed, despite uncertainty in the input matrix B.

Simulation has been done for other types of uncertainties. These results show that controller is capable of suppressing the LCOs in closed-loop system. It is noted that a differential game-based controller [27] has been developed for this aeroelastic model with gust load, but for that design it is essential to know the system parameters exactly. Unlike [27], the controller designed here achieves finite-time stabilization, despite parametric uncertainties and gust loads.

5   Conclusion

In this paper, control of a multi-input multi-output aeroelastic system for the finite-time stabilization in the presence of gust load and parametric uncertainties was considered. The control system included a primary feedback loop designed for the trajectory control for the nominal model. This nonlinear control law yielded a nominal homogeneous system which achieved finite-time stabilization. Then a discontinuous control law was developed based on a second-order sliding mode control scheme for eliminating the effect of uncertainties in the model.

In the complete closed-loop system stabilization of the state vector to the origin was accomplished. Simulation results were presented which validated finite-time robust suppression of the limit cycle oscillations despite uncertainties and gust load.

Appendix

Table 1. System parameters

Parameter value	Parameter value	Parameter value
$a = -0.4$	$b = 0.1064$ [m]	$s_p = 0.6$ [m]
$\rho_a = 1.225$ [kg/m^3]	$r_{cg} = (0.82 \times b - b - a \times b)$ [m]	$x_\alpha = r_{cg}/b$
$c_h = 27.43$ [kg/s]	$c_\alpha = 0.0360$ [N·s]	$k_h = 2844$ [N/m]
$r_c = 1.1936$	$m_c = 0.718$ [kg]	$m_w = 1.662$ [kg]
$m_t = 12.0$ [kg]	$I_{EA} = 0.04325 + m_{wing} \cdot r_{cg}^2$ [kg ·m^2]	$c_{m_\gamma} = -0.1005$ [rad^{-1}]
$c_{l_\alpha} = 6.757$ [rad^{-1}]	$c_{l\beta} = 3.774$ [rad^{-1}]	$c_{l\gamma} = -0.1566$ [rad^{-1}]
$c_{m_\alpha} = 0$ [rad^{-1}]	$c_{m_\beta} = -0.6719$ [rad^{-1}]	
$k_\alpha = 6.861422(1+1.1437925\alpha+96.669627\alpha^2 +9.513399\alpha^3 -727.664120\alpha^4)$ [N·m/rad]		

References

1. Mukhopadhyay, V.: Historical perspective on analysis and control of aeroelastic responses. J. Guidance Control Dyn. **26**, 673–684 (2003)
2. Thomas, J.P., Dowell, E.H., Hall, K.C.: Nonlinear inviscid aerodynamic effects on transonic divergence, flutter, and limit-cycle oscillations. AIAA J. **40**, 638–646 (2002)
3. Marzocca, P., Librescu, L., Silva, W.A.: Flutter, postflutter, and control of a supersonic wing section. J. Guidance Control Dyn. **25**, 962–970 (2002)
4. Mabey, D.G.: Physical phenomena associated with unsteady transonic flows. In: Nixon, D. (ed.) Unsteady Transonic Aerodynamics. AIAA Progress in Aeronautics & Astronautics, vol. 120, Chap. 1, pp. 1–55 (1989)
5. Waszak, M.R.: Robust multivariable flutter suppression for benchmark active control technology wind-tunnel model. J. Guidance Control Dyn. **24**, 147–153 (2001)
6. Mukhopadhyay, V.: Transonic flutter suppression control law design and wind-tunnel test results. J. Guidance Control Dyn. **23**(5), 930–937 (2000)
7. Ko, J., Kurdila, A.J., Strganac, T.W.: Nonlinear control of a prototypical wing section with torsional nonlinearity. J. Guidance Control Dyn. **20**, 1181–1189 (1997)
8. Sheta, E.F., Harrand, V.J., Thompson, D.E., Strganac, T.W.: Computational and experimental investigation of limit cycle oscillations of nonlinear aeroelastic systems. J. Aircr. **39**, 133–141 (2002)
9. Block, J.J., Strganac, T.W.: Applied active control for a nonlinear aeroelastic structure. J. Guidance Control Dyn. **21**, 838–845 (1998)

10. Platanitis, G., Strganac, T.W.: Control of a nonlinear wing section using leading- and trailing-edge surfaces. J. Guidance Control Dyn. **27**, 52–58 (2004)
11. Lee, K.W., Singh, S.N.: Global robust control of an aeroelastic system using output feedback. J. Guidance Control Dyn. **30**, 271–275 (2007)
12. Lee, K.W., Ghorawat, P., Singh, S.N.: Wing rock control by finite form adaptation. J. Vib. Control **22**, 2687–2703 (2016). doi:10.1177/1077546314550222
13. Xing, W., Singh, S.N.: Adaptive output feedback control of a nonlinear aeroelastic structure. J. Guidance Control Dyn. **23**, 1109–1116 (2000)
14. Behal, A., Marzocca, P., Rao, V.M., Gnann, A.: Nonlinear adaptive control of an aeroelastic two-dimensional lifting surface. J. Guidance Control Dyn. **29**, 382–390 (2006)
15. Gujjula, S., Singh, S.N., Yim, W.: Adaptive and neural control of a wing section using leading- and trailing-edge surfaces. Aerosp. Sci. Technol. **9**, 161–171 (2005)
16. Behal, A., Rao, V.M., Marzocca, P., Kamaludeen, M.: Adaptive control for a nonlinear wing section with multiple flaps. J. Guidance Control Dyn. **29**, 744–749 (2006)
17. Reddy, K.K., Chen, J., Behal, A., Marzocca, P.: Multi-input/multi-output adaptive output feedback control design for aeroelastic vibration suppression. J. Guidace Control Dyn. **30**, 1040–1048 (2007)
18. Wang, Z., Behal, A., Marzocca, P.: Model-free control design for multi-input multi-output aeroelastic system subject to external disturbance. J. Guidance Control Dyn. **34**, 446–458 (2011)
19. Lee, K.W., Singh, S.N.: Multi-input noncertainty-equivalent adaptive control of an aeroelastic system. J. Guidance Control Dyn. **33**, 1451–1460 (2010)
20. Mannarino, A., Mantegazza, P.: Multifidelity control of aeroelastic systems: an immersion and invariance approach. J. Guidance Control Dyn. **37**, 1568–1582 (2014)
21. Lee, K.W., Singh, S.N.: L_1 adaptive control of a nonlinear aeroelastic system despite gust load. J. Vib. Control **19**, 1807–1821 (2013)
22. Wang, Z., Behal, A., Marzocca, P.: Continuous robust control for two-dimensional airfoils with leading- and trailing-edge flaps. J. Guidance Control Dyn. **35**, 510–519 (2012)
23. Bhat, S.P., Bernstein, D.S.: Finite-time stability of continuous autonomous systems. SIAM J. Control Optim. **38**, 751–766 (2000)
24. Bhat, S.P., Bernstein, D.S.: Geometric homogeneity with applications to finite-time stability. Math. Control Signals Syst. **17**, 101–127 (2005)
25. Defoort, M., Floquet, T., Kokosy, A., Perruquetti, W.: A novel higher order sliding mode control scheme. Syst. Control Lett. **58**, 102–108 (2009)
26. Lee, K.W., Singh, S.N.: Robust higher-order sliding-mode finite-time control of aeroelastic systems. J. Guidance Control Dyn. **37**, 1664–1671 (2014)
27. Ghorawat, P., Lee, K.W., Singh, S.N.: Differential game-based control law for stabilization of aeroelastic system with gust load. In: AIAA Guidance, Navigation, and Control Conference, (Paper AIAA 2016–1866), San Diego, CA, January 2016

Resource Allocation in Self-managing Networks Under Uncertainty

Dariusz Gąsior[(⊠)]

Faculty of Computer Science and Management,
Wroclaw University of Science and Technology,
Wybrzeze Wyspianskiego 27, 50-370 Wroclaw, Poland
Dariusz.Gasior@pwr.edu.pl

Abstract. Nowadays, the increasing complexity of the computer networks causes that most resource management tasks have to be performed in automatic manner. The paper presents the approach to the capacity allocation problem in self-managing networks with uncertain parameters. The network operation is described using Network Utility Maximization framework. The uncertainty is modeled with uncertain variable formalism and the aforementioned problem is formulated in terms of the game theory. The algorithm finding pure Nash Equilibrium for the defined game is introduced.

Keywords: Computer network · Game theory · Optimization · Self-management · Uncertainty · Uncertain variables

1 Introduction

Recently, the operational costs of computer networks maintenance started to grow rapidly. Since, the new technologies, protocols, frameworks have been developed, the administration of such environments became extremely difficult. On the other hand, the management tasks have to be executed quickly and correctly. This is the reason that new mechanisms enabling automation of the management processes in the computer networks have become the main topic of interest. One of the first ideas was presented by IBM and called autonomic networking [1]. Then, the concept of the self-organizing networks (SON) was introduced [2]. This approach was successfully applied especially for the mobile networks [3]. In [4] authors argued that existing protocols like well known Transmission Control Protocol (TCP) or Open Shortest Path First (OSPF) may be also viewed as the beginnings of autonomic networking.

A survey of the current results in autonomic network management may be found in [5, 6]. The most common approaches adapt control theoretic approach [7], biological inspired mechanisms [8] and game theory [9].

However, most of the proposed algorithms were developed for the deterministic cases. This means that despite they enable distributed managing it is assumed that the values of the local network parameters have to be precise. Unfortunately, it has been shown that due to many reasons, it is unachievable [10].

In this paper, the efficient self-managing algorithm for capacity allocation under uncertainty basing on the game theory is presented.

© Springer International Publishing AG 2017
J. Świątek and J.M. Tomczak (eds.), *Advances in Systems Science*, Advances in Intelligent Systems and Computing, 539, DOI 10.1007/978-3-319-48944-5_30

2   Uncertain Variables

The formalism of uncertain variables is based on the paradigm of soft property $\phi(\lambda)$, which means that for the fixed λ the logic value fulfills condition $v[\phi(\lambda)] \in [0, 1]$. The following soft properties are used in the definition of the uncertain variable $\bar{\lambda}$:

- "$\bar{\lambda} \cong \lambda$" which means "$\bar{\lambda}$ is approximately equal to λ,"
- "$\bar{\lambda} \tilde{\in} D_\lambda$" which means "$\bar{\lambda}$ approximately belongs to the set D_λ".

The uncertain variable $\bar{\lambda}$ is defined by a set of values Λ (the real number vector space) and the function $h_\lambda(\lambda) = v(\bar{\lambda} \cong \lambda)$ (i.e. the certainty index that $\bar{\lambda} \cong \lambda$, given by an expert). For the properties of uncertain variables and further details see [11, 12].

For non-empty set D_λ, the certainty index that "$\bar{\lambda}$ approximately belongs to the set D_λ" is given by $v(\bar{\lambda} \tilde{\in} D_\lambda) = \max_{\lambda \in D_\lambda} h_\lambda(\lambda)$.

Usually, a certainty distribution has a form of

$$h_\lambda(\lambda) = \begin{cases} \bar{h}_\lambda(\lambda) & \text{for} & \hat{\lambda} - d_\lambda \leq \lambda \leq \hat{\lambda}, \\ \underline{h}_\lambda(\lambda) & \text{for} & \hat{\lambda} < \lambda \leq \hat{\lambda} + d_\lambda, Z \\ 0, & \text{otherwise.} \end{cases}$$

where $\bar{h}_\lambda(\lambda)$ is the increasing function, $\underline{h}_\lambda(\lambda)$ is the decreasing function, and $\bar{h}_\lambda(\hat{\lambda} - d_\lambda) = 0, \bar{h}_\lambda(\hat{\lambda}) = \underline{h}_\lambda(\hat{\lambda}) = 1, \underline{h}_\lambda(\hat{\lambda} + d_\lambda) = 0$. The values $\hat{\lambda}$ and d_λ indicate respectively the most certain value of the unknown parameter according to the expert and the range of possible values of the unknown parameter (i.e., $[\hat{\lambda} - d_\lambda, \hat{\lambda} + d_\lambda]$).

3   Capacity Allocation Problem in Computer Networks

3.1   Deterministic Case

Let us assume that the computer network consists of L links. The lth link has a capacity C_l. There are R traffic flows in the network. The routing matrix for traffic flows is given by binary matrix a_{rl}. Let us denote by x_{rl} capacity allocated to the rth traffic flow on the lth link. So, the transmission rate of such flow is given by $\bar{x}_r = \min_{l:a_{rl}=1} x_{rl}$. The higher is the transmission rate, the more money the flow's owner is willing to pay. This dependence is clearly explained in terms of the network utility [13, 14]. So, we assume that there is given the utility function f_r for each transmission flow which reflect the possible income for serving the particular transmission with the given rate $\bar{x}_r$. Usually, the utility function is given in the form of:

$$f_r(\bar{x}_r) = w_r \varphi(\bar{x}_r)$$

where

$$\varphi(\bar{x}_r) = \begin{cases} (1-\rho)^{-1}(\bar{x}_r)^{(1-\rho)} & \rho > 0, \rho \neq 1, \\ ln\bar{x}_r & \rho = 1, \end{cases}$$

and w_r is a parameter characterizing the value of the transmission flow (user's willingness-to-pay).

Once we assume that all networks are the property of the same owner, adapting the network utility maximization framework, we may formulate the following capacity allocation problem:

Given: $R, L, C_l, a_{rl}, f_r, w_r,$

Find:

$$x^* = argmax \sum_{r=1}^{R} w_r \varphi(\bar{x}_r) \tag{1}$$

such that:

$$\forall_l \quad \sum_{r=1}^{R} a_{rl} x_{rl} \leq C_l,$$

$$\bar{x}_r = \min_{l:a_{rl}=1} x_{rl},$$

$$\forall_l \forall_r \quad x_{rl} \geq 0,$$

where $x = [x_{rl}]_{r=1,2,...,R_k; l=1,2,...,L}.$

3.2   Uncertain Case

As it was stated, e.g. in [10, 14], the exact values of the network parameters usually are not precisely known in advance. Furthermore, it is justify using non-probabilistic approaches in such cases [15]. In this paper, we consider the case when the objective function parameters w_r are uncertain and there is only expert's (e.g. network operator's) knowledge concerning the possible values of these parameters and express it in terms of the certainty distributions (denoted by $h_r(w_r)$). Since w_r are not crisp values, the precise value of the objective in (1) cannot be determined. Instead, we can only say what is the certainty index that the objective is approximately not less than the given value α. So, the capacity allocation problem under uncertainty consists in finding such an allocation that this certainty index is maximal, i.e.:

Given: $R, L, C_l, a_{rl}, f_r, h_r, \alpha$

Find:

$$x^* = argmaxv \left[\sum_{r=1}^{R} w_r \varphi(\bar{x}_r) \gtrsim \alpha \right] \tag{2}$$

such that:

$$\forall_l \ \sum_{r=1}^{R} a_{rl} x_{rl} \leq C_l,$$

$$\bar{x}_r = \min_{l:a_{rl}=1} x_{rl},$$

$$\forall_l \forall_r \ \ x_{rl} \geq 0.$$

There are some algorithms, which enable finding optimal solution for such an allocation problem in for deterministic case, e.g. in [16, 17] and also for uncertain case [18]. However, the proposed methods require collaboration and an information ex-change between the networks. In this paper it is assumed that each network device in a network should work independently enabling self-managing feature.

4   Capacity Allocation Game for Network Self-Managing

The problem of capacity allocation in self-managed environment may be seen as the game. The networks devices (usually corresponding to the network links) may be seen as the players. Their goal is allocate their resources (e.g. links' capacities), so their local profit is maximized. However, their income depends on the other devices' resource allocations. Thus, in this paper, the appropriate game is proposed to model such a situation.

4.1   Deterministic Case

The deterministic capacity allocation game was introduced in [19] and may be summatized as follows. It is assumed that every link in the network is a player (i.e. there are L players). Each player decides how to allocate his capacity among all traversing it traffic flows.

For lth link player, the permissible pure strategy is any vector of the allocations $s_l = [x_1, x_2, \ldots, x_R]^{\mathrm{T}} \in S_l$, where the set of feasible strategies is defined as follows:

$$S_l = \left\{ s_l : \sum_{r=1}^{R} a_{rl} x_{rl} \leq C_l \wedge \forall_r x_{rl} \geq 0 \right\}. \tag{3}$$

The payoff of the lth link player is given as follows:

$$Q_l = \sum_{r=1}^{R} a_{rl} b_r w_r \varphi \left(\min_{l:a_{rl}=1} x_{rl} \right),$$

where $b_r = \dfrac{1}{\sum_{l=1}^{L} a_{rl}}$.

One must notice that for such defined payoff functions, the social welfare [20] is given by:

$$SW = \sum_{l=1}^{L} Q_l = \sum_{l=1}^{L} \sum_{r=1}^{R} a_{rl} \frac{1}{\sum_{l=1}^{L} a_{rl}} w_r \varphi\left(\min_{l:a_n=1} x_{rl}\right) = \sum_{r=1}^{R} w_r \varphi\left(\min_{l:a_n=1} x_{rl}\right)$$

which is identical to the objective in (1).

4.2   Uncertain Case

In this paper, the new capacity allocation game is introduced. Unlike in previous one, the uncertainty is taken into consideration now. It is assumed that the values of the wr parameters are not known in advance, but each link has its own expert (network operator), who expresses his knowledge about possible values of these parameters in terms of certainty distribution. Notice that for each link certainty distribution corresponding to the particular flow may be different. Thus, let us $h_{rl}(w_r)$ denote the certainty distribution of parameter w_r given by an expert corresponding to the lth link.

Furthermore, unlike in centralized problem (2), each link must have its own satisfactory threshold of objective (payoff). Let us denote it by α_l.

Now we can define the game as follows. All links are players. Each link has to allocate its capacity among all flows traversing it. The set of feasible strategies is defined by (3). The payoff for the lth player is given by:

$$v_l = v\left[\sum_{r=1}^{R} a_{rl} b_r \bar{w}_{rl} \varphi\left(\min_{l:a_n=1} x_{rl}\right) \gtrsim \alpha_l\right]. \tag{4}$$

In this paper, the further considerations will be limited to the case when all the certainty distributions have the shape of the equilateral triangle with the following range of possible values: $[w^*_{rl} - \gamma\, w_{rl}, w^*_{rl} + \gamma\, w_{rl}]$.

5   The Algorithm for Pure Nash Equilibrium

The Nash Equilibriums (NE) are import strategy profiles of the proposed game. They correspond to such a situation when no player is willing to change their allocation while they cannot increase their incomes in such a way. Regarding the self-managed networks, the most payable algorithm for any device is the one, which finds a strategy in Nash Equilibrium strategy profile.

The algorithm finding the pure NE has been introduced in [19]. This paper is focused only on the uncertain case.

From the lth player's point of view, the game consist in solving the following optimization problem:

Given: R, C_l, a_{rl}, f_r, h_{rl}, α

Find:

$$s_l^* = argmaxv \left[\sum_{r=1}^{R} a_{rl} \bar{w}_{rl} \varphi(min\{x_{rl}, \min_{k:a_{rk}=1} x_{rk}\}) \tilde{\geq} \alpha_l \right] \qquad (5)$$

such that:

$$\sum_{r=1}^{R} a_{rl} x_{rl} \leq C_l, \qquad (6)$$

$$\forall_r \quad x_{rl} \geq 0. \qquad (7)$$

The certainty index in (4) for the triangular certainty distributions may be given as follows:

$$v_l = \frac{\sum_{r=1}^{R} a_{rl} w_{rl}^* (1+\gamma)\varphi(\bar{x}_r) - \alpha_l}{\sum_{r=1}^{R} a_{rl}\gamma w_{rl}^* \varphi(\bar{x}_r)}$$

Let us denote by R_l' set of flows that are local, i.e. $R_l' = \left\{ r : a_{rl} = 1 \wedge \sum_{l=1}^{L} a_{rl} = 1 \right\}$ and let us denote by $\bar{R}_l$ set of flows that are global, i.e. $\bar{R}_l = \left\{ r : a_{rl} = 1 \wedge \sum_{l=1}^{L} a_{rl} > 1 \right\}$.

Let us also denote the final solution $\bar{x}^* = [\bar{x}_1^*, \bar{x}_2^*, \ldots, \bar{x}_R^*]$, since for NE the allocated capacities should be same for all links that particular flow traverses.

Finally, the algorithm calculating pure NE for the uncertain case is as follows:

1. Let us introduce auxiliary variables $\hat{L}$ and $\hat{R}$.
2. Let us initiate $\hat{L} := \{1.2\ldots, L\}$ and $\hat{R} := \{1.2\ldots, R\}$
3. For each link $l \in \hat{L}$ solve the following optimization problem:
   Find:

$$\hat{s}_l(\bar{v}_l) = [\hat{x}_{1l}(\bar{v}_l), \hat{x}_{2l}(\bar{v}_l), \ldots, \hat{x}_{Rl}(\bar{v}_l)]^T$$

$$= argmax \sum_{r=1}^{R} a_{rl} \bar{w}_{rl} (1 + (1 - \bar{v}_l)\gamma)\varphi(x_{rl})$$

such that

$$\forall_{r\in\hat{R}} \quad x_{rl} = \bar{x}_r^* \qquad (8)$$

and (6) and (7) are fulfilled. Such a solution is parameterized with $\bar{v}_l$.
4. For each link $l \in \hat{L}$, find $\bar{v}_l$ solving the following equation:

$$\sum_{r=1}^{R} a_{rl}\bar{w}_{rl}(1 + (1 - \bar{v}_l)\gamma)\varphi(\hat{x}_{rl}(\bar{v}_l)) = \alpha_l$$

5. Calculate $\bar{x}_r = \min_{l:a_{rl}=1} \hat{x}_{rl}(\bar{v}_l)$.

6. For each link $l \in \hat{L}$ solve the following optimization problem:
   Find:

$$s'_l(v'_l) = [x'_{1l}(v'_l), x'_{2l}(v'_l), \ldots, x'_{Rl}(v'_l)]^{\mathsf{T}}$$

$$= argmax \sum_{r=1}^{R} a_{rl}\bar{w}_{rl}(1 + (1 - v'_l)\gamma)\varphi(x_{rl})$$

such that

$$\forall_{r \in \bar{R}_l \setminus \hat{R}} \quad \bar{x}_r = x_{rl}$$

and (6), (7) and (8) are fulfilled. Such a solution is parameterized with v'_l.

7. For each link $l \in \hat{L}$, find v'_l solving the following equation:

$$\sum_{r=1}^{R} a_{rl}\bar{w}_{rl}(1 + (1 - v'_l)\gamma)\varphi(x'_{rl}(v'_l)) = \alpha_l.$$

If the solution does not exist, let $v'_l = 0$.

8. Find $l' \in \hat{L}$ for which the value of $v'_{l'}$ has the greatest value.
9. The final solution for the $r \in (R'_l \cup \bar{R}_l) \setminus \hat{R}$ is $\bar{x}^*_r = x_{rl}(v'_l)$.
10. $\hat{L} := \hat{L} \setminus \{l'\}$
11. $\hat{R} := \hat{R} \cup R'_l \cup \bar{R}_l$
12. If $\hat{L}$ is not empty then go to the step 3, stop otherwise.

The introduced algorithm finds pure NE. However, due to the space limitation, the formal proof is omitted. Nevertheless, the idea of the proof is analogous to the one that algorithm given in [23] finds the pure NE for the deterministic game.

6   Final Remarks and Future Works

In this paper, the game modeling for uncertain capacity allocation problem in self-managed networks environment is considered. The algorithm finding pure Nash Equilibrium is given and illustrated.

Obviously, elaborated solution need to be further investigated and compared with other approaches to the capacity allocation in computer network under uncertainty.

It is also promising to examine cases, when the computer networks are quality of service networks or content aware networks (CAN). While some results concerning self-managed CAN networks for deterministic case have been already obtained [21], especially with game-theoretical approach [22].

References

1. Kephart, J.O., Chess, D.M.: The vision of autonomic computing. IEEE Comput. **36**(1), 41–50 (2003)
2. Bettstetter, C., Prehofer, C.: Self-organisation in communication networks: Principles and design paradigms. IEEE Commun. Mag. **43**(7), 78–85 (2005)
3. Hu, H., Zhang, J., Zheng, X., Yang, Y., Wu, P.: Self-configuration and selfoptimization for lte networks. IEEE Commun. Mag. **48**(2), 94–100 (2010)
4. Mortier, R., Kiciman, E.: Autonomic network management: some pragmatic considerationsl. In: Proceedings of the ACM SIGCOMM Workshop on Internet Network Management, pp. 89–93 (2006)
5. Dobson, S., Denazis, S., Fernández, A., Gaïti, D., Gelenbe, E., Massacci, F., Nixon, P., Saffre, F., Schmidt, N., Zambonelli, F.: A survey of autonomic communications. ACM Trans. Auton. Adapt. Syst. **1**(2), 223–259 (2006)
6. Agoulmine, N.: Autonomic Network Management Principles. From Components to Applications. Elsevier, New York (2011)
7. Diao, Y., Hellerstein, J.L., Parekh, S., Griffith, R., Kaiser, G., Phung. D.: Selfmanaging systems: a control theory foundation. In: Proceedings of 12th IEEE International Conference and Workshops on the Engineering of Computer-Based Systems, pp. 441–448 (2005)
8. Balasubramaniam, S., Botvich, D., Donnelly, W., Foghlú, M.Ó., Strassner, J.: Biologically inspired self-governance and self-organisation for autonomic networks. In: Proceedings of the 1st International Conference on Bio-Inspired Models of Network, Information and Computing Systems (2006)
9. MacKenzie, A.B., Wicker, S.B.: Game theory and the design of self-configuring, adaptive wireless networks. IEEE Commun. Mag. **39**(11), 126–131 (2001)
10. Guerin, R.A., Orda, A.: QoS Routing in Networks with Inaccurate Information: Theory and Algorithms. IEEE Trans. Netw. **7**(3), 350–364 (1999)
11. Bubnicki, Z.: Analysis and Decision Making in Uncertain Systems. Springer-Verlag, London (2004)
12. Bubnicki, Z.: Modern Control Theory. Springer-Verlag, London (2005)
13. Kelly, F.P., Maulloo, A.K., Tan, D.K.H.: Rate control for communication networks: shadow prices, proportional fairness and stability. J. Oper. Res. Soc. **49**(3), 237–252 (1998)
14. Gasior, D.: QoS rate allocation in computer networks under uncertainty. Kybernetes **37**(5), 693–712 (2008)
15. Gąsior, D., Orski, D.: On the rate allocation problem under co-existence of uncertain utility function parameters and uncertain link capacities. J. Oper. Res. Soc. **65**(10), 1562–1570 (2014)
16. He, J., Zhang-Shen, R., Li, Y., Lee, C.Y., Rexford, J., Chiang, M.: DaVinci: dynamically adaptive virtual networks for a customized internet. In: Proceedings of the 2008 ACM CONEXT Conference (2008)
17. Drwal, M., Gasior, D.: Utility-based rate control and capacity allocation in virtual networks. In: Proceedings of the 1st European Teletraffic Seminar (2011)
18. Gasior, D.: Capacity allocation in multilevel virtual networks under uncertainty. In: XVth International IEEE Telecommunications Network Strategy and Planning Symposium (2012)
19. Gasior, D., Drwal, M.: Pareto-optimal Nash equilibrium in capacity allocation game for self-managed networks. Comput. Netw. **57**(14), 2817–2832 (2013)
20. Koutsoupias, E., Papadimitriou, C.: Worst-case equilibria. In: Meinel, C., Tison, S. (eds.) STACS 1999. LNCS, vol. 1563, pp. 404–413. Springer, Heidelberg (1999). doi:10.1007/3-540-49116-3_38

21. Gasior, D., Drwal, M.: Decentralized algorithm for joint data placement and rate allocation in content-aware networks. In: Kwiecień, A., Gaj, P., Stera, P. (eds.) Computer Networks. Springer, Heidelberg (2012)

22. Gasior, D., Drwal, M.: Caching and capacity allocation game in self-managed content provider networks. In: 16th International IEEE Telecommunications Network Strategy and Planning Symposium (2014)

Author Index

© Springer International Publishing AG 2017   339
J. Świątek and J.M. Tomczak (eds.), *Advances in Systems Science*, Advances in Intelligent
Systems and Computing, 539, DOI 10.1007/978-3-319-48944-5

Printed in the United States
By Bookmasters